ELECTRONICS

PACKAGING

FORUM

VOLUME TWO

EDITED BY

JAMES E. MORRIS

State University of New York
Binghamton

VNR VAN NOSTRAND REINHOLD
New York

Manufactured in the United States of America

Published by Van Nostrand Reinhold
115 Fifth Avenue
New York, New York 10003

Chapman and Hall
2-6 Boundary Row
London, SE 1 8HN

Thomas Nelson Australia
102 Dodds Street
South Melbourne 3205
Victoria, Australia

Nelson Canada
1120 Birchmount Road
Scarborough, Ontario M1K 5G4, Canada

16 15 14 13 12 11 10 9 8 7 6 5 4 3 2 1

Library of Congress Cataloging-in-Publication Data

Electronics packaging forum.

    Based on the annual Electronics Packaging Symposium,
which is run at the State University of New York at
Binghamton by the Continuing Education Division of the
T.J. Watson School of Engineering, Applied Science, and
Technology.
    1. Electronic packaging--Congresses. I. Morris,
James E., 1944-   . II. State University of New York
at Binghamton. T.J. Watson School of Engineering,
Applied Science, and Technology. Continuing Education
Division.
TK7870.15.E43   1990   621.381'046   90-12156

ISBN 0-442-00178-9 (v. 1)
ISBN 0-442-00476-1 (v. 2)

PREFACE

Each May, the Continuing Education Division of the T.J.Watson School of Engineering, Applied Science and Technology at the State University of New York at Binghamton sponsors an Annual Symposium in Electronics Packaging in cooperation with local professional societies (IEEE, ASME, SME, IEPS) and UnIPEG (the University-Industry Partnership for Economic Growth.) Each volume of this Electronics Packaging Forum series is based on the the preceding Symposium, with Volume Two based on the 1990 presentations.

The Preface to Volume One included a brief definition of the broad scope of the electronics packaging field with some comments on why it has recently assumed such a more prominent priority for research and development. Those remarks will not be repeated here; at this point it is assumed that the reader is a professional in the packaging field, or possibly a student of one of the many academic disciplines which contribute to it. It is worthwhile repeating the series objectives, however, so the reader will be clear as to what might be expected by way of content and level of each chapter.

It is intended that each chapter address the "mature" theoretical basis to its specific topic area, pitched at a level which will be accessible to a non-specialist reader in the particular topic. The reader should have a general appreciation for the technical issues, however, due to some involvement in one of the other, different packaging disciplines. In this way the series directly addresses the specific and widely recognized problem presented by the extraordinarily multi-disciplinary nature of the field and its practical problems. So any chapter found here in the reader's own area of expertise should not be expected to provide the latest research results in that area, although it may prove to be valuable for other reasons, (e.g. as a convenient tutorial for one's colleagues.) The book's value to the reader will lie in the chapters outside his or her area of specialization.

As with the preceding volume, Volume Two covers a diverse range of topics, balancing academic research with industrial applications, and including material from the electrical, mechanical, thermal, materials, manufacturing and systems fields. The book opens in Chapter 1 with a comprehensive review of one specific technology, in this case tape-automated bonding (TAB), with a most extensive bibliography for further research purposes. A feature of this review is the recent material on thermo-elasticity. The next two chapters address the mechanics of circuit cards, Chapter 2 further developing some of the stress analysis introduced in the previous volume, and Chapter 3 considering vibrational modes. Board-level power distribution architectures are changing rapidly with the widespread adoption of switched-mode power supply (SMPS) designs, and the effects of these developments, which are traced out in Chapter 4 will undoubtedly spread out to affect the complete system design. One of the driving forces to SMPS adoption is the potential for reduced power dissipation, so the topic leads naturally into the next one of thermal management. Chapter 5 reviews the field at the system level, while Chapter 6 focuses on the issues behind heat-sink design. To improve heat conduction from circuit modules, one needs a good thermal conductor, which unfortunately must also usually be an insulator. This brings us in Chapter 7 to diamond films, the leading candidate to meet these normally conflicting goals. The materials research area continues in Chapter 8; high speed propagation requires low dielectric constant materials, and gains are being made with the development of new polymers. Since Chapter 8 relates directly to the problem of high speed electrical transmission systems, it is appropriate to follow it with consideration of optical signal transmission in Chapter 9. Chapter 10 seems at first glance to be one of the most peripheral chapters to the goal of the book, but that could not be further from the truth; it is, in fact, the first of the sequence of chapters covering manufacturing issues. In modern electronics assembly, there is considerable awareness of the dangers of electrostatic damage (ESD) to CMOS parts in particular. The work reported in Chapter 10 identified distinctly different damage mechanisms in Si and GaAs devices, but more significantly to the assembly industry, identified the "walking wounded" -- those chips which will pass test, even though damaged, to fail later in the field. Another issue of major significance to the assembly industry today is the drive towards environmentally safe processing materials; one aspect of the environmental issue is addressed in Chapter 11. In a sense, the topic of Chapter 12 is also environmental, but the problem is now turned around, with the issue now being one of environmental damage to electrical performance. Chapter 13 is another where it just so happens that the work described is very recent; in fact, the Symposium presentation on which it is based included data obtained just the day before! It provides an excellent example of industrial research. Finally, Chapter 14 maintains prior "tradition" and introduces some possibly controversial ideas on the national competitiveness of the industry. The particular point being made here is that the different branches of the packaging industry (computer, military, consumer, automotive, etc.,) must learn from each other, and that in some cases the problems of one are largely solved already in another.

This is a contributed volume, and I must offer my thanks to the contributors for their hard work in preparing manuscripts, diagrams, etc. I am also grateful to Gary Arnold (the Director of Continuing Education), whose organization of the Symposium is fundamental to the whole endeavor, and to the various members of the Symposium's organizing committee. At Van Nostrand Reinhold, the project is handled by Steve Chapman to whom I am grateful for both advice and assistance. Closer to home, I must also acknowledge the assistance of the Electrical Engineering Department secretary Mrs. Mary-Lou Curry with word-processing, print-outs and day-to-day crises. I also appreciate my wife and family's patience with the inevitable disruptions to domestic routine.

James E. Morris

CONTENTS

AN INTRODUCTION TO

TAPE AUTOMATED BONDING

TECHNOLOGY

JOHN H. LAU

STEVE J. ERASMUS

DONALD W. RICE

1.1 Introduction

Figure 1.1 shows a Tape Automated Bonding (TAB) assembly with 544 I/Os at 0.008" (0.203 mm) outer-lead spacing (courtesy of Hewlett-Packard Electronics Packaging Laboratory). It consists of three major parts, the silicon chip, the copper beam leads, and the epoxy/glass FR-4 printed circuit board. In this chapter, an introduction to TAB for fine pitch, high I/O, high performance, high yield, high volume, and high reliability is presented. Emphasis is placed upon a new understanding of the key elements (e.g., tapes, bumps, inner lead bonding, testing and burn-in on tape-with-chip, encapsulation, outer lead bonding, thermal management, reliability, etc.) of this rapidly moving technology.

TAB was introduced in the late 1960s by the General Electric Company for Small Scale Integration (SSI) devices. It was developed as a possible replacement for wire bonding technology and was widely perceived as one of the most promising approaches to electronics packaging. During the 1970s, this technology was the topic of discussion and consideration. There was little industrial acceptance, because high speed automatic wire bonders met the chip to package interconnect needs.

The intense development of surface mount technology (SMT) in the early 1980s for light weight, high density, and automated electronic assemblies has increased the interest in TAB (a natural extension of SMT). Furthermore, due to the advance of Integrated Circuit (IC) technology and the requirement of high-density I/Os and high-speed circuitry, TAB has been applied to the more demanding field of Very Large Scale Integration (VLSI) packaging for a variety of consumer, medical, security, computer, peripheral, telecommunication, automotive, and aerospace products.

TAB is a technique allowing full automation of the bonding of one end of the etched copper beam lead (beam tape or lead frame) to a semiconductor device and the other end of the lead to a conventional package or printed circuit board (PCB). TAB offers many advantages over conventional wire bonding. The more prominent benefits of TAB are listed below (though not in order of importance) [1-312]:
   * Smaller bonding pad on chip.
   * Smaller bonding pitch on chip.
   * Less gold required.
   * Elimination of wire loop (low-profile assembly).
   * More precise geometry.
   * Opportunity for gang bonding.
   * Stronger and more uniform inner lead bond pull strength.

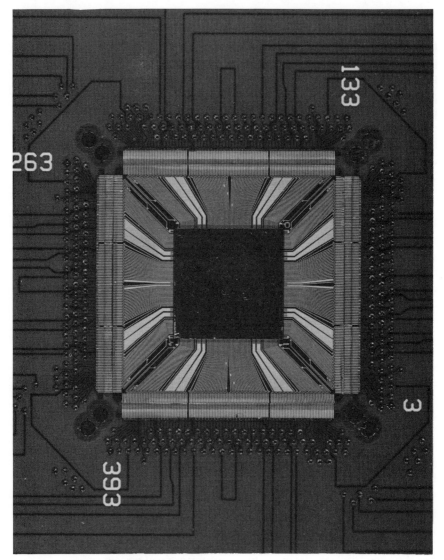

Figure 1.1 544 I/O TAB assembly.

* Lower molding cost.

* Like wire bonding chip on board (COB), one less interconnect point required from chip to PCB.

* Like wire bonding COB, direct mounting of chip onto a multichip substrate without a package.

* Improved conduction heat transfer.

* Better electrical performance.
* In some cases, better high frequency performance.
* Ability to electrically test and "burn-in" the tape-with-chip prior to assembly.
* Reduced labor costs.
* Higher I/Os.
* Lighter weight.
* Less PCB surface area required.
* Chip can be attached in either a face-up or a face-down configuration.
* The passivation opening is hermetically sealed by the bump metallurgy such that no portions of the monolithic circuit are exposed.
* For very high I/Os, TAB packages cost less than wire bonding packages.

However, as with any new technology, TAB is not without its problems. At this point in the development of TAB, the following issues must be noted and understood in order to obtain all of its benefits (not in order of importance) [1-312]:
* Inflexible process (special tape design and retooling for each different chip design).
* Bonds are more difficult or impossible (tape to device) to repair.
* Production equipment is difficult to obtain.
* Requires extra wafer processing to provide bumped bond sites.
* Larger capital investment.
* General use of TAB requires advances in tape technology.
* General use of TAB requires advances in PCB technology.
* Requires extensive engineering development.
* Requires special materials and equipment.
* TAB design and manufacturing talents are still rare.
* Entrenched position of conventional SMT technology.
* Resistance to change.
* Assembly rework techniques are more difficult.
* Coplanarity of beam leads with chip and substrate.
* Overall thermal management problems are greater.
* System testability.
* TAB standards.
* Tape carriers are not economically available in small quantities.
* Lack of commercially available "bumped" tapes or wafers.
* Beam lead reliability is more critical because of its small thickness.

In the present chapter, the literature of TAB will be reviewed. The purpose is to bring past work together into one source which is easily accessible to engineers and researchers interested in TAB. This review of the literature is by no means comprehensive, but the references selected [1-312] are representative of the state-of-the-art. The review is limited primarily to English language publications.

1.2 Types of TAB

There are at least four popular kinds of TAB assembly technology, bumped chip TAB, bumped tape TAB, area TAB, and multichip packaging with TAB, [8-9, 12, 15, 24-25, 105-106, 122, 169-174, 182-186, 201, 204, 234-236, 249-250, 266-267, 277-278, 287-290, 296].

1.2.1 Bumped chip (planar lead) TAB

The primary characteristic of TAB technology is a bonding projection, or bump between the chip and the beam lead. This bump provides both the necessary bonding metallurgy for inner lead bonding and a physical standoff to prevent lead-chip shorting. Conventional TAB uses specially processed bumped wafers which may be supplied by the semi-conductor manufacturer, or by electronic companies producing "in-house" TAB devices; Figure 1.2, [189-190]. However, because of the current nature of TAB technology, bumped wafers from semi-conductor manufacturers are not readily available.

1.2.2 Bumped tape (planar chip) TAB

Bumped tape TAB (BTAB) was developed in the late 1970s, [24-25, 122, 218, 249-250, 277-278, 287-290]. BTAB puts the bump on the inner end of the tape rather than on the chip; Figure 1.3, [189-190].

Recently, Hatada and his associates [93-98] at Matsushita of Japan developed a cost effective TAB process called "transferred bump TAB". In this process, bumps are first formed on a glass or ceramic substrate by means of gold plating and are then transferred (bonded) onto copper beam leads before being bonded to the aluminum electrodes on the chip; Figure 1.4. At the present stage this process can only work for tin-plated copper beam leads.

1.2.3 Area (array) TAB

Both bumped chip and bumped tape TABs are also called peripheral TABs, that is, with leads along the periphery of the chip. The third kind of TAB is called area TAB, for example [12,15,269], where a multilayer tape provides an interconnection between the grid of solder bumps of a

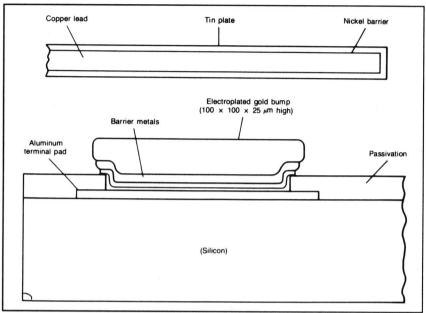

Figure 1.2 Bumped chip (planar lead) TAB. [189,190]

VLSI chip and the bonding pads of a ceramic package; Figure 1.5, [169-174].

1.2.4 Multichip packaging with TAB

Figure 1.6 is an example of multichip packaging with TAB [84,151]. It consists of a ceramic substrate with copper thick-film interconnects. The chips are nine complementary metal-oxide semiconductor (CMOS) 20K Gate Arrays, TAB bonded to the thick film. There are eight decoupling capacitors to reduce power supply noise. A metal cover hermetically seals the active devices and a heat sink is attached to the back of the substrate for added cooling. The unit is pressure mounted in the connector with the connector clamp, and the connector is soldered onto the PCB. Some advantages of multichip TAB are an increase in I/Os, density, performance, and pre-testability prior to mounting components into the module.

1.3 Types of Tape

Basically, there are three different beam tape constructions: one, two and

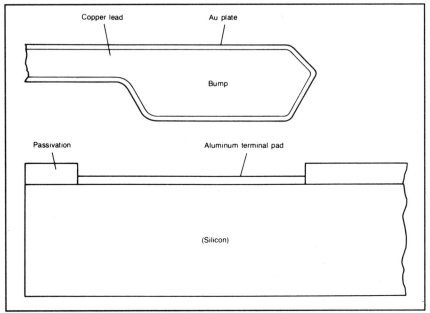

Figure 1.3 Bumped tape (planar chip) TAB. [189,190]

three layers, [8-9, 12, 29, 43, 45, 63, 67, 104, 109-111, 141, 165, 201, 234-236, 260, 268, 275, 310-311]. Color examples of different tapes can be found in [182-186, 204]. Each of these tape constructions has a unique application area, in which its use is preferred over the other two. Figure 1.7 summarizes the differences between the three construction types, and Figure 1.8 compares typical fabrication processes used in construction of the three tapes [29].

1.3.1 One-layer tape

One-layer tape is made using a metal foil with a series of lead patterns photo-etched into it. A few metals shown in Table 1.1 [268] can be used for tape construction. However, the metal used today is typically a 2 ounce rolled, annealed copper, i.e., 0.0028" (0.07 mm) thick. The tape can be bumped in the etching process for use with a planar chip, or may be planar for use with a bumped chip. This tape is ideally suited for low-cost assembly and is very precise for high-speed inner lead bonding because its sprocket holes are etched simultaneously with its bonding leads. Because the tape is all metal and ties all leads together, electrical test and burn-in in tape is impossible (except for National Semiconductor's Tape-Pack, which requires an additional manufacturing

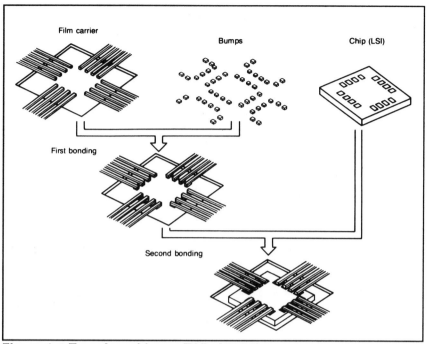

Figure 1.4 Transferred bump TAB. [93-98]

process). Also, unsupported long leads are more likely to suffer mechanical damage.

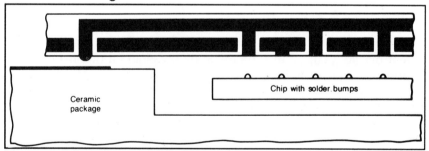

Figure 1.5 Area (array) TAB. [169-174]

1.3.2 Two-layer tape

Two-layer tape is typically made of 1 ounce copper (0.0014" or 0.035 mm thick) with material properties given in Table 1.1, and 0.0005"-0.002"

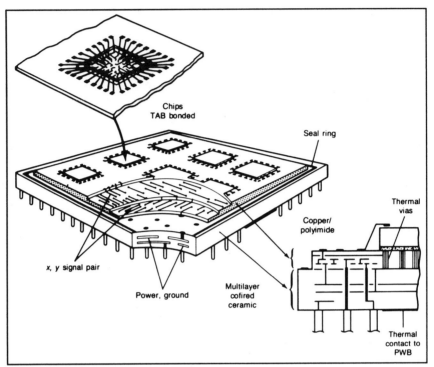

Figure 1.6 Multi-chip TAB. [84]

(0.013-0.051 mm) plastic film. Tables 1.2(a-d) list the material properties of 13 candidate plastic film carriers. ("Melting point" in Figure 1.2(c) means "zero-strength temperature" in some cases). Examination of these properties indicates that polyimide is the best all-around material for the carrier plastic [268]. It may be made by screening a polyimide film on copper, by casting a liquid polyimide film onto copper, or by plating copper onto a polyimide film. Sprocket and window patterns in the polyimide film are formed by a chemical etching process, while the lead and sprocket patterns in the copper foil are formed by the normal photo-patterning and etching process. The major advantages of the two-layer type are the design freedom allowed and testability. Furthermore, it is stronger than the one-layer type because of the polyimide film. However, two-layer tape is more expensive than one-layer tape.

1.3.3 Three-Layer Tape

A typical three-layer tape starts with 0.003"-0.005" (0.076-0.152 mm) thick polyimide film coated with 0.0005"-0.001" (0.013-0.025 mm) of an

Tape type	Construction	Advantages	Disadvantages
One-layer	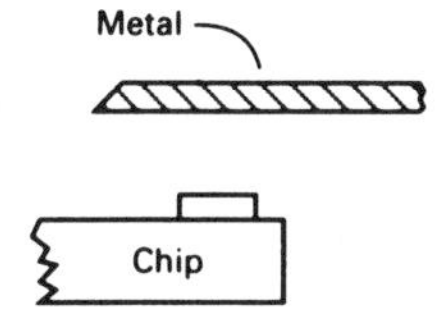 Metal / Chip	Excellent for high-temperature applications Low cost	Bonded chips cannot be tested on tape Unsupported leads are more likely to incur mechanical damage
Two-layer	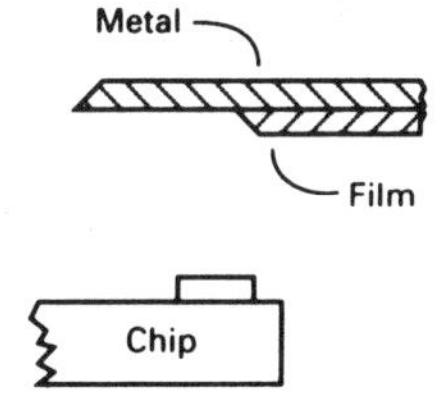 Metal / Film / Chip	Very good for high-temperature applications Bonded chips can be tested on tape Makes excellent use of available design area	High cost (compared to one-layer tapes) Poor dimensional stability on wide tape widths (for example, 35 mm) Long leads may not lie flat if not given enough film support Marked tendency to curl
Three-layer	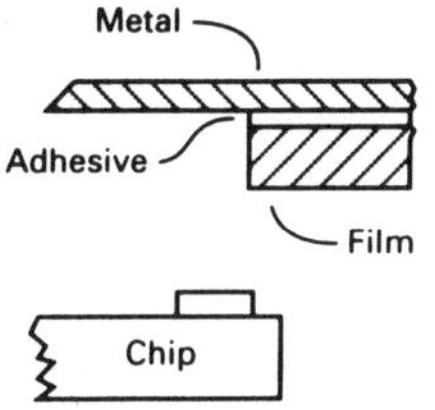 Metal / Adhesive / Film / Chip	Bonded chips can be tested on tape Excellent for large chips needing tape widths of 35 or 70 mm Long leads lie flat for good alignment	High cost (compared to one-layer tapes) Adhesive stability can be a problem

Figure 1.7 Comparison of beam tape carrier constructions. Adapted from [29]

adhesive. The physical properties of some commercial available adhesives are given in Table 1.3, [268]. Sprocket holes are mechanically fabricated (by die punch or laser cut, etc.) in the film, which is then laminated to 0.0014" (0.035 mm) thick copper foil. As with the two-layer tape, the coated polyimide film acts as a dielectric carrier and electrically isolates the beam leads to allow functional testing before the IC is committed to its final connection.

Both two- and three-layer tapes have nonconductive backings and, consequently, potentially offer a method of support for long, delicate TAB beam leads. In the case of two-layer tape, where patterns are etched into both copper and polyimide layers, a support structure of virtually any shape, including free-floating, is possible. On the other hand, three-layer tape has had a more limited range of support structures available because of its process. Until fairly recently, rings had to be supported, typically by using corner tie bars, Figure 1.9(a). Now,

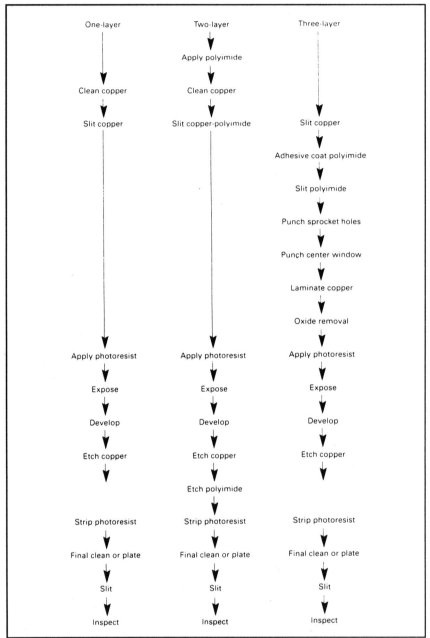

Figure 1.8 Comparison of typical tape fabrication processes. [29]

fortunately, a process has been developed which creates such a ring

Table 1.1 Physical properties of metal foils. [268]

Property	Rolled, Annealed Cu	Electro-deposited Cu	Al	Rolled Ni	Steel	Stainless Steel
Specific Gravity	8.89	8.94	2.70	8.90	7.8	7.9
Resistivity Ω.cm x10^{-6}	1.72	1.77	2.83	9.50	13-22	74
TCR* Ω.cm/K x10^{-6}	.00393	.00382	.00390	.00470	.00160	.....
CTE** 10^{-6}/K	16.6	16.6	23.6	13.3	15.1	17.3
Thermal Conductivity W/m.K (W/in.C)	392 (10.0)	392 (10.0)	155 (3.9)	125 (3.2)	47 (1.2)	12.5 (.32)
Tensile Strength MPa (ksi)	235 (34)	310 (45)	83 (12)	495 (72)	305 (44)	620 (90)
Elongation %	20.0	12.0	18.0	40.0	36.0	50.0
Modulus of Elasticity Gpa (10^6psi)	117 (17.0)	110 (16.0)	70 (10.0)	205 (29.6)	200 (29.0)	195 (28.0)
Thickness range mm (mils)	0.018-0.36 (0.7-14)	0.018-0.36 (0.7-14)	0.025-0.50 (1.0-20)	0.013-0.38 (0.5-15)	0.038-0.25 (1.5-10)	0.013-0.25 (0.5-10)
Yield ft^2/lb/mil	22.0	22.0	71.2	21.7	25.0	25.0

* TCR: temperature coefficient of resistivity
** CTE: coefficient of thermal expansion

(Figure 1.9(b)) to give full design freedom for the copper lead pattern and to allow simple excision tooling [109-110]. The three-layer tape is more expensive than the one-layer tape.

Table 1.2(a) Physical properties of plastic films: Mechanical. [268]

Plastic Film	Moisture Absorption %	Ultimate Elongation %	Tensile Strength	
			Mpa	ksi
Polyimide (Kapton)	4.0	70	170	25
Polyester (Mylar)	<0.8	120	170	25
TFE (Teflon)	<0.01	350	28	4
FEP (Teflon)	<0.01	300	21	3
Polyamide (Nomex)	3.0	10	75	11
Polyvinyl Chloride	<0.5	130	35	5
Polyvinyl Fluoride (Tedlar)	<0.05	110-300	70-130	10-19
Polyethylene	<0.01	>300	21	3
Polypropylene	<.005	250	170	25
Polycarbonate	0.35	110	62	9
Polysulfone	0.22	95	68	9.9
Polyparabanic acid	1.8	10	97	14
Polyether sulfone	...	...	84	12.2

1.3.4 Special Considerations of Tape

Recently, multiconductor (multimetal) tapes have been in great demand. Two-metal layers to provide signal lines and ground plane are being made in production quantities. Five metal layers to provide signal, ground, and power isolation, as well as controlled impedance are being researched and developed. Although, multiconductor tapes can be

Table 1.2(b) Physical properties of plastic films: Electrical. [268]

Plastic Film	Dissipation Factor at 1kHz	Dielectric Constant at 1kHz	Dielectric Strength	
			MV/m	kV/mil
Polyimide (Kapton)	0.003	3.5	280	7.0
Polyester (Mylar)	0.005	3.25	300	7.5
TFE (Teflon)	0.0002	2.0	17	0.43
FEP (Teflon)	0.0002	2.0	255	6.5
Polyamide (Nomex)	0.007	2.0	18	0.45
Polyvinyl Chloride	0.009	3.0	40	1.0
Polyvinyl Fluoride (Tedlar)	0.02	8.5-10.5	140	3.5
Polyethylene	0.0003	2.2	20	0.5
Polypropylene	0.0003	2.1	160	4.0
Polycarbonate	1.000	3.2	16	0.4
Polysulfone	0.001	3.1	295	7.5
Polyparabanic acid	0.004	3.8	235	6.0
Polyether sulfone	0.001	3.5	...	...

produced in a manner similar to that used for conventional tapes, a host of problems involving materials, registration, and processing need to be solved.

In general, the interconnection surfaces on copper beam leads of tape are either electroplated gold (0.00064-0.0015 mm), electroplated tin (0.00064 mm), or electroless tin. In the case of gold-plated leads, a nickel underlayer (0.00025-0.001 mm) is sometimes used. In the case of

Table 1.2(c) Physical properties of plastic films: Thermal and economic. [268]

Plastic Film	Melting Point[*]		Coeff of Linear Expansion 10^{-6}/K	Yield ft^2/lb/mil	Cost Factor
	°C	°F			
Polyimide (Kapton)	>600	>1110	36	136	17.0
Polyester (Mylar)	180	355	31	140	1.0
TFE (Teflon)	328	620	12.2	84	9.2
FEP (Teflon)	280	535	9.7	90	16.0
Polyamide (Nomex)	482	895	38-154	240	1.9
Polyvinyl Chloride	163	325	63	140	0.36
Polyvinyl Fluoride (Tedlar)	299	570	50	140	3.5
Polyethylene	121	250	198	205	0.17
Polypropylene	204	400	104-184	215	0.48
Polycarbonate	132	270	68	140	1.32
Polysulfone	190	375	56	155	3.33
Polyparabanic acid	299	570	51	...	8.0
Polyether sulfone	203	395	...	...	4.5

tin leads, incoming inspection and proper storage of the tapes are very important because of both copper migration and tin oxidation that can reduce lead wettability, and the whisker effect, that can short adjacent leads.

Three widths of tapes have been registered with the Joint

Table 1.2(d) Physical properties of plastic films: Comments. [268]

Plastic Film	General Comments
Polyimide (Kapton)	Expensive
Polyester (Mylar)	Low temperature
TFE (Teflon)	Too weak; linear coefficient of thermal expansion (CTE) too high
FEP (Teflon)	Weak; CTE too high; expensive
Polyamide (Nomex)	CTE too high; porous structure; only in paper form
Polyvinyl Chloride	Weak; low temperature
Polyvinyl Fluoride (Tedlar)	CTE too high; dielectric constant higher for DIP use
Polyethylene	Low temperature
Polypropylene	CTE much too high
Polycarbonate	Low temperature; CTE high
Polysulfone	Low temperature
Polyparabanic acid	Brittle; not yet in production
Polyether sulfone	Medium temperature; unproven

Electron Device Engineering Council (JEDEC): 35 mm, 48 mm, and 70 mm. The standards for TAB tape carriers that are compatible with outer lead bonding pitches of 0.02" (0.5 mm), 0.015" (0.37 mm), and 0.01" (0.25 mm) have also been approved. These tapes can be stored either on reels or in slide carriers.

Table 1.3 Physical properties of adhesives. [268]

Adhesives	Type	Nip Roll Temperature		Maximum Solder Temperature		Minimum Peel Strength	
		°C	°F	°C	°F	N/mm	lbf/in
Rogers 8145	Epoxy	204	400	316	600	0.53	3
DuPont WA	Acrylic	188	370	260	500	1.75	10
CMC 1477	Epoxy	190	375	288	550	1.58	9
CMCX-1496	Polyester	190	375	232	450	1.05	6

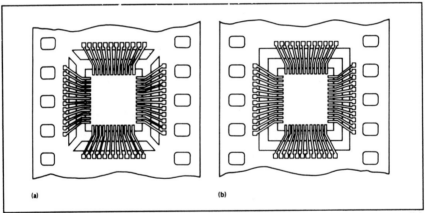

(a)   (b)

Figure 1.9 (a) Annular ring supported by tie bars at each corner. (b) Free-floating annular ring supports leads. [110]

1.4 Wafer Bumping Technology

For bumped tape (planar chip) TAB, no wafer bumping is necessary, and the wafer preparation of the aluminum bond pads is simple; that is, open up the passivation in the bond pad areas in the usual way and provide a stable, bondable interface (like wire bonding technology).

For bumped chip (planar lead) TAB, one of the key processes is wafer bumping, [6, 8-9, 13, 31, 34, 54-58, 69, 88-91, 93-98, 114, 134, 139,

143, 161-164, 192-196, 201, 243, 264-265, 275, 300]. This process is to make thick metal bumps on the chip bond pads for inner lead bonding. These bumps are made of various metals of which the most commonly used are gold and solder. It should be pointed out that the cost per wafer of bumping is borne by the surviving good chips.

1.4.1 Gold bumps

Extensive work on gold wafer bumping has been done by Oswald, Montante, Liu, Rodrigues de Miranda, and Zipperlin [161-164, 206-207, 195-196] at Honeywell Inc. Their metallization process for bumping wafers is shown in Figure 1.10. This process makes 0.004" (0.1 mm) square and 0.001" (0.025 mm) high gold bumps on an aluminum bond pad after the passivation over the bond pads has been opened.

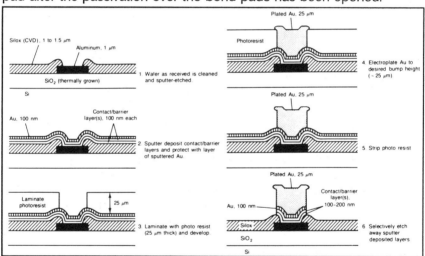

Figure 1.10 Metallization process for bumping wafers. [161-164, 206-207, 195-196]

Barrier metallization in the form of titanium, palladium, and gold is deposited on the wafer. The reason for this metallization system is to provide a diffusion barrier between the original aluminum bonding pads and the gold bumps. Low contact resistance is required through the barrier, as is good adhesion between all layers. The barrier layers also act as a metal substrate during the selective electroplating of the bumps.

After barrier layer metallization, the surface of the wafer is laminated with a dry film photoresist. Dry film was selected for its uniformity and thickness. The 0.025 mm thick photoresist lends itself precisely to the 0.025 mm height selected for the finished bump. After the photoresist has been developed and inspected for proper hole size and positioning, the wafer is ready for electroplating. The plated areas, located at each pad and via site will form the gold bumps, as shown in Figure 1.10 [162, 195-196, 206-207].

Bumps can be effectively formed using dry photoresist if certain precautions are taken [162]. Before plating, the gold surface of the wafer must be activated. The plating bath must produce a soft gold (99.99%) to be compatible with thermocompression bonding. Accurate measurement of the area to be plated must be performed to adjust plating current properly.

After plating the bump areas, the selective etching of the unwanted evaporated or sputtered layers is a very important step in preserving yield. Care must be taken not to let the etchant penetrate defects in the passivation and dissolve the easily corroded aluminum interconnect metallurgy. Several processes, such as plasma stripping, may avoid this yield loss.

Other barrier layer metallizations currently employed include Ti-Pt-Au, Cr-Ni-Au, Ti-W-Au, etc. Refer to [162] for more details. Furthermore, bumps can also be formed using a liquid photoresist anisotropic etch process which may lead to a more vertical bump walls.

1.4.2 Solder Bumps.

Solder bumps can also be used for wafer bumping, especially for area TAB [269]. The bonding pressure and planarity requirements for inner lead bonding are lower with solder bumps than with gold bumps. However, the bond pull strength and ductility with solder bumps can be smaller than with gold bumps.

1.5 Inner Lead Bonding

Inner lead bonding (ILB) is the process of attaching the inner leads on the tape to the semiconductor device. The purpose of this process is to form a strong metallurgical and reliable electrical bond between the beam lead and the device. The pull strength of an inner lead bond typically exceeds 50 grams instead of the 5 to 10 grams for the traditional 0.001" (0.025 mm) diameter gold flying lead wire bond. TAB leads may be bonded one at a time with a single-point bonder or bonded simultaneously with a gang (mass) bonder. Bonding parameters (time, temperature, pressure, and energy) will vary based on chip size, pad size, tape characteristics and metallurgy, [6, 8-10, 16, 24-26, 31, 33, 39, 41, 54-58, 88-91, 99-100, 116, 120, 123, 125-127, 130, 144-148, 150, 156,

159-164, 201, 206-207, 211-217, 232-236, 243, 254, 266-268, 270, 275, 282-283, 287-291, 294, 297, 307-309, 311].

It should be pointed out that depending on the level of test (component test, burn-in, etc.) prior to the assembly of IC and TAB, the costs of assembling "poor chip" must be shared by the final yielded components.

1.5.1 Single-Point Bonding.

As with wire bonding technology, TAB inner leads can be bonded by methods including thermocompression bonding, ultrasonic bonding, thermosonic bonding, laser bonding, and solder reflow. The disadvantage of single-point is the speed. The advantages of single-point bonding are:

* no planarity problem
* minimal set up time
* no thickness (both tape and bump) problem
* consistent bonding pressure on each lead
* easy to remove individual leads, and
* very similar to wire bonding.

The disadvantage is the speed.

1.5.2 Gang (Mass) Bonding.

Unlike wire bonding technology, TAB beam leads can be bonded simultaneously by methods including thermocompression bonding, dynamic alloy formation, and solder reflow. For high rate mass production operations, thermocompression bonding is often preferred. Thermocompression gang bonding does increase throughput, but subject to planarity (missing bonds), bump/lead height (cracking chip) and alignment tolerance (poor bonds) problems. All of these problems may lead to inconsistent bond quality.

(a) Thermocompression bonding process. Figure 1.11 shows a thermocompression inner lead bonding sequence [125-127] for a bumped chip TAB. It can be seen that the thermode is the reference and chips are automatically pre-aligned by the precision XY-coordinate table (Figure 1.11(a)). During bonding (Figure 1.11(b)) the chip carrier rises to the bond level and the thermode drops to apply heat and pressure through metallic fingers on the tape to the bumps on the chip. At the completion of the preset bond cycle (Figure 1.11(c)) the thermode retracts. As soon as it is clear of the tape, the chip carrier drops to provide clearance for the tape to index and carry the finished part away. The tape then

indexes to position a fresh tape frame for the next bond and the coordinate table indexes to position a new chip (Figure 1.11(d)). The chip
carrier then rises to the "align" position. It should be pointed out that the
sorting process is simultaneous with the bonding process. In some
cases, it may be advantageous to bond the chips into tape from a
previously sorted array [275].

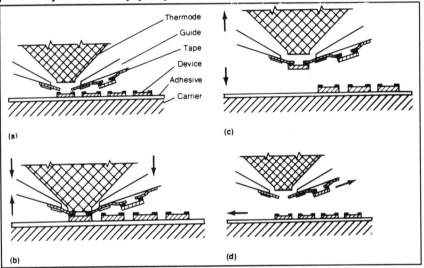

**Figure 1.11 Inner lead bonding sequence. (a) Precision x-y coordinate
table. (b) Bonding. (c) Pre-set bond cycle. (d) Coordinate table
indexes to position a new chip.** [125-127]

(b) Characterization of thermocompression bonding. The fundamental
parameters that influence bond formation have been studied by Ahmed
and Svitak [6] and others [211-217]. By using the models shown in
Figures 1.12(a) and 1.12(b) and the theory of plasticity, the normal
stresses, shear stresses, displacements and strains, were determined at
the interface. Correlating theoretical and experimental studies, Ahmed
and Svitak [6] observed the following:

        * There is a readily predicted central dead zone, where bonding
is difficult.
        * The magnitude of the normal stress, above that required for
contact, is unimportant.
        * Mutual extension of two nominally clean surfaces in contact is
sufficient for bond formation (after some time at temperature).
        * The required surface extension is considerably reduced by the
simultaneous application of an interfacial shear stress, though the

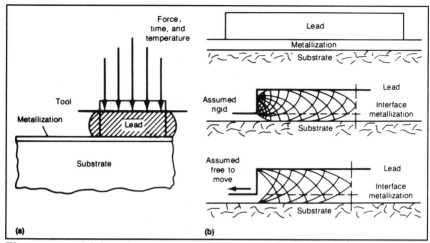

Figure 1.12 (a) Lead attachment by thermocompression bonding. (b) Cross-section through lead being bonded (top) with corresponding slip-line fields for the surrounding rigid metallization (center) and the free-flow model (bottom). [6]

magnitude of the shear stress above some threshold value is not very important.

* Neither sliding nor high interfacial shear stresses result in bonding unless both surfaces are deformed.

(c) Gold bump to tin-plated copper lead (Au/Sn). The metallurgical interactions at the interface between the sputtered or plated gold bumps and the tin plated copper leads have been studied by Liu and Fraenkel [161, 164] among others. Their experimental results indicate that interaction of liquid tin with solid gold takes place during thermo-compression bonding, forming a gold-rich Au-Sn eutectic structure. Figure 1.13(a) depicts line scan displays for tin across gold bumps, and Figure 1.13(b) shows the corresponding percent tin in the gold bump. Locations probed in this microsection are marked on the insert sketch. It can be seen that approximately 20wt%Sn has been dissolved into the gold bump (80wt%Au). The resultant metallurgical structure favors a gold rich Au-Sn eutectic [161, 164].

Aging experiments have also been conducted by Liu [161]. Figure 1.14(a) shows a photomicrograph of a cross-section of an ILB with a plated gold bump aged at 75°C for 1,000 hours in air. Tin distribution in the vicinity of ILB is shown as tin concentration contours in Figure 1.14(b). In Figure 1.14(b), the values above the bump-lead interface do not reflect the actual tin concentration; they represent only a relative

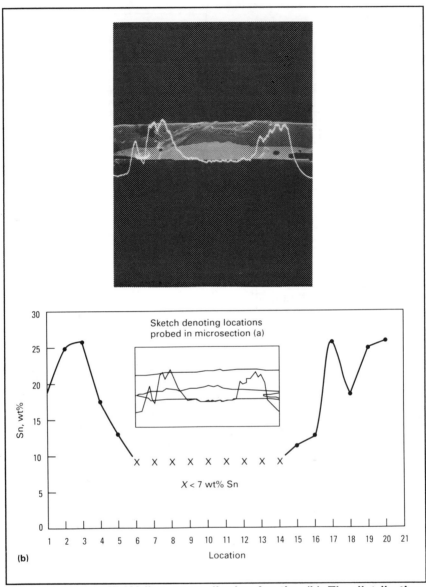

Figure 1.13 (a) EDAX line scan display for tin. (b) Tin distribution. [161-164]

distribution of tin. It can be seen that aging results indicate little change in the tin distribution at ILB compared with the as-bonded condition.

All the experimental results [161, 164] indicate deeper penetration

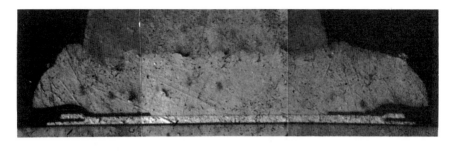

Figure 1.14 (a) Photomicrograph of an ILB section (plated bump), 75°C (165°F) 1000 hrs 600x. (b) Concentration contours of Sn. [161-164]

of tin into gold bumps at higher bonding temperatures. However, no difference in tin distribution is indicated for bonding pressure.

<u>(d) Gold bump to gold-plated copper lead (Au/Au).</u> Thermocompression bonding of gold-plated copper leads to gold bumps has been studied extensively [88-91, 211-217, 294]. Most of the work deals with the effect of bond pressure, bond temperature, bond time, and stage temperature on bond strength. Very recently, Vaz, Iverson, and Lynch [294] have done a fractional factorial experiment and proposed empirical equations for the bond strength and the lead lifts.

The circuit used in the experiment [294] was a 64-pad device (180 mils square) with straight wall bumps consisting of electroplated gold on top of a sputtered Au and Ti/W diffusion barrier. Bump dimensions were 114 microns square, 27 microns high. The hardness of the gold bumps ranged from 40 to 60 on the Vickers scale. The three-layer tape used consisted of one ounce, gold plated, electro-deposited copper which is bonded to Kapton by means of an adhesive. The hardness of the copper foil ranged from 90 to 115 on the Vickers scale.

The bond pull test configuration is shown in Figure 1.15, [294]. It can be seen that the chip is held rigidly by a vacuum chuck and the inner leads are clamped in place at the Kapton ring to reduce the stress being induced on adjacent inner leads being pulled. A hook, placed under the lead, is pulled at a constant rate and the resultant force to failure is measured.

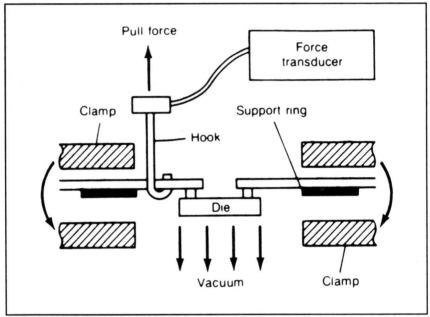

Figure 1.15 The bond pull test configuration. [294]

Figure 1.16 (at 95% confidence level) shows the bond strength as a function of temperature and force. It can be seen that compressive force had a stronger effect on bond strength than did thermode temperature.

Figure 1.17 (at 95% confidence level) shows the number of expected lead lift failures as a function of force and temperature. It can be seen that the bond temperature was a more dominant factor in determining lead failures than was the bond force. Also, when compared to the bond strength curve (Figure 1.16), the lead lift curve (Figure 1.17) was not as linear in nature. It should be noted that there is an error in Equation (2) of [294]. The correct equation was provided by Oscar Vaz and is shown in Figure 1.17. In [294], they also showed that the bond time and stage temperature are not as important as the bond force and bond pressure.

The kinetic and diffusion effects of thermocompression bonding have been studied extensively by Panousis, Hall, Morabito, Menzel, and Kershner [88-91, 128, 211-217]. They found that the rate of diffusion is strongly dependent on temperature. The effect of pressure on the rate of diffusion is not as great as that of temperature.

Thermocompression bondability windows for as-fired thick-film Au and thin-film TiPdAu have been provided by Panousis and Kershner [215]. The results, Figure 1.18, showed that it was possible to get

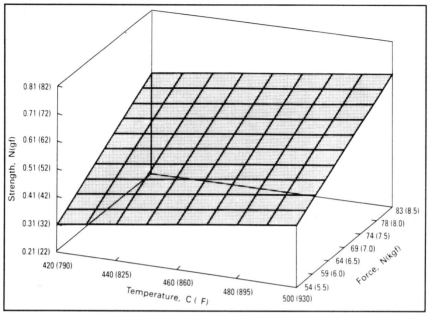

Figure 1.16 Three-dimensional plot showing the relationship of bond strength (S) to temperature (T) and force (F) where:
$$S = 0.022TF - 1.25F - 10.32 \ [294]$$

acceptable initial bondability to the thick-film gold, but that the bondability window was significantly smaller than that for thin-film gold. This large difference in bondability is due to the thick films being less dense, rougher, and having more surface contaminants.

1.6 TAB Testing and Burn-In

One of the advantages of TAB is that, after ILB, the component can be tested and burned-in prior to assembly, [37-38, 40, 61, 201, 251, 263].

1.6.1 Component testing.

The purposes of component testing are to identify assembly problems and to ensure electrical functionality. Testing of a tape-with-chip requires electrical contact between the probe and the test pads on the tape that are connected to the beam leads. Testing can be performed in reel-to-reel mode or by utilizing slide carriers that hold individual tape frames.

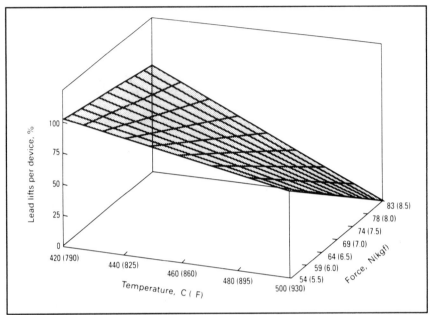

Figure 1.17 Three-dimensional plot showing the relationship of number of lead lifts (N) to temperature (T) and force (F), where: N = 1.112T + 79.2F - 0.198TF - 379.4 [294]

1.6.2 Component burn-in

The purposes of component burn-in are to offer an effective screen to produce highly reliable components and to eliminate components that might fail prematurely prior to the next level of the package system. Current TAB component burn-in technologies, oven chamber (thermal management), connector, and control and power circuitry, can only be done in a slide carrier and socket configuration.

Recently, the thermal behavior of TAB device burn-in in air was studied by Chung [40]. It was found that the thermal resistance between the chip junction and the ambient can be on the order of 10°C per watt under single sided air impingement during burn-in, which is about one third of the thermal resistance of the non-impingement condition. This can be important for high power chips.

1.7 Encapsulation

After the beam leads of a tape are fully bonded on the chip and then

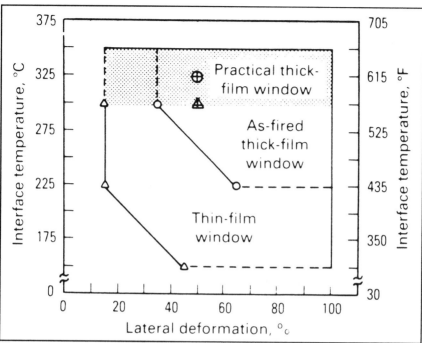

Figure 1.18 Bondability windows for thick-film Au and thin-film Ti-Pd-Au. [215]

tested and burned-in, the next operation (encapsulation) is to make a protective coating on the top and/or bottom surface of the ILB chip (Figure 1.19). Encapsulation can be accomplished by dispensing a liquid polymer onto the chip and curing the polymer to a state in which its physical and mechanical properties are compatible with the chip, bump, and lead [18, 23, 74, 77, 103, 117, 201, 224-226].

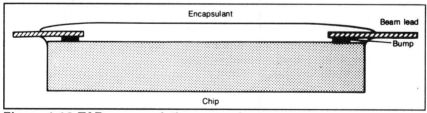

Figure 1.19 TAB encapsulation example.

1.7.1 Characteristics of encapsulation.

The purpose of an encapsulant is to protect the chip, bumps, and leads. Protection from mechanical loads, shock and vibration, and edge shorts is of utmost importance. The material of the encapsulant should not create unusual stresses and strains during functional cycling, nor should it cause corrosion. Handling and processing characteristics are also important.

Desirable physical characteristics of encapsulants are: high glass transition temperature, low coefficient of thermal expansion, void-free cure, low dielectric constant, low alpha particle emission, and high alpha particle absorption. The state of cure is the most important factor to achieve these desirable properties. To optimize the cure profile of a thermoset material, the following parameters are necessary: cure initiation temperature, temperature of cure, presence of volatile components, gel time and temperature, and properties of the fully cured polymer.

1.7.2 Epoxy encapsulants.

Epoxies are chemically polar materials which provide excellent adhesive properties and good thermal and chemical resistance. Table 1.4 [226] summarizes the physical and mechanical properties of epoxy encapsulants. It can be seen that epoxy encapsulants can develop high modulus (stiff) structures during cure. This often leads to high stress in the chip surface and cracking of the encapsulant, the bump, or the chip. Most commercial epoxy formulations are mixtures of diglycidyl ether of bisphenol A and cycloaliphatic epoxies with acid anhydride crosslinking agents. They are heavily filled (to 75%) with fumed silica and/or crushed quartz and may contain a carboxylated elastomeric polymer to provide additional stress relief.

Using microdielectric spectroscopy and thermal analysis techniques, Pennisi, Nounou, and Machuga [226] found that a cure time of 20-25 minutes at 160-170°C should be sufficient to fully cure the epoxy encapsulant (Table 1.5), although variations among suppliers do limit the value of a generic characterization. Their results, Table 1.6, also indicate that none of the physical properties were sacrificed when the cure time was reduced from the three hours recommended by the vendor to 25 minutes.

1.7.3 Silicone encapsulants.

Silicone encapsulants are polymers with very low Young's modulus (Table 1.4) and with very high coefficient of thermal expansion (more than

Table 1.4 Material properties of epoxy and silicone encapsulants.
[224-226]

Property	Material	
	Epoxy	Silicone
Glass transition temperature, °C (°F)	100-175 (212-350)	<-20 (<-4)
Coefficient of thermal expansion, 10^{-6}/K	>20	>200
Dielectric constant	3.5	3.0
Dielectric loss constant	0.003	0.001
Room temperature modulus, Pa (psi)	10^8 (14,5000)	10 (0.00145)
Ionics, ppm	<30	<5
Thermal stability, °C (°F)	250 (480)	250 (480)
Freon resistance	Good	Poor

200 ppm/°C). They are virtually ion- free and provide excellent low temperature properties. Moisture resistance and electrical properties are generally quite good. However, low mechanical strength (100 dynes/sq.cm), poor solvent resistance, and limited protection from handling damage are the major drawbacks of these materials. The silicone encapsulants cure through the addition of hydride-containing silicones to vinyl silicones. Halogenated platinum compounds are employed to catalyze this polymerization reaction.

1.8 Outer Lead Bonding

The final step of the TAB assembly process is outer lead bonding (OLB). This operation transfers the chip with leads in place to the next-level

Table 1.5 Results for epoxy encapsulant curing. [225-226]

Curing Parameters	Conditions
Cure initiation temperature °C (°F)	138 (280)
Cure temperature °C (°F)	161 (320)
Weight loss %	2.9
Gel time at 150°C (300°F), min	7

Cure time:
  DMA, min
  MDS, min
	16
	15

package or PCB. The sites where the outer leads are to be bonded to the PCB are called land patterns. OLB is a four-step process, Figure 1.20, usually done sequentially on the same piece of equipment. These four steps are: excising, lead forming, transporting, and welding or soldering, [7-9, 14, 24-25, 31, 33, 39, 41, 54-58, 99, 128, 136, 144-148, 150, 177, 198, 203, 205-206, 232-236, 238-243, 248, 253-254, 262, 266-268, 275, 287-291, 304-305].

1.8.1 Land pattern [115].

The design of land patterns in TAB is very important. Design affects the strength and toughness of the bonds and impacts the defects, cleanability, testability, and repair/rework.

As mentioned earlier, the JEDEC standards for OLB pitches are: 0.02" (0.5mm), 0.015" (0.37 mm), and 0.01" (0.25 mm). Consequently, IPC/ASTM/EIA recommended the principal land pattern dimensions for these OLB pitches. Table 1.7 is for 0.5 mm pitch, Table 1.8 for 0.37 mm pitch, and Table 1.9 is for 0.25 mm pitch. The variables in Tables 7-9 are defined in Figure 1.21. Note that the tables refer to "Package Body Size", which is equivalent to the inner dimensions of the OLB window or, conversely, the outer dimension of the polyimide support ring.

1.8.2 Excising.

The first step in the OLB process is to excise (punch out) the chip with

Table 1.6 Physical properties of cured epoxy encapsulants. [225-226]

Property	3 hour cure	25 min cure
Glass transition temp, °C (°F)		
DMA	174(345)	174(345)
DSC	172(342)	170(338)
TMA	168(334)	165(325)
Coeff of thermal expansion, 10^{-6}/K	29	31
Thermal stability, °C (°F)	300(570)	300(570)
G' at 50°C (120°F), GPa (10^6psi)	3(0.44)	3(0.44)
G' at 200°C (390°F), GPa (10^6psi)	.1(.015)	.1(.015)
E' at 50°C (120°F), 1Hz	3.7	3.5
E" at 50°C (120°F), 1Hz	0.029	0.059
Moisture absorption, %	0.44	0.48

leads in place from the tape carrier. This is shown schematically in Figure 1.20(a).

1.8.3 Lead forming.

The second step in the OLB process is to form (bend) the beam leads, Figure 1.20(b). The purpose of the lead forming operation is three fold [238-242]: (1) bring the area of the lead designated to accept the OLB from a plane at the topside of the chip into a plane at the level of the topside of the substrate, and (2) provide a flat area sufficiently large to accept the thermode (bonding tool), and (3) increase compliance.

There are at least two basic ways in which lead forming can be accomplished [238-242], "fan-out on level" and "fan-out on slope". In the first approach the beam lead is designed such that fan-out from bump separation to bond pad separation occurs on a level plane through the top surface of the chip, Figure 1.22. At the point that each lead is lined up with its respective bonding pad, a steep step downward followed by a sharp bend to the horizontal plane completes the form trajectory. This

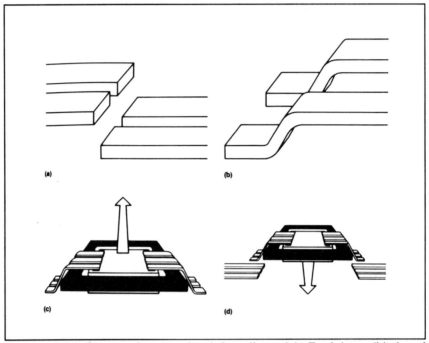

Figure 1.20 Steps of outer lead bonding. (a) Excising. (b) Lead forming. (c) Transporting. (d) Welding or soldering.

forming requires simple dual action tools, but large real estate on the substrate.

The second approach uses a more complex beam lead design where the fan-out from bump separation distance to pad separation distance takes place on the descent slope, Figure 1.23. Thus, the leads move both outward and downward and take on complex shapes. The advantage of this design is that the outer lead bonds can be made close to the chip, saving real estate on the substrate. The drawbacks include the twisting, lifting, and misalignment of the leads.

The design of lead forming is unlimited. However, the fundamental principle is to bend the beam leads into curved shapes so that they can withstand mechanical and thermal stresses and strains.

1.8.4 Transporting and placement.

After the beam leads have been formed into curved shapes to increase compliance, the component is then transferred to the PCB in a way that precisely maintains the original relationship of the chip with respect to its

Table 1.7 Land patterns for 0.5mm (0.020in.) pitch TAB. Land width X=0.33mm (0.013in.) (a) Metric. [115]

Package Body Size mm	Tape Format mm	A/A1 mm	B/B1 mm	C/C1 mm	D/D1 mm	Number of Lands
10.41	S35	10.82	15.14	12.98	10.16	84
15.49	S35	15.90	20.22	18.06	12.19	100
20.57	S35	20.98	25.30	23.14	17.27	140
20.57x 41.15	S35	20.98x 41.55	25.30x 45.88	23.01x 43.72	17.27x 37.59	220
26.65	S48	26.06	30.38	28.22	22.35	180
33.27	S48	33.68	38.00	35.84	30.48	244
38.35	S70	38.76	43.08	40.92	35.56	284
43.43	S70	43.84	48.16	46.00	40.64	324
48.51	S70	48.92	53.23	51.08	45.72	364
53.39	S70	54.00	58.32	56.16	50.80	404

tape sprocket holes or to a vision system that positions leads to lands. Thus the component will be placed on the substrate in a position that is accurately known (Figure 1.20(c)).

1.8.5 Welding or soldering.

Welding or soldering is the most important step in OLB assembly processes (Figure 1.20(d)). This bonding process attaches the outer leads to the lands of the PCB by methods including single-point bonding, gang (mass) bonding, vapor phase or infrared reflow, conductive adhesive techniques, etc.

(a) Single-point bonding. As with ILB, OLB can be done by ultrasonic

Table 1.7 Land patterns for 0.5mm (0.020in.) pitch TAB. Land width X=0.33mm (0.013in.) (b) British. [115]

Package Body Size in.	Tape Format mm	A/A1 in.	B/B1 in.	C/C1 in.	D/D1 in.	Number of Lands
0.410	S35	0.426	0.596	0.511	0.400	84
0.610	S35	0.626	0.796	0.711	0.480	100
0.810	S35	0.826	0.996	0.911	0.680	140
0.810x 1.620	S35	0.826x 1.636	0.996x 1.806	0.911x 1.721	0.680x 1.480	220
1.010	S48	1.026	1.196	1.111	0.880	180
1.310	S48	1.326	1.496	1.411	1.200	244
1.510	S70	1.526	1.696	1.611	1.400	284
1.710	S70	1.726	1.896	1.811	1.600	324
1.910	S70	1.926	2.096	2.011	1.800	364
2.110	S70	2.126	2.296	2.211	2.000	404

bonding, thermosonic bonding, laser bonding, etc. Ultrasonic bonding is done at room temperature by vibrating the beam leads against the bond pad of the PCB. Thermosonic bonding can be four to five times faster than ultrasonic bonding due mainly to the addition of heat during the bonding process. Because the thermode and the substrate are heated at 150 to 200°C, and since the glass transition temperature of epoxy FR-4 falls around 125°C [154], this prohibits the possible use of thermosonic bonding with FR-4.

Laser bonding uses laser energy to melt the solder for a plated PCB, or to weld the lead to the bond pad for a bare PCB. However, holding the lead down to the pad during laser bonding is a difficult task. It is not uncommon to observe that the copper lead melts before the bond is formed.

Table 1.8 Land patterns for 0.38mm (0.015in.) pitch TAB. Land width X=0.23mm (0.009in.) (a) Metric. [115]

Package Body Size mm	Tape Format mm	A/A1 mm	B/B1 mm	C/C1 mm	D/D1 mm	Number of Lands
10.41	S35	10.82	15.14	12.98	7.62	84
15.49	S35	15.90	20.22	18.06	12.19	132
20.57	S35	20.98	25.30	23.14	17.53	188
20.57x 41.15	S35	20.98x 41.55	25.30x 45.88	23.01x 43.72	17.53x 38.10	296
26.65	S48	26.06	30.38	28.22	22.86	244
33.27	S48	33.68	38.00	35.84	27.94	324
38.35	S70	38.76	43.08	40.92	35.05	372
43.43	S70	43.84	48.16	46.00	40.39	428
48.51	S70	48.92	53.23	51.08	45.72	484
53.39	S70	54.00	58.32	56.16	50.29	532

(b) Gang (mass) bonding. Unlike ILB, thermocompression bonding is seldom used for OLB with conventional PCB laminates. This is because of the high temperature and pressure requirements for thermocompression bonding, which will deform the PCB laminate locally at the bonding site.

The most widely used gang bonding method for OLB is the pulse-thermode reflow soldering method. The temperature and pressure required by this method are much lower than those in the thermocompression method.

The principle of this method is to heat (by conduction) the solder at the bond site until it melts, and to stabilize (by pressure) the bond until it re-solidifies.

The procedure of this method is to slightly press (1.4-1.7 Mpa)

Table 1.8 Land patterns for 0.38mm (0.015in.) pitch TAB. Land width X=0.23mm (0.009in.) (b) British. [115]

Package Body Size in.	Tape Format mm	A/A1 in.	B/B1 in.	C/C1 in.	D/D1 in.	Number of Lands
0.410	S35	0.426	0.596	0.511	0.300	84
0.610	S35	0.626	0.796	0.711	0.480	132
0.810	S35	0.826	0.996	0.911	0.690	188
0.810x 1.620	S35	0.826x 1.636	0.996x 1.806	0.911x 1.721	0.690x 1.500	296
1.010	S48	1.026	1.196	1.111	0.900	244
1.310	S48	1.326	1.496	1.411	1.100	324
1.510	S70	1.526	1.696	1.611	1.380	372
1.710	S70	1.726	1.896	1.811	1.590	428
1.910	S70	1.926	2.096	2.011	1.800	484
2.110	S70	2.126	2.296	2.211	1.980	532

the pre-heated (150°C) bonding thermode to the beam lead. Then the thermode receives a short current pulse to a resistive heater which quickly increases the thermode temperature to 275-325°C. Before the solder reaches this temperature, the pulse is removed, the solder has reflowed and the thermode begins to cool back down to 150°C and is lifted from the component.

The outer lead pulse-thermode reflow soldering sequence is shown in Figure 1.24 [125-127]. It can be seen that precision feed mechanisms advance both tape and lead frame (A). During the punch and bond sequence (B) an accurately fitted punch with a spring-loaded pressure pad cuts the device and its leads from the tape and carries it up through the die plate until it contacts the lead frame. The thermode simultaneously drops and applies heat and pressure to form the bond

Table 1.9 Land patterns for 0.25mm (0.010in.) pitch TAB. Land width X=0.17mm (0.007in.) (a) Metric. [115]

Package Body Size mm	Tape Format mm	A/A1 mm	B/B1 mm	C/C1 mm	D/D1 mm	Number of Lands
10.41	S35	10.82	15.14	12.98	7.62	124
15.49	S35	15.90	20.22	18.06	12.70	204
20.57	S35	20.98	25.30	23.14	17.78	284
20.57x 41.15	S35	20.98x 41.55	25.30x 45.88	23.01x 43.72	17.78x 38.10	444
26.65	S48	26.06	30.38	28.22	22.86	364
33.27	S48	33.68	38.00	35.84	30.48	484
8.35	S70	38.76	43.08	40.92	35.56	564
43.43	S70	43.84	48.16	46.00	40.64	644
48.51	S70	48.92	53.23	51.08	45.72	724
53.39	S70	54.00	58.32	56.16	50.80	804

through the lead frame (using the punch as the bond anvil). At (C) the punch, pressure pad and thermode retract for the part index.

(c) Vapor phase/infrared reflow soldering. Traditional SMT assembly processes, such as vapor phase or infrared reflow soldering, have been applied to OLB. More than 1 million 170-lead TABs have been assembled into Hewlett-Packard's calculators [7, 304-306] by these SMT-compatible processes. The defect rates have been slightly lower than the previous best yielding conventional SMT assembly process. The advantage of this process is that it is simple and low cost, and no new equipment for OLB is necessary. However, it may be very difficult when the pitch of the outer leads is smaller than 0.015" (0.37 mm).

Table 1.9 Land patterns for 0.25mm (0.010in.) pitch TAB. Land width X=0.17mm (0.007in.) (b) British. [115]

Package Body Size in.	Tape Format mm	A/A1 in.	B/B1 in.	C/C1 in.	D/D1 in.	Number of Lands
0.410	S35	0.426	0.596	0.511	0.300	124
0.610	S35	0.626	0.796	0.711	0.500	204
0.810	S35	0.826	0.996	0.911	0.700	284
0.810x 1.620	S35	0.826x 1.636	0.996x 1.806	0.911x 1.721	0.700x 1.500	444
1.010	S48	1.026	1.196	1.111	0.900	364
1.310	S48	1.326	1.496	1.411	1.200	484
1.510	S70	1.526	1.696	1.611	1.400	564
1.710	S70	1.726	1.896	1.811	1.600	644
1.910	S70	1.926	2.096	2.011	1.800	724
2.110	S70	2.126	2.296	2.211	2.000	804

(d) Special considerations of gang bonding. Instead of solder alloy, chemical materials such as conductive epoxy can also be used for OLB joints. One of the drawbacks of this class of materials is its especially low strength and toughness.

As with inner lead gang bonding, outer lead gang bonding can give high throughput but is subjected to planarity, lead-height, and alignment tolerance problems. Consistent bond quality remains a challenge for gang bonding.

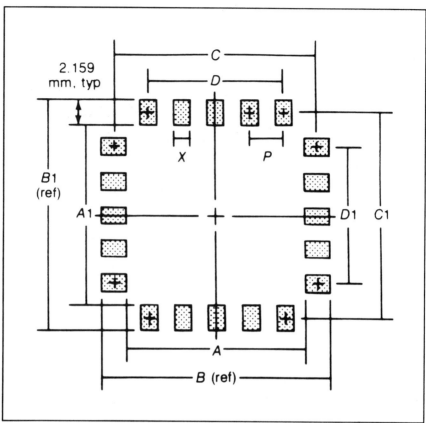

Figure 1.21 Land pattern principal dimensions. [115]

1.9 TAB Thermal Management

With the trends towards high power density at ambient temperature and wide electrical bandwidth (required by VLSI) comes the need for cost effective thermal management at device level packaging, e.g. TAB. Consequently, fully understanding and adequately defining and determining the thermal requirements of the TAB package are essential, [21, 26-27, 37-38, 40, 84, 125-127, 137, 179, 201, 295, 301-302].

1.9.1 Thermoelasticity for electronics packaging.

Let us consider a linear elastic TAB package (with no heat source in the

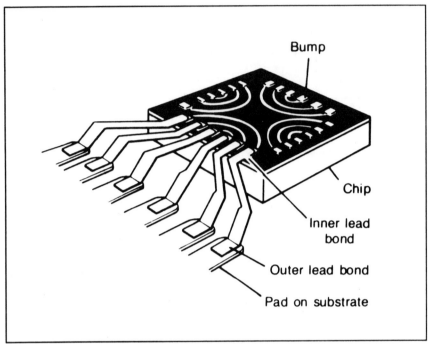

Figure 1.22 Fan-out on level. [238-242]

material) subjected to heating and external forces. We assume that the package is stress-free at a uniform temperature T_0 when all external forces are removed. The stress-free state will be referred to as the reference state, and the temperature T_0 as the reference temperature. Furthermore, the displacement and velocity of every particle of the package in the instantaneous state from its position in the reference state will be assumed to be small (i.e. infinitesimal strains and the material derivative is the same as the partial derivative with respect to time). Then the fundamental equations of thermoelasticity for the package are:

Infinitesimal strain-displacement relations

$$\epsilon_{ij} = \frac{1}{2}\left(\frac{\partial u_i}{\partial x_j} + \frac{\partial u_j}{\partial x_i}\right)$$

[1.1]

Constitutive equation (Duhamel-Neumann law)

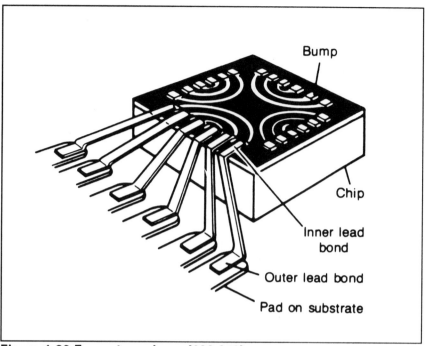

Figure 1.23 Fan-out on slope. [238-242]

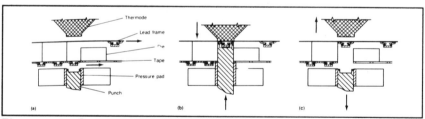

Figure 1.24 Outer lead bonding sequence. (a) Feed. (b) Punch and bond. (c) Retract. [125-127]

$$\sigma_{ij} = C_{ijkl}e_{kl} - \beta_{ij}(T-T_0)$$

[1.2]

Conservation of Mass (Continuity equation)

$$\frac{\partial \rho}{\partial t} + \frac{\partial \rho v_i}{\partial x_i} = 0$$

[1.3]

Conservation of Momentum (Newton's law)

$$\rho \frac{\partial v_i}{\partial t} = \frac{\partial \sigma_{ij}}{\partial x_j} + X_i$$

[1.4]

Conservation of Energy

$$\frac{\partial \varepsilon}{\partial t} = T \frac{\partial \varphi}{\partial t} + \frac{1}{2\rho} \sigma_{ij} \left(\frac{\partial v_i}{\partial x_j} + \frac{\partial v_j}{\partial x_i} \right)$$

[1.5]

Rate of Change of Entropy (if no heat source in the material)

$$\rho \frac{\partial \varphi}{\partial t} = -\frac{1}{T} \frac{\partial h_i}{\partial x_i} = -\frac{\partial}{\partial x_i} \left(\frac{h_i}{T} \right) - \frac{h_i}{T^2} \frac{\partial T}{\partial x_i}$$

[1.6]

Heat Conduction (Fourier's law)

$$h_i = -k_{ij} \frac{\partial T}{\partial x_j}$$

[1.7]

Definition of Specific Heat (if $\frac{\partial e_{ij}}{\partial t} = 0$)

$$-\frac{\partial h_i}{\partial x_i} = \rho C_v \frac{\partial T}{\partial t}$$

[1.8]

where

T is the instantaneous absolute temperature,
x_i are Cartesian coordinates,
u_i are components of the displacement vector,
C_v is the heat capacity per unit mass,
T_0 is the reference temperature,
X_i are body force components per unit volume,
β_{ij} are thermal moduli,
σ_{ij} are components of the stress tensor,
e_{ij} are components of the strain tensor,
k_{ij} are heat conduction coefficients,
C_{ijkl} are elastic moduli,
ρ is the mass density,
ε is the internal energy per unit mass,
φ is the entropy per unit mass,
h_i are components of the heat flux vector, and
v_i are components of the velocity vector.

In the present chapter, all the indices range over 1,2,3 and the summation convention for repeated indices is used.

By combining Equations 1.1-1.8, we have the following Coupled Equations of Thermoelasticity for the package:

$$\frac{\partial}{\partial x_i}\left(k_{ij}\frac{\partial T}{\partial x_j}\right) = \rho C_v \frac{\partial T}{\partial t} + \frac{1}{2}T\beta_{ij}\frac{\partial}{\partial t}\left(\frac{\partial u_i}{\partial x_j}+\frac{\partial u_j}{\partial x_i}\right)$$

[1.9]

$$\frac{1}{2}\frac{\partial}{\partial x_j}\left[C_{ijkl}\left(\frac{\partial u_k}{\partial x_l}+\frac{\partial u_l}{\partial x_k}\right)\right] = -X_i + \rho\frac{\partial^2 u_i}{\partial t^2} + \frac{\partial}{\partial x_j}(\beta_{ij}T)$$

[1.10]

It can be seen that the heat conduction equation, Equation 1.9 contains,

besides the temperature T, the rate of strains $(\frac{\partial}{\partial t}(\frac{\partial u_i}{\partial x_j}+\frac{\partial u_j}{\partial x_i}))$, whereas

the equations of motion, Equation 1.10 contain, besides the displacement u_i, the temperature increase T.

In principle, every non-stationary problem of thermoelasticity is a dynamic problem. The inertial forces must be taken into account if the temperature field undergoes a sudden change with time, e.g. in cases of sudden cooling or heating of a structure. For small variations of temperature with time, however, the inertial terms may be neglected in the equations of motion, Equation 1.10, i.e.

$$\rho \frac{\partial^2 u_i}{\partial t^2} = 0$$

[1.11]

then we have the Coupled, Quasi-Static Thermoelasticity problem

$$\frac{\partial}{\partial x_i}\left(k_{ij}\frac{\partial T}{\partial x_j}\right) = \rho\, C_v \frac{\partial T}{\partial t} + \frac{1}{2}\, T\beta_{ij}\frac{\partial}{\partial t}\left(\frac{\partial u_i}{\partial x_j} + \frac{\partial u_j}{\partial x_i}\right)$$

[1.12]

$$\frac{1}{2}\frac{\partial}{\partial x_j}\left[C_{ijkl}\left(\frac{\partial u_k}{\partial x_l} + \frac{\partial u_l}{\partial x_k}\right)\right] = -X_i + \frac{\partial}{\partial x_j}(\beta_{ij}\,T)$$

[1.13]

Boundary-value problems involving Equations 1.12 and 1.13 are rather complex and difficult to solve. Fortunately, in most engineering applications it is possible to neglect the mechanical coupling term in Equation 1.12 without significant error, i.e.

$$\frac{1}{2}\, T\beta_{ij}\frac{\partial}{\partial t}\left(\frac{\partial u_i}{\partial x_j} + \frac{\partial u_j}{\partial x_i}\right) = 0$$

[1.14]

Consequently, Equations 1.12 and 1.13 become

$$\frac{\partial}{\partial x_i}\left(k_{ij}\frac{\partial T}{\partial x_j}\right) = \rho\, C_v \frac{\partial T}{\partial t}$$

[1.15]

$$\frac{1}{2}\frac{\partial}{\partial x_j}\left[C_{ijkl}\left(\frac{\partial u_k}{\partial x_l} + \frac{\partial u_l}{\partial x_k}\right)\right] = -X_i + \frac{\partial}{\partial x_j}(\beta_{ij}\,T)$$

[1.16]

Equations 1.15 and 1.16 are called the equations of the Uncoupled, Quasi-Static Thermoelasticity or the Theory of Thermal Stresses for electronics packaging.

The physical meaning of Equation 1.14 is that the interaction between strain and temperature is ignored and the effects of elasticity (change in dimensions of the package) on the temperature distribution are negligible. For example, the change in dimension of a package is of the order of the product of the linear dimension of the package L, the temperature rise T, and the coefficient of thermal expansion α. If L=1in, $T-T_0=100°C$, $\alpha=10^{-6}/°C$, the change in dimension is 10^{-4}in., which is negligible for electronics packaging problems of heat conduction.

A plausible example of Equation 1.11 is as follows. If the temperature rise from 0 to 100°C were achieved in a time interval of 0.1 second, then the acceleration is of the order $10^{-4}/0.1^2=0.01$in/sec^2. The change of stress due to this acceleration may be estimated from the equation of equilibrium, $\Delta\sigma_{xx} \propto \Delta x\rho\frac{d^2u}{dt^2}$. If the specific gravity of the material is ten and the material is 1 inch thick, we have $\Delta\sigma_{xx}=(1/12)(10)(62.4)(1/32.2)(0.01/12)(1/144)\approx10^{-5}$lb/sq in. This stress is negligible in most electronics packaging problems in which the magnitude of the stresses concerned are of the order of the strength (yielding stress or ultimate stress) and fracture toughness of the packaging material.

Fundamentally speaking, most electronics packaging materials are anisotropic. For practical applications, however, it is possible to gain some insights of the electronics packaging problems by assuming the materials to be isotropic. In that case, the governing equations of the Theory of Isotropic Thermal Stresses for electronics packaging are

$$k\delta_{ij}\frac{\partial}{\partial x_i}\left(\frac{\partial T}{\partial x_j}\right) = \rho C_v\frac{\partial T}{\partial t}$$

[1.17]

$$Gu_{i,\mu\mu}+(\lambda+G)u_{\mu,\mu i} = -X_i+\beta\frac{\partial T}{\partial x_i}$$

[1.18]

or

$$\nabla^2 T = \frac{\rho C_v}{k}\frac{\partial T}{\partial t}$$

[1.19]

$$G\nabla^2 U+(\lambda+G)\nabla(\nabla\cdot U)+X = \beta\nabla(T)$$

[1.20]

where

$$\sigma_{ij} = \lambda e_{\mu\mu}\delta_{ij} + 2Ge_{ij} - \beta\delta_{ij}(T-T_0)$$

[1.21]

$$\beta = \frac{\alpha E}{1-2v}$$

[1.22]

$$\lambda = \frac{Ev}{(1+v)(1-2v)}$$

[1.23]

$$G = \frac{E}{2(1+v)}$$

[1.24]

U = displacement vector,
x = body force per unit volume,
α = thermal coefficient of linear expansion,
k = heat conductivity,
E = Young's modulus,
v = Poisson's ratio,
λ = Lame's constant,
G = shear modulus,
T_0 = reference temperature,
∇ = del operator,
∇^2 = Laplace operator,
σ_{ij} = components of the stress tensor,
e_{ij} = components of strain tensor, and
δ_{ij} = Kronecker delta.

For the special case of steady heat flow, i.e. $\frac{\partial T}{\partial t} = 0$, we have

$$\nabla^2 T = 0$$

[1.25]

$$G\nabla^2 U + (\lambda + G)\nabla(\nabla\cdot U) + x = \beta\nabla(T)$$

[1.26]

These are the equations for Isotropic Thermal Stresses with Steady Heat Flow.

Heat transfer and thermal stress in electronics packaging are usually applied in two stages, transient (power on/off) and steady state (during operation). In both cases, the temperature distribution $T(x_i, t)$ in the package is calculated by solving the heat conduction equation (Equation 1.19) with the prescribed initial and boundary conditions. The stress and strain distributions everywhere inside the package are then determined by Equations 1.20 with the prescribed stress/displacement boundary conditions and with the calculated temperature distribution as an imposed boundary condition. This temperature distribution is mathematically shown at the right-hand side of Equation 1.20 and is a known function.

1.9.2 Thermal management analysis.

Because of the geometry and material construction of most electronic packages, exact solutions of Equations 1.19 and 1.20 are very difficult to obtain. Therefore, numerical schemes such as the finite element methods are usually adopted. The basic concept of the finite element method is that a structure can be decomposed into a finite number of elements. For each element, trial function approximations of temperature/ displacement are used in conjunction with the variational principle or Galerkin's approximation and matrix methods to transform the boundary-value problem, Equations 1.19 and 1.20 and the initial/boundary conditions, into a system of simultaneous algebraic equations. These equations can then be easily solved by a computer for the temperature/ displacement/strain/stress distributions within the electronic packages. The thermal stresses in a TAB assembly will be determined by the finite method in the next section.

A very simple 1-D heat dissipation model for a TAB package is given by Mahalingam and Andrews [179], Figure 1.25. It can be seen that the dissipation of heat from the TAB package is not only through the back of the chip but also through the interconnections (leads) of the package. Unlike wire bonding technology, more than 80% of the heat can be dissipated through the leads of the TAB package [295]. Thus, thermal resistance through the leads has to be considered.

1.9.3 Thermal management design.

An example of TAB thermal management design for achieving low thermal resistance is given by Winkler [301-302], Figure 1.26. The design utilizes an attached aluminum heat sink plate on the back of the board,

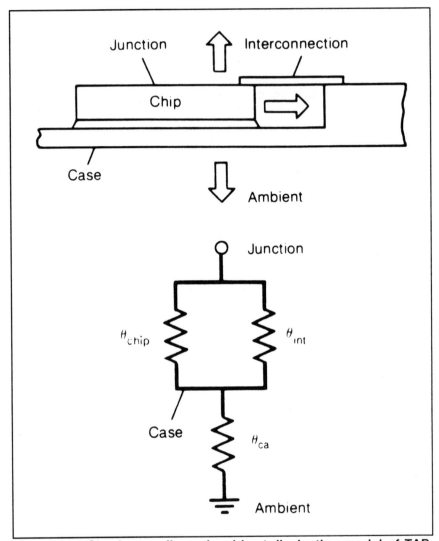

Figure 1.25 Simple one-dimensional heat-dissipation model of TAB, where θ = thermal resistance. [179]

with epoxy bonding, a PCB with a copper heat transfer feedthrough under the chip bond and conductive epoxy for the chip bond and the bond between heat sink and feedthrough. Heat sink cooling requires forced convective cooling fluid.

Design dimensions, and thermal conductivities are included for the thermal resistance calculations in Table 1.10. These approximate results, using one dimensional heat transfer models, are applied to each layer in turn as indicated in the table [302]. The calculation assumes sufficient cooling fluid flow to dissipate the heat from the highly conducting heat sink plate. If required, the heat sink plate may be designed to have fins. Figure 1.27 schematically shows a TAB system on a PCB with 18 million devices and 3.6 million gates (assuming 5 devices/gate on average).

As another example, Figure 1.28(a) shows the cross-section arrangement of an indirect liquid cooling system (without heat sinks) for a multichip package with TAB [135]. It can be seen that very fine coolant channels having a cross-section of less than 1mm (0.0015in.) were formed into the lower portion and between the buried via holes of a multi-layered alumina substrate. The coolant is supplied from the inlet port and distributed to each channel by the coolant distributor, thus ensuring a uniform velocity profile. The heat generated in the chips is conducted to the coolant channels by way of the alumina substrate, where it is transferred to the flowing liquid. The warmed coolant is subsequently collected by the collector after flowing through the coolant channels. It then flows out through the outlet port. The high-density stack-packaged system is shown in Figure 1.28(b). The upper and lower stack packages are connected in series by piping. One concern with this method of cooling is the potential of fouling of the cooling channels.

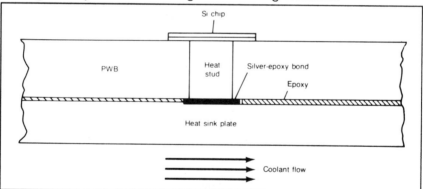

Figure 1.26 Schematic drawing of the TAB system and thermal management design. [301-302]

1.10 TAB Reliability

Reliability of TAB assembly, R(t), can be defined as the probability that a TAB product will perform its intended function for a specified period of

Table 1.10 Thermal resistance estimates for advanced thermal management design. [301-302]

Layer→	Silicon	Ag-epoxy	Copper feedthrough	Ag-epoxy

Thickness, t
  mm
  in
	380
	15
		100
		4
			1525 60	100
				4

Thermal conductivity K
  W/m.K
  W/in.°C
	83.9
	2.13
		2.03
		0.0515
			354
			9.0
				2.03
				0.0515

Size, L=W
  mm
  in
	9.1
	0.36
		9.1
		0.36
			6.9
			0.27
				6.9
				0.27

Thermal resistance
  °C/W
  °F/W
	0.05*
	0.09*
		1.06*
		1.91*
			0.09*
			0.16*
				1.03**
				1.85**

* Calculated as t/KL^2
** Calculated as $t/K(L+2t)$

time, under a given operation condition, without failure, i.e. R(t)=1-P(f). P(f) is the probability of failure over the life of the TAB product which can be determined by first identifying the failure modes and then calculating the failure rate by accelerated testing and the stresses/strains to cause failure, [15, 21, 26-27, 30-32, 39, 46-49, 64, 67, 73, 75, 81, 88-92, 100, 107, 120, 129, 133, 158, 201, 203, 210-217, 244, 251, 253, 255-256, 276, 300, 308, 311-312].

1.10.1 Modes of Mechanical Failure [46-49].

Many mechanical failure modes are possible with TAB assemblies. Seven different failure modes have been observed during various temperature cycle experiments and are illustrated in Figure 1.29.

  * Silicon Pullout: This results from excessive pressures applied during ILB, or improper photomask to wafer alignment during bump plating. It is recommended that the bond pressure not exceed 210 MPa.

  * Chip-to-Bump Separation: This can occur at any of the

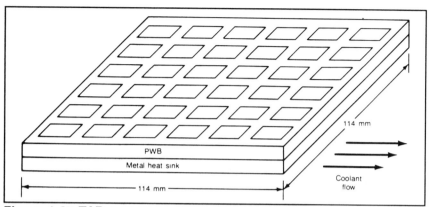

Figure 1.27 TAB system on a board with 18 million devices and 3.6 million gates. [301-302]

interfaces.   Care must be taken during chip processing to eliminate contaminants and other causes of poor adhesion.

* Beam-to-Bump Separation: Control of beam-to-bump separation requires clean Au/Au or Au/Sn interfaces and the use of suitable ILB parameters such as those previously noted.

* Beam-to-Substrate Separation: This can occur at the beam-to-film interface, at an intrafilm (Au-Pd or Pd-Ti) interface, or at the film-to-ceramic interface.   Control of the intra-film and film-to-ceramic separation requires care to eliminate contaminants and other causes of poor adhesion.   Controlling beam-to-film separation requires clean Au-Au or Au-Sn interfaces.

* Inner and Outer Bond Heel Breaks and Beam Breaks: These are the normal wearout modes in temperature cycling.   Care must be taken to insure that heel breaks do not occur prematurely, the known causes of which are poorly dressed bonding thermodes and excessive deformation.   On the other hand, inadequate deformation increases vulnerability to bond interface separation.

1.10.2 Thermal Stresses in TAB Assemblies

The stresses and strains in TAB assemblies have been studied by Lau, Rice, and Harkins [153] using a nonlinear three-dimensional finite element method.   Figure 1.30 shows half of a typical TAB assembly section.   It consists of six major parts: the silicon chip, the gold bump, the polyimide ring, the copper beam lead, the Sn/Pb solder joint, and the FR-4 printed circuit board.   The physical and mechanical properties of

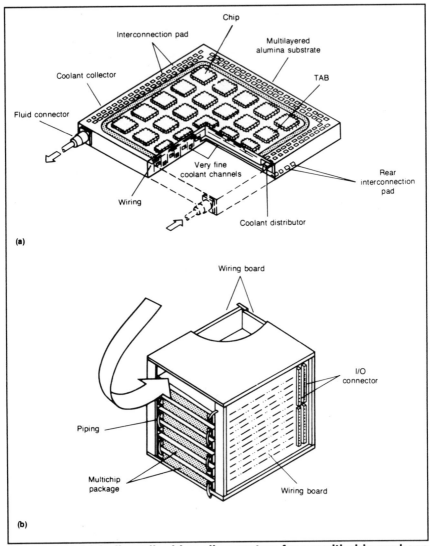

Figure 1.28 (a) Indirect liquid cooling system for a multi-chip package with TAB. (b) High-density stack-packaged system. [135]

the TAB assembly are shown in Table 1.11. The Sn/Pb solder and the copper beam leads are treated as elasto-plastic materials with strain-hardening. The boundary-value problem was to calculate the thermal stresses and strains generated in this assembly when it was subjected to a temperature cycle between -55°C and +125°C.

Figure 1.31 shows the Mises stress contours in the copper beam

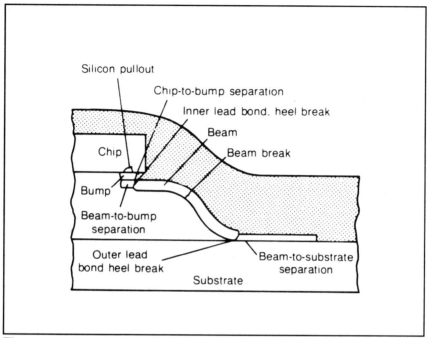

Figure 1.29 Possible modes of mechanical failure of TAB assemblies. [46]

lead (Point-A on Figure 1.30). It can be seen that the maximum stress (9,092 psi or 62.6 MPa) occurs near the intersection between the solder joint and the beam lead. Consequently, any fatigue cracking of the lead should initiate there and propagate through the thickness of the lead from its extreme fibers.

The Mises stress contours in the portion of the copper beam lead directly over the polyimide ring (Point-B on Figure 1.30) are shown in Figure 1.32. It can be seen that the maximum stress (7,686 psi or 53 MPa) occurs near the intersection between the polyimide ring and the beam lead and is large enough to cause failure of the lead.

Figure 1.33 shows a 48-lead TAB device mounted on a ceramic PCB subjected to a temperature cycling of (-55°C,+125°C) [158]. After 750 temperature cycles, cracks were found at A (beam lead near the solder joint) and at B (beam lead near the polyimide ring), Figure 1.34. These observed failure locations agreed very well with the finite element prediction. It should be noted that, unlike conventional SMT assemblies, the beam leads in TAB assemblies are more likely to fail than the solder joints.

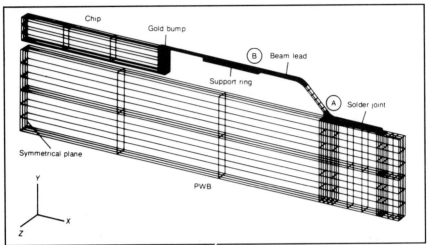

Figure 1.30 Finite-element model for a TAB assembly. [153]

1.11 TAB Rework

The use of statistical process control (SPC) and total quality control (TQC), coupled with the ability to test and burn in prior to TAB assembly, has minimized the need for rework of less dense assemblies. However, as the number of chip I/Os increases, the probability of a high yield is reduced for fine-pitch TAB. Consequently, TAB components must be removed, replaced, repaired and realigned. In the following two paragraphs, the rework of solder-reflow TAB will be briefly mentioned.

1.11.1 Component removal and replacement.

TAB components with broken, cracked or severely bent leads should be removed and discarded. They should be removed from the board using a stream of hot air that is directed (through nozzles) onto the component leads only. The hot air should be time and temperature controlled to minimize damage to the sensitive lands and adjacent parts. Because the hot air is a potential source of electrostatic discharge damage, precautions such as grounded work stations and ionized air sources should be used. A rework station with a vacuum nozzle that gently raises the removed component from the PCB is desirable. In addition, for more reliable solder joints, the solder should be removed from the PCB before a component is re-soldered onto the PCB.

Table 1.11 Material properties of TAB assembly. [153]

Material	Young's Modulus		Poisson's Ratio	TCE* 10^{-6}/K
	GPa	10^6psi		
Solder	10	1.5	0.40	21
Copper	120	17.5	0.35	17
Silicon	130	19.1	0.30	2.9
Polyimide	4	0.6	0.30	45
Gold	78	11.3	0.30	15
FR-4	11	1.6	0.28	15

* Thermal coefficient of linear expansion

1.11.2 Component repair and realignment.

TAB components with minor lead bends, planarity and misalignment problems can be repaired by heating the leads with forced hot air, a hot bar or a fine-tip soldering iron, and repositioning the lead with the aid of a sharpened dental pick and a microscope.

1.11.3 Special considerations of rework.

It should be emphasized that rework is one of the most important processes of TAB and needs to be performed properly. If not, rework could be the most thermally severe process that the TAB component and PCB will encounter.

Machine vision and computers will play important roles in the design of rework systems. A microprocessor can monitor and adjust the rework variables (time, temperature, gas, flow, pressure, and so on). Machine vision can ensure fast and accurate placement of the TAB components.

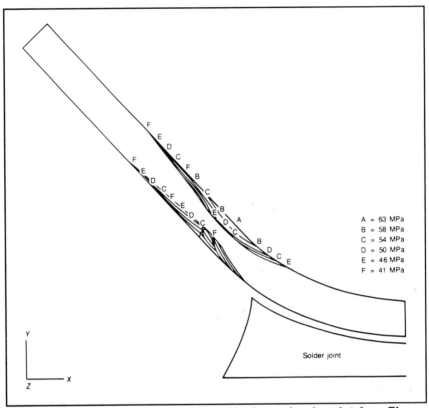

Figure 1.31 Mises stress contours in the beam lead: point A on Figure 1.30. [153]

1.12 Summary

A review of state-of-the-art technology has been presented for fine pitch, high I/O, high performance, high yield, high volume, and high reliability TAB assemblies. An understanding of the key elements of this rapidly moving technology has also been provided.

From a technical point of view, TAB is an advanced and difficult technology. In order for TAB to succeed, further understanding of the economic, material, process, equipment, quality and reliability issues of tape, bump, ILB, testing and burn-in, encapsulation, OLB, PCB, thermal management, and rework is a must.

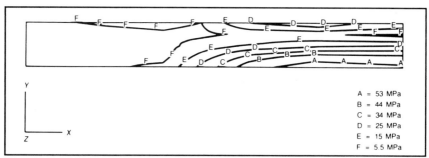

A = 53 MPa
B = 44 MPa
C = 34 MPa
D = 25 MPa
E = 15 MPa
F = 5.5 MPa

Figure 1.32 Mises stress contours in the beam lead: point B on Figure 1.30. [153]

1.13 Acknowledgements

The authors would like to thank Larry Moresco, Marcos Karnezos, Clinton Chao, Glen Leinbach, Jim Baker, Hans Obermaier, Albert Jeans, Tom Chung, Brett Sharenow, and Robert Howard for their constructive suggestions and useful comments. They also would like to thank Girvin Harkins for his support during the early phase of the present work, and Teresa T. Lau and Judy M. Lau for their effective help in collecting the references.

1.14 Recommendations

TAB is not yet a mature packaging technology, so the following recommendations for research and development are provided:

(1) BUMPING: Fundamental analysis of the bump interconnect is needed, and new materials and processes for reliable bumps should be investigated.

(2) TAPE: Advanced tapes with very fine pitch and high lead count are needed, often in the form of multiconductor tapes with controlled impedance signal and ground planes. New materials and new processes for their manufacture are required to increase yields and lower costs. The S-N-P curves for coppers are needed for optimal design of beam leads.

(3) BONDING: Optimal process windows need to be defined for both ILB and OLB which provide acceptable bond strength and fracture toughness, whether these be single-point bonding processes or gang-bonding processes. For ILB, the need for minimum damage to chip pad barrier metallization requires development of multi-parameter controlled

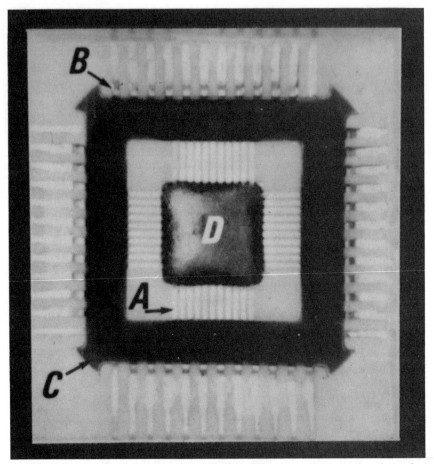

Figure 1.33 Thermal cycling of a TAB assembly. (A: a 48-lead device mounted on a ceramic PCB. B: outer lead. C: Kapton ring. D: epoxy coated chip.) [158]

equipment (pressure, temperature, planarity, pattern recognition, stability) to do the job. For the OLB, similar precision equipment needs development for traditional solder reflow, but new materials (conductive polymer, amalgams) offer research opportunities.

(4) ENCAPSULANTS: New materials providing a higher glass transition temperature, lower coefficients of thermal expansion, void-free curing, and low dielectric constant should be synthesized and developed for TAB encapsulants.

(5) PRINTED CIRCUIT BOARDS: Higher performance TAB technology requires the development of finer pitch, controlled-impedance

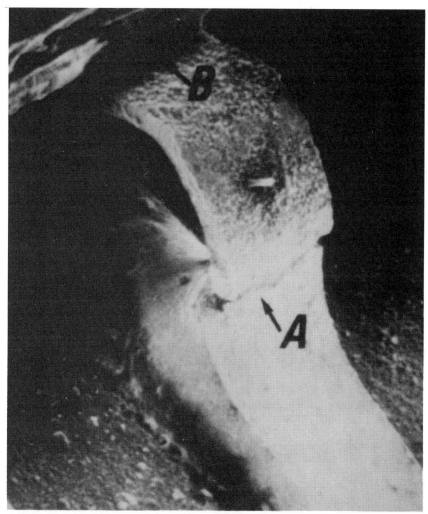

Figure 1.34 Failure mechanisms in the thermal cycling TAB assembly. [158]

printed circuit boards. Many of the process tools in this area could be leveraged from TAB tape development.

(6) RELIABILITY AND REWORK: Reliability modeling and testing methods for very high density and high lead count TAB should be developed. One aspect of this topic is the 3-D design of TAB thermal management systems using finite element modeling. A second aspect is the connector design and thermal stressing procedure for TAB component burn-in and testing. Advanced inspection systems are also

needed to detect the occasional TAB bonding failure. Simultaneously more research and development is required to provide rework systems that do not degrade the overall mechanical and electrical performance of TAB components during the repair process.

(7) ADVANCED TAB: Progress in tape and bumping technology will speed up area TAB development. Meanwhile much more research should be done on multi-chip packaging with TAB, which offers the promise of improving performance and reliability while relaxing some printed circuit board requirements.

1.15 References

1   J.R. Adams and H.L. Floyd, "Evaluation of the Mechanical Integrity of Beam Lead Devices and Bonds Using Thermomechanical Stress Waves," Proceedings of IEEE Reliability Physics Symposium, pp. 187-194, 1971.

2   J.R. Adams and H.B. Bonham, "Analysis and Development of a Thermocompression Bond Schedule for Beam Lead Devices," IEEE Transactions on Parts, Hybrids, and Packaging, Vol. PHP-8, No. 3, pp. 22-26, September 1972.

3   D.G. Aeschliman, "Multiconductor TAB Tape Materials", Proceedings of the 2nd International TAB Symposium, pp. 250-257, 1990.

4   A.H. Agajanian, "A Bibliography on Electronic Packaging, II," Solid State Technology, pp. 56-63, October 1975.

5   A.H. Agajanian, "A Bibliography on Electronic Packaging-III," Solid State Technology, pp. 87-101, September 1976.

6   N. Ahmed and J.J. Svitak, "Characterization of Gold-Gold Thermocompression Bonding," Solid State Technology, pp. 25-32, November 1975. Altendorf, J. M., "SMT-Compatible, High-Yield TAB Outer Lead Bonding Process," Proceedings of NEPCON West, March 1989.

7   J. M. Altendorf, "SMT-Compatible, High-Yield TAB Outer Lead Bonding Process," Proceedings of NEPCON West, March 1989.

8   T.L. Angelucci, "Gang Lead Bonding Integrated Circuits," Solid State Technology, pp. 21-25, July 1976.

9   Angelucci, T. L., "Gang Lead Bonding Equipment, Materials and Technology," Solid State Technology, pp. 65-68, March 1978.

10   K. Atsumi, N. Kashima and Y. Maehara, "Inner Lead Bonding Techniques for 500 Lead Dies Having a 90 μm Lead Pitch", Proceedings of the 39th IEEE Electronic Components and Technology Conference, pp. 171-176, 1989.

11   W.V. Ballard and V.S. Cardashian, "Semi-Conductor Packaging Featuring Copper/Polyimide Multilayer Tape," Proceedings of IEEE Electronic Components Conference, pp. 556-559, 1986.

12   D.A. Behm and B. Parizek, "Beam Tape Carriers - A New Design Guide," Proceeding of ISHM International Symposium on Microelectronics, pp. 384-390, 1985.

13   C. Bernadotte, "Development of the TAB-Process at SAAB-SCANIA," Proceedings of IEEE 28th Electronic Components Conference, pp. 151-158, 1978.

14   M. Bertram, "Repair Method for Solder Reflow TAB OLB", Proceedings of the 2nd International TAB Symposium, 147-163, 1990.

15   E.C. Blackburn, "Non-Destructive Evaluation of Tape Automated Bonds- A Serious Reliability Need'" Proceedings of ISHM International Symposium on Microelectronics, pp. 311-313, 1986.

16   R.J. Blazek and W.A. Piper, "The Optimization of Lead Frame Bond Parameters for Production of Reliable Thermocompression Bonds," Proceedings of IEEE Electronic Components Conference, pp. 373-379, 1978.

17   A. Boetti, J.T. Lynch, J.P. McCarthy and M.R. Hepher, "Investigation of Micropackaging Techniques for Hi-Rel Active Chips," Presented at the 3rd European Hybrid Microelectronics Conference, Avignon, France, May 20-22, 1981. Also, ESA Journal, Vol. 5, pp. 117-125, 1981.

18   J.Boutillier and R. Roche, "Chip Protection on Tape Automated Bonding," International Journal of Hybrid Microelectronics, Vol. 4, No. 2, pp. 320-325, 1981.

19   R. Bowlby, "The DIP May Take Its Final Bows," IEEE SPECTRUM, pp. 37-42, June 1985.

20   M. Boyd, "TAB Packaging At Hewlett-Packard", Proceedings of the 39th IEEE Electronic Components and Technology Conference, pp. 184-186, 1989.

21   M.F. Bregman and C.A. Kovac, "Plastic Packaging for VLSI-Based Computers," Solid State Technology, pp. 75-80, June 1988.

22   D.B. Brown and M.G. Freedman, "Is There a Future for TAB?," Solid State Technology, pp. 173-175, September 1985.

23   A. Burkhart, "Considerations for Choosing Chip-On-Board Encapsulants," ELECTRIONICS, pp. 67-69, September 1985.

24   C.D. Burns and J.W. Kanz, "Gang Bonding to Standard Aluminized Integrated Circuits with Bumped Interconnect Tape," Solid State Technology, pp. 79-81, September 1978.

25   C.D. Burns, "Trends in Tape Bonding," Semiconductor International, pp. 25-30, April 1979.

26   C. Burns, A. Keizer and M. Tonner, "Beam Tape Automated Assembly of Dips," Proceedings of International Microelectronics Conference, pp. 99-103, 1975.

27   C. Burns, "Improved Reliability for Molded DIP IC's Using Microinterconnect Bonding," Proceedings of Automated Microcircuit Interconnections Symposium, pp. 105-111, Philadelphia, February 1976.

28   M. Buschbom, S. Johannesmeyer, T. Poul, D. Mohr and P. Wallace, "The CAD/CAE Challenge for Semiconductor Multi-Chip Module Design", Proceedings of NEPCON West, pp. 178-193, 1989.

29   R.L. Cain, "Beam Tape Carriers-A Design Guide," Solid State Technology, pp. 53-58, March 1978.

30   J.J. Camarda, "Reliability Study of "TAB-On-Board" Packaged Devices for Multi-Chip Module Assembly," Proceedings of NEPCON West, pp. 576-586, February 1988.

31   B. Chaffin, "Reliability of Tab Products," Solid State Technology, pp. 136-138, September 1981.

32   N.J. Chaplin and A.J. Massesa, "Reliability of Epoxy and Silicone Molded Tape-Carrier Silicon Integrated Circuits with Various Chip-Protective Coatings," Proceedings of IEEE Reliability Physics Symposium, pp. 187-193, 1978.

33   J-M. Cheype, "Development of T.A.B. Processes for V.L.S.I Devices," Proceedings of IEEE Electronic Components Conference, pp. 65-69, 1986.

34   Y. Chikawa, K. Mori, S. Sasaki, T. Maeda and M. Hayakawa, "Bumping Technology for Tape-Automated Bonding," Proceedings of the 1st IEEE CHMT Symposium, pp. 227-232, 1984.

35   S. Chin, "Chip-on-board Technology Chugs Along," Electronic Products, pp. 37-40, March 1987.

36   T. Chung, R.T. Smith, J. Chang and T. Hunter, "A Substrate Connector Design Utilizing TAB Technology", Proceedings of NEPCON West, pp. 929-943, 1989.

37   T. Chung and R.T. Smith, "Design and Analysis of High Performance High Density Multilayer Tape Interconnect," Proceedings of NEPCON West, pp. 757-774, February 1988.

38   T. Chung and R.T. Smith, "High Performance High Density Connectors Utilizing Multiple Layer Metal/Polymer Construction," Proceedings of IEEE Electronic Components Conference, pp. 11-18, May 1987.

39   T.C. Chung and H.A. Moore, "Analysis and Evaluation of TAB Bonds", Proceedings of the NEPCON East, pp. 746-762, 1989.

40   T. Chung, "Exploration of The Thermal Management and Interconnection Problems on Tape Automated Bonding Device Burn-In," Proceedings of International Electronic Packaging Symposium, pp. 50-67, November 1986.

41   C.G. Cleveland, "Practical Design Considerations for TAB Packaging", Proceedings

of NEPCON West, pp. 877-893, 1989.

42   A.D. Close, "Initial Standardization Activities on Tape Carrier Packaging," Proceedings of ISHM International Symposium on Microelectronics, pp. 22-27, 1977. Compass Program, "Is TAB Ready for high Growth?," BPA (Technology and Management) Ltd., 1986.

43   Compass Program, "Is TAB Ready for High Growth?" BPA (Technology and Management) Ltd., 1986.

44   A. Coucoulas, "Compliant Bonding," Proceedings of IEEE Electronic Components Conference, pp. 380-389, 1970

45   G.S. Cox, J.R. Dorfman, C.A. Fedetz, T.D. Lantzer, R.S. Lawton and J.W. Lott, "DuPont Process for High Performance Two Conductor TAB", Proceedings of the 2nd International TAB Symposium, pp. 222-229, 1990.

46   J.L. Dais and F.L. Howland, "Fatigue Failure of Encapsulated Gold-Beam Lead and TAB Devices," IEEE Transactions on Components, Hybrids, and Manufacturing Technology, Vol. CHMT-1, No. 2, pp. 158-166, June 1978.

47   J.L. Dais, J.S. Erich and D. Jaffe, "Face-Down TAB for Hybrids," IEEE Transactions on Components, Hybrids, and Manufacturing Technology, Vol. CHMT-3, No. 4, pp. 623-634, December 1980.

48   J.L. Dais, "The Mechanics of Gold Beam Leads During Thermocompression Bonding," IEEE Transactions on Parts, Hybrids, and Packaging, Vol. PHP-12, No. 3, pp. 241-250, September 1976.

49   J.L. Dais and F.L. Howland, "Beam Fatigue as A Failure Mechanism of Gold Beam Lead and TAB Encapsulated Devices," Proceedings of ISHM International Symposium on Microelectronics, pp. 100-104, 1977.

50   F.J. Dance, "Chip-On-Board Has Designs on High-Density Packaging," Electronic Packaging and Production, pp. 70-75, October 1985.

51   F.J. Dance, "Printed Wiring Boards: A Five Year Plan," Circuits Manufacturing, pp. 23-36, June 1986.

52   H. Danielsson, "Different Chip Interconnecting Techniques," Electrocomponent Science and Technology, Vol. 7, pp. 149-154, 1980.

53   C.J. Dawes, "An Evaluation of Techniques for Bonding Beam-Lead Devices to Gold Thick Films," Solid State Technology, pp. 23-28, March 1976.

54   G. Dehaine and K. Kurzweil, "Tape Automated Bonding Moving into Production," Solid State Technology, pp. 46-52, October 1975.

55   G. Dehaine and M. Leclercq, "Tape Automated Bonding, A New Multichip Module Assembly Technique," Proceedings of IEEE 23rd Electronic Components Conference, pp. 69-74, 1973.

56   G. Dehaine, "Ceramic Package Lead Frame Replaced by TAB Carrier", Proceedings of the 2nd International TAB Symposium, pp. 54-61, 1990.

57   G. Dehaine and K. Kurzweil, "T.A.B. Automation in Component Attachment on Substrates," Proceedings of International Conference on Manufacturing and Packaging Techniques for Hybrid Circuits, pp. 107-121, April 7-8, 1976.

58   L.A. DelMonte, "Fabrication of Gold Bumps for Integrated Circuit Terminal Contact," Proceedings of IEEE Electronic Components Conference, pp. 21-25, 1973.

59   S. Denda, "Hybrid IC Technologies Satisfy High Density and High Function Requirements," Journal of Electronic Engineering, Vol. 23, No. 232, pp. 35-38, April 1986.

60   D. Devitt and J. George, "Beam Tape plus Automated Handling Cuts IC Manufacturing Costs," ELECTRONICS, pp. 116-119, July 1978.

61   L. DiFrancesco, "High Speed TAB Testing", Proceedings of NEPCON West, pp. 1123-1123, 1989.

62   L. DiFrancesco, "TAB Implementation: A Military User's Viewpoint", Proceedings of the IEEE International Electronic Manufacturing Technology Symposium, pp. 234-238, 1989.

63   T. Dixon, "TAB Technology Tackles High Density Interconnections," Electronic Packaging and Production, pp. 34-39, December 1984.

64   G.A. Dodson, "Analysis of Accelerated Temperature Cycle Test Data Containing Different Failure Modes," IEEE International Reliability Physics Symposium, pp. 238-246, 1979.

65   K. Doss, "Materials Issues for The TAB Technology of '90s", Proceedings of NEPCON West, pp. 944-956, 1989.

66   L. Dries, "Deformation and Structural Changes Occurring at The Interface During Gold-Gold Thermocompression Bonding," Master Thesis, Lehigh University, 1973.

67   P.M. Duncan, D.L. Fett, A.H. Mones and T.R. Poulin, "Modeling and Analysis of Two Layer TAB", Proceedings of the 2nd International TAB Symposium, pp. 165-171, 1990.

68   P.M. Duncan, D.L. Fett, A.H. Mones and T.R. Poulin, "Noise Characteristics of High Density TAB Configurations", Proceedings of NEPCON West, pp. 638-649, 1989.

69   O. Ehrmann, G. Engelmann, J. Simon and H. Reichl, "A Bumping Technology for Reduced Pitch", Proceedings of the 2nd International TAB Symposium, pp. 41-48, 1990.

70   M. El Refaie, "Interconnect Substrate for Advanced Electronic Systems," Electronic Engineering, pp. 133-141, September 1982.

71   M. El Refaie, "Chip-Package Substrate Cushions Dense, High-Speed Circuitries," Electronics, pp. 135-141, July 1982.

72   T.S. Ellington, "Lead Frame Bonding," Proceedings of IEEE Electronic Components Conference, pp. 357-361, 1972.

73   A. Emamjomeh and M.J. Sandor, "Evaluation of Material and Geometric Parameters in a TAB-On-Board Packages Using Finite Element Analysis", Proceedings of the 2nd International TAB Symposium, pp. 172-198, 1990.

74   J. Erlewein and H. Rachner, "Chip on Glass Technology for Large Scale Automotive Displays," Proceedings of Electronic Displays and Information Systems Conferences, pp. 31-35, 1984.

75   L.G. Feinstein, "Do Flame Retardants Affect the Reliability of Molded Plastic Packages?," Microelectron Reliability, Vol. 21, No. 4, pp. 533-541, 1981.

76   E. Fuchs, "Chip-On-Board: An Economical Packaging Solution," Electronic Packaging and Production, pp.182-185, January 1985.

77   K. Fujita, T. Onishi, S. Wakamoto, T. Maeda and M. Hayakawa, "Chip-Size plastic Encapsulation on Tape Carrier Package," The International Journal for Hybrid Microelectronics, Vol. 8, Number 2, pp. 9-15, June 1985.

78   R. Gardner, "Packaging Trends and Considerations: A Global View from a Custom Manufacturer," WESCON Conference Record, pp. 3/3/1-3, 1983.

79   G.L. Ginsberg, "Chip and Wire Technology: The Ultimate in Surface Mounting," Electronic Packaging and Production, pp.78-83, August 1985.

80   G.L. Ginsberg, "Chip-On-Board Profits From TAB and Flip-Chip Technology," Electronic Packaging and Production, pp. 140-143, September 1985.

81   B.E. Goblish, R.G. Arno and R.V. Nicolaides, "Reliability Study on Environmentally-Protected/Tape-Automated-Bonded Integrated Circuits", Proceedings of the 2nd International TAB Symposium, pp. 103-123, 1990.

82   C.T. Goddard, "The Role of Hybrids in LSI Systems," IEEE Transactions on Components, Hybrids, and Manufacturing Technology, Vol. CHMT-2, No. 4, pp. 367-371, December 1979.

83   J.W. Greig, P.J. Moneika and R.A. Brown, "High-Density High-Reliability Multichip Hybrid Packaging with Thin Films and Beam Leads," Proceedings of IEEE Electronic Components Conference, pp. 146-150, 1978.

84   N. Griffin, "High Lead Count TAB in A Multi-Chip Environment," Proceedings of NEPCON West, pp. 569-575, February 1988.

85   S.E. Grossman, "Film-Carrier Technique Automates the Packaging of IC Chips," ELECTRONICS, pp. 89-95, May 1974.

86   M.P. Hagen, P.L. Li and R.T. Ogan, "A Hybrid Oscillator Circuit for the 30 KG, Non-Hermetic Environment," Proceedings of ISHM International Symposium on Microelectronics, pp. 331-335, 1979.

87   M.P. Hagen, "Methods for Measuring Plating Thicknesses on TAB Lead Frames," Proceedings of NASA/ISHM Microelectronics Conference, pp. 111-121, Alabama, 1977.

88   P.M. Hall, N.T. Panousis and P.R. Menzel, "Strength of Gold-Plated Copper Leads on Thin Film Circuits Under Accelerated Aging," IEEE Transactions on Parts, Hybrids, and Packaging, Vol. PHP-11, No. 3, pp. 202-205, September 1975.

89   P.M. Hall, J.M. Morabito and N.T. Panousis, "Interdiffusion in the Cu-Au Thin Film System at 25 C to 250 C," Thin Solid Films, Vol. 41, pp. 341-361, 1977.

90   P.M. Hall and J.M. Morabito, "Thin Film Phenomena-Interfaces and interactions," Surface Science, Vol 67, pp. 373-392, 1977.

91   P.M. Hall and J.M. Morabito, "A Formalism for Determining Grain Boundary Diffusion Coefficients Using Surface Analysis," Surface Science, Vol. 59, pp. 624-630, 1976.

92   G.G. Harman, "Acoustic-Emission-Monitored Tests for TAB Inner Lead Bond Quality," IEEE Transactions on Components, Hybrids, and Manufacturing Technology, Vol. CHMT-5, No. 4, pp. 445-453, December 1982.

93   K. Hatada, H. Fujimoto and K. Matsunaga, "New Film-Carrier-Assembly Technology "Transferred Bump TAB"," Proceedings of IEEE International Electronic Manufacturing Technology Symposium, pp. 122-127, 1986.

94   K. Hatada, H. Fujimoto and T. Kawakita, "Application to the Electronic Instrument by Transferred Bump-TAB Technology," Proceedings of the 1987 International Symposium on Microelectronics, pp. 649-653.

95   K. Hatada, H. Fujimoto and T. Kawakita, "Application To The Electronic Instrument by Transferred Bump-TAB Technology," Proceedings of IEEE International Electronic Manufacturing Technology Symposium, pp. 81-86, September 1987.

96   K. Hatada, H. Fujimoto and T. Kawakita, "A Direct Mounting of a 4000-pin LSI Chip Using a Photo-Hardening Resin," Nikkei Microdevices, pp. 107-115, September 1987.

97   K. Hatada, "The Development of the High Density Mounting System," Internepcon-Japan/Semiconductor, January 1987.

98   K. Hatada, H. Fujimoto and T. Kawakita, "A New LSI Bonding Technology: Micron Bump Bonding Assembly Technology," Proceedings of IEEE International Electronic Manufacturing Technology Symposium, pp. 23-27, October, 1988.

99   M. Hayakawa, T. Maeda, M. Kumura, R.H. Holly and T.A. Gielow, "Film Carrier Assembly Process," Solid State Technology, pp. 52-55, March 1979.

100   P. Hedemaim, L-G Liljestrand and H. Bernhoff, "Quality and Reliability of TAB Inner Lead Bond", Proceedings of the 2nd International TAB Symposium, pp. 88-102, 1990.

101   R.W. Helda, "Spider Bonding Technique for I/C's," Proceedings of Wescon Technical Paper, pp. 2/2/1-2, 1969.

102   D. Herrell and D. Carey, "High-Frequency Performance of TAB," IEEE Transactions on Components, Hybrids, and Manufacturing Technology, Vol. CHMT-10, No. 2, pp. 199-203, June 1987.

103   N.J. Ho and J. Kratochvil, "Encapsulation of Polymeric Membrane-Based Ion-Selective Field Effect Transistors," Sensors and Actuators, Vol. 4, No. 3, pp. 413-421, 1983.

104   P. Hoffman, "Design, Manufacturing, and Reliability Aspects of Two Metal Layer TAB Tape", Proceedings of the 2nd International TAB Symposium, pp. 124-146, 1990.

105   P. Hoffman, "TAB Implementation and Trends," Solid State Technology," pp. 85-88, June 1988.

106   P. Hoffman, "An Overview of TAB," Proceedings of EXPO SMT, pp. 257-261, Nov. 1988.

107   V.R. Holalkere, "Finite Element Analysis of TAB Packages", Proceedings of the NEPCON East, pp. 762-771, 1989.

108   R.E. Holmes, "VLSI Packaging and Interconnect Technologies," Proceedings of IEEE Custom Integrated Circuits Conference, pp. 132-134, 1983.

109   S.T. Holzinger, "TAB Types and Material Choices," Proceedings of EXPO SMT, pp. 247-255, Nov. 1988.

110   S.T. Holzinger and B. Sharenow, "Advantages of A Floating Annular Ring in Three-Layer TAB Assembly," IEEE Transactions on Components, Hybrids, and Manufacturing Technology, pp. 332-334, Vol. CHMT-10, No. 3, September 1987.

111   S.T. Holzinger, "Conductor and Substrate Concerns in TAB Material Selection", Proceedings of the NEPCON East, pp. 351-357, 1990.

112   B.W. Hueners, "Wire Bonding and Alternative Technologies for Hybrid Circuits," Semiconductor International, pp. 120-121, September 1986.

113   C.E. Huwen, "Tape Automated Bonding (TAB), With Special Emphasis on Surface Mounting," WESCON Conferences Record, pp. 15/1, 1-3, 1984.

114   M. Inaba, K. Yamakawa and N. Iwase, "Solder Bump Formation Using Electroless Plating and Ultrasonic Soldering," Proceedings of IEEE International Electronic Manufacturing Technology Symposium, pp. 13-17, October 1988.

115   IPC, "An introduction to Tape Automated Bonding and Fine Pitch Technology," Technical Report, The Institute for Interconnecting and Packaging Electronic Circuits, January 1989.

116   V. Iyer and R. Pendse, "A New Inner Lead Bonding Scheme for Tape Automated Bonding (TAB)", Proceedings of the 40th IEEE Electronic Components and Technology Conference, pp. 754-756, 1990.

117   D. Jaffe, "Encapsulation of Integrated Circuits Containing Beam Leaded Devices with a Silicone RTV Dispersion," IEEE Transactions on Parts, Hybrids, and Packaging, Vol. PHP-12, No. 3, pp. 182-187, September 1976.

118   L.J. Jardine, "Worldwide Tape Automated Bonding Marketplace," Proceedings of NEPCON West, pp. 119-122, February 1987.

119   J. Johansson, "Beam Lead Bonding Investigation," Proceedings of ISHM Symposium, pp. 340-347, 1975.

120   Y. Jee and M. Andrews, "Failure Mode Analysis for TAB Inner Lead Bonding", Proceedings of the 39th IEEE Electronic Components and Technology Conference, pp. 325-334, 1989.

121   D.R. Johnson and D.L. Willyard, "The Influence of Lead Frame Thickness on The Flexure Resistance and Peel Strength of Thermocompression Bonds," Proceedings of IEEE Electronic Components Conference, pp. 80-85, 1976.

122   J.W. Kanz, G.W. Braun and R.F. Unger, "Bumped Tape Automated Bonding (BTAB) Practical Application Guidelines," IEEE Transactions on Components, Hybrids, and Manufacturing Technology, Vol. CHMT-2, No. 3, pp. 301-308, September 1979.

123   T. Kawanobe, K. Miyamoto and M. Hirano, "Tape Automated Bonding Process for High Lead Count LSI," Proceedings of IEEE Electronic Components Conference, pp. 221-226, 1983.

124   R. Keeler, "Chip-On-Board Alters the Landscape of PC Boards," Electronic Packaging and Production, pp.62-67, July 1985.

125   A. Keizer, "Mass Outerlead Bounding and Applications to Open Packages," International Microelectronic Conferences, pp. 37-46, May 1977.

126   A. Keizer and D. Brown, "Bonding Systems for Microinterconnect Tape Technology," Solid State Technology, pp. 59-64, March 1978.

127   A. Keizer and J. Thompson, "Tape Automated Bonding Systems," New Electronics, Vol. 13, No. 11, pp. 122/125/128, May 27, 1980.

128   R.C. Kershner and N.T. Panousis, Diamond-Tipped and Other New Thermodes for Device Bonding," IEEE Transactions on Components, Hybrids, and Manufacturing Technology, Vol. CHMT-2, No. 3, pp. 283-288, September 1979.

129   L.W. Kessler, "Acoustic Microscopy: A Nondestructive Tool for Bond Evaluation on TAB Interconnections," Proceedings of ISHM International Symposium on Microelectronics, pp. 79-84, 1984.

130   L.W. Kessler, J.E. Semmens and F.T. Cichanski, "Nondestructive Evaluation of TAB Bonds by Acoustic Microscopy", Proceedings of the IEEE International Electronic Manufacturing Technology Symposium, pp. 330-331, 1989.

131   S. Khadpe, "Worldwide TAB Status -- 1990", Proceedings of the 2nd International TAB Symposium, pp. 1-6, 1990.

132   S. Khadpe, "Automated Wire Bonding Versus Tape Automated Bonding: What Are The Tradeoffs?," Proceedings of IEEE 28th Electronic Components Conference, pp. 402-408, 1978.

133   H. Khajezadeh and A.S. Rose, "Reliability Evaluation of Hermetic Integrated Circuit Chips in Plastic Packages," Proceedings of IEEE International Reliability Physics Symposium, pp. 87-92, 1975.

134   D. Killam, "Development of Tape Automated Bonding for High Leadcount Integrated Circuits," Professional Technical Record 19/4, 14, Southcon, March 1986.

135   T. Kishimoto and T. Ohsaki, "VLSI Packaging Technique Using Liquid-Cooled Channels", IEEE Transactions on Components, Hybrids, and Manufacturing Technology, Vol. CHMT-9, No. 4, pp. 328-335, December 1986.

136   T. Kleiner, "TAB Excise and From Tooling Principles", Proceedings of the 2nd International TAB Symposium, pp. 62-76, 1990.

137   M. Kohara, S. Nakao and K. Tsutsumi, H. Shibata and H. Nakata, "High-Thermal Conduction Module," Proceedings of IEEE Electronic Components Conference, pp. 180-186, 1985.

138   T. Kon, S. Sasaki and R. Konno, "A New Technique for Testing TAB-LSIs in Gigahertz Frequency Range", Proceedings of the 40th IEEE Electronic Components and Technology Conference, pp. 949-952, 1990.

139   K. Kopejtko and J. Vilim, "An Alternative Way of Making Bumps for Tape Automated Bonding," Electrocomponent Science and Technology, Vol. 6, pp. 185-187, 1980.

140   J.L. Kowalski, "Individual Carriers for TAB Integrated Circuit Assembly," Proceedings of IEEE Electronic Components Conference, pp. 315-318, 1979.

141   J.L. Kowalski, "Multichip Hybrid Assembly Using TAB Components," Proceedings of International Microelectronics Conference, pp. 28-33, 1978.

142   H. Kroger, D. Carey, T. Chung, D.C. Duane, L. Gilg, D.J. Herrell, W. Mulholland, B.H. Nelson, D. Nelson, R. Nelson, T. Pan, C.N. Potter, K. Ramsey, D. Walshak, M. Wesling, B.H. Whalen and P. Winberg, "Tape Automated Bonding for High

Lead-Count, High Performance Devices," IEEE International Reliability Physics Symposium, Tutorial Notes, pp. 2.1-2.39, 1987.

143   Y-X Kuang and L. Liu, "Tape Bump Forming and Bonding in BTAB", Proceedings of the 40th IEEE Electronic Components and Technology Conference, pp. 943-947, 1990.

144   K. Kurzweil, "TAB: Now in Factory," Proceedings of European Hybrid Microelectronics Conference, pp. 367-377, Ghent, May 1979.

145   K. Kurzweil, "Tape Automated Bonding in Volume Production," Proceedings of IEEE 30th Electronic Components Conference, pp. 500-503, 1980.

146   K. Kurzweil, and G. Dehaine, "Density Upgrading in Tape Automated Bonding," Electrocomponent Science and Technology, Vol. 10, pp. 51-55, 1982.

147   K. Kurzweil, "Tape Automated Bonding for High Density Packaging, " Electrocomponent Science and Technology, Vol. 8, pp. 15-19, 1981.

148   K. Kurzweil, "An Installed Tape Automated Bonding Unit," Electrocomponent Science and Technology, Vol. 6, pp. 159-163, 1980.

149   M. Lancaster, "Tape Automated Bonding - The Ultimate in Surface Mount Density?," Proceedings of NEPCON WEST, pp. 503-515, 1987.

150   R. Landis, "Alternative Bonding Methods for Chip-On-Board Technology," Proceedings of IEEE Electronic Components Conference, pp. 53-58, 1986.

151   T.A. Lane, F.J. Belcourt and R.J. Jensen, "Electrical Characteristics of Copper/Polyimide Thin Film Multilayer Interconnects," Proceedings of IEEE Electronic Components Conference, pp. 614-622, 1987.

152   J.H. Lau, S.J. Erasmus and D.W. Rice, "Overview of Tape Automated Bonding Technology", Circuit World, Vol. 16, No. 2, pp. 5-24, 1990.

153   J.H. Lau, D.W. Rice and G. Harkins, "Thermal Stress Analysis of TAB Packagings and Interconnections," IEEE Transactions on Components, Hybrids, and Manufacturing Technology, Vol. 13, No. 1, pp. 183-188, March 1990.

154   J.H. Lau and D.W. Rice, "Solder Joint Fatigue in Surface Mount Technology: State of the Art," Solid State Technology, Vol. 28, pp. 91-104, October 1985.

155   M.P. Lepselter, "Beam-Lead Technology," The Bell System Technical Journal, Vol. XLV, No. 2, pp. 233-253, February 1966.

156   L. Levine and M. Sheaffer, "Optimizing The Single Point TAB Inner Lead Bonding Process", Proceedings of the 2nd International TAB Symposium, pp. 16-24, 1990.

157   E.T. Lewis, "Interconnection Design Considerations for VLSI Multichip Packaging," Proceedings of ISHM International Symposium on Microelectronics, pp. 722-729, 1986.

158   L.-G. Liljestrand, "Bond Strengths of Inner and Outer Leads on TAB Devices," Hybrid Circuits, No. 10, pp. 42-48, May 1986.

159   F.A. Lindberg, "Bumped Tape Processing and Application," Semiconductor Processing, ASTM STP 850, pp. 512-530, 1984.

160   F.A. Lindberg, "Hybrid Tape Bonding With Fast Turnaround, Standard Cell, Reusable Tape," Proceedings of International Microelectronic Conference, pp. 117-125, 1980.

161   T.S. Liu, "Aspects of Gold-Tin Bump-Lead Interconnection Metallurgy," Proceedings of ISHM International Symposium on Microelectronics, pp. 120-126, 1977.

162   T.S. Liu, W.R. Rodrigues de Miranda and P.R. Zipperlin, "A Review of Wafer Bumping for Tape Automated Bonding," Solid State Technology, pp. 71-76, March 1980.

163   T.S. Liu, D.E. Pitkanen and C.H. McIver, "Surface Treatments and Bondability of Copper Thick Film Circuits," Proceedings of IEEE 31st Electronic Components Conference, pp. 9-17, 1981.

164   T.S. Liu and H.S. Fraenkel, "Metallurgical Considerations in Tin-Gold Inner Lead Bonding Technology," International Journal for Hybrid Microelectronics, Vol. 1(2), pp. 69-76, 1978.

165   S.C. Lockard, J.M. Hansen and G.H. Nelson, "Multimetal Layer TAB for High Performance Digital Applications", Proceedings of NEPCON West, pp. 1113-1122, 1989.

166   S.C. Lockard, M. Hansen and G.H. Nelson, "High Lead Count Multimetal Layer TAB Circuits", Proceedings of the 2nd International TAB Symposium, pp. 214-221, 1990.

167   J. Loughran and K. Kurzweil, "Economic Considerations in Multilayer Thick Film Hybrids," IEEE Transactions on Parts, Hybrids, and Packaging, Vol. PHP-10, No. 2, pp. 120-131, June 1974.

168   D.P. Ludwig, "Chip-In-Tape--Its Role in Hybrid Integrated Circuit Manufacture," Proceedings of International Microelectronics Conference, pp. 22-28, 1977.

169   J. Lyman, "Technology Update: Packaging and Production,", Electronics, pp. 153-156, October 1982.

170   J. Lyman, "Special Report: Film Carriers Star in High-Volume IC Production," Electronics, pp. 61-68, December 1975.

171   J. Lyman, "Tape Automated Bonding Meets VLSI Challenge," Electronics, pp. 100-105, December 1980.

172   J. Lyman, "Is Tom Angelucci's Big Gamble Finally Paying Off?," Electronics, pp.

38-39, February 1986.

173   J. Lyman, "Film Carriers Win Productivity Prize," Electronics, pp. 122-125, October 1975.

174   J. Lyman, "Growing Pin Count is Forcing LSI Package Changes," Electronics, pp. 81-91, March 1977.

175   R.E. MacDougall and N. Grossman, "Properties of Gold and Nickel Beam Lead Bonds," Proceedings of IEEE Electronic Components Conference, pp. 347-356, 1972.

176   S. Machuga, R. Pennisi and D. Wilburn, "The Relationship Between Polymerization Reaction Kinetics and the Optimum Physical Properties and Processing of Epoxy-Based Encapsulating Systems," Proceedings of IEEE International Electronic Manufacturing Technology Symposium, pp. 113-116, October 1988.

177   D. MacIntyre, "TAB To Board Attach Techniques", Proceedings of NEPCON West, pp. 1124-1132, 1989.

178   R.B. Maciolek, E.D. Pisacich and C.J. Speerschneider, "Thermal Aging Characteristics of In-Pb Solder Bonds to Gold," Proceedings of ISHM International Symposium on Microelectronics, pp. 209-212, 1977.

179   M. Mahalingam and J.A. Andrews, "TAB vs. Wire Bond- Relative Thermal Performance," IEEE Transactions on Components, Hybrids, and Manufacturing Technology, Vol. CHMT-8, No. 4, pp. 490-499, December 1985.

180   P.P. Marcoux, "Overview of TAB Assembly Process", Proceedings of the NEPCON East, pp. 737-745, 1989.

181   P.P. Marcoux, "The Benefits of Surface Mount Technology," WESCON Conferences Record, pp. 20/2-1-5, 1984.

182   H.W. Markstein, "Automatic Wire Bonding and TAB for Hybrids," Electronic Packaging and Production, pp. 63-70, February 1979.

183   H.W. Markstein, "TAB Rebounds as I/Os Increase," Electronic Packaging and Production, pp. 42-45, August 1986.

184   H.W. Markstein, "TAB Tames High-Density Chip I/Os," Electronic Packaging and Production, pp. 42-45, December 1988.

185   J.F. Marshall and R.P. Sheppard, "New Applications of Tape Bonging for High Lead Count Devices," Semiconductor Processing, ASTM STP 850, pp. 500-511, 1984.

186   J. Marshall and J. Hullmann, "Encapsulated Chip Packaged on Tape," Semiconductor International, pp. 170-173, August 1985.

187   T.J. Matcovich, "Trends in Bonding Technology," Proceedings of IEEE 31st

Electronic Components Conference, pp. 24-30, 1981.

188   C.H. McIver, "Flip TAB, Copper Thick Films Create the Micropackage," Electronics, pp. 96-99, November 1982.

189   Mesa Technology, "FASTRACK Design Guidelines," Olin Interconnect Technologies Company, March 1987.

190   Mesa Technology, "Tape Automated Bonding Engineering/Reliability Study," Olin Interconnect Technology Company, May 1987.

191   G. Messner and C. Lassen, "Substrates For Surface Mounted Components," Proceedings of The Printed Circuit World Convention II, Vol. 1, pp. 415-427, 1981.

192   D. Meyer, A. Kohli, H. Firth and H. Reis, "Metallurgy of Ti-W/Au/Cu System for TAB Assembly," Journal of Vacuum Society Technology, pp. 772-776, May/June 1985.

193   D.E. Meyer, "Product Applications for Tape Automated Bonding," Electro and Mini/Micro Northeast Conferences Record, pp. 6/3-1-5, 1986.

194   J.C. Miller, "VLSI Packaging Trends - An Update", Proceedings of the 2nd International TAB Symposium, pp. 7-15, 1990.

195   J.M. Montante, W.R. Rodrigues de Miranda and P. Zipperlin, "Storage and Life Tests of Thermocompression Tape Automated Bonded Leads to Various Gold Metallizations," Proceedings of ISHM International Symposium on Microelectronics, pp. 169-175, 1978.

196   J.M. Montante and R.G. Oswald, "Wafer Bumping for Tape Automated Bonding," Proceedings of ISHM International Symposium on Microelectronics, pp. 115-119, 1977.

197   J.M. Morabito and C.T. Hartwig, "Indium Gold Alloy Bonding," Proceedings of ISHM International Symposium on Microelectronics, pp. 73-79, 1976.

198   S. Nakabu and T. Nukii, R. Miyake and K. Awaue, "Development of a Pocket-Sized and High Resolution Liquid Crystal Display Unit Using a Press-Formed MO Substrate (Metal-Based Organic Film Substrate)," Proceedings of ISHM International Symposium on Microelectronics, pp. 318-324, 1984.

199   H. Nakahara, "Japan's Swing to Chip-On-Board," Electronic Packaging and Production, pp. 38-41, December 1986.

200   S. Nakao, T. Hashimoto and Y. Nemoto, "New Packaging Technology for Smart Card by TAB," Proceedings of IEEE Electronic Components Conference, pp. 41-46, 1987.

201   B. Nelson, B. Hargis, M. Bertram, D. Duane, L. Gilg, W., Hanna, H.S. Hashemi, G. Ogden, K. Pitts, D. Schulze, D. Walshak and M. Wesling, "Performance and Reliability of Tape Automated Bonding of High Lead-Count, High Performance Devices," IPC-TP-682, 1988.

202   C.A. Neugebauer, R.O. Carlson, R. A. Fillion and T.R. Haller, "High Performance Interconnections Between VLSI Chips," Solid State Technology, pp. 93-98, June 1988.

203   M. Nitta, T. Tsuge, Y. Hiki and R. Kato, "Process and Accelerated Aging Test of Gang Bonding for Gold-Plated TAB Outer Leads," Electrocomponent Science and Technology, Vol. 8, pp. 27-36, 1981.

204   T.G. O'Neill, "The Status of Tape Automated Bonding," Semiconductor International, pp.33-51, February 1981.

205   T.G. O'Neill, "VLSI Packaging Requirements and Trends," Semiconductor International, pp. 43-60, March 1981.

206   R.G. Oswald, J.M. Montante and W.R. Rodrigues de Miranda, "Automated Tape Carrier Bonding for Hybrids," Solid State Technology, pp. 39-48, March 1978.

207   R.G. Oswald and W.R. Rodrigues de Miranda, "Application of Tape Automated Bonding Technology to Hybrid Microcircuits," Solid State Technology, pp. 33-38, March 1977.

208   J.A. Owczarek, F.L. Howland and D. Jaffe, "Formation of Voids in Silicone RTV Dispersion Under Beam-Leaded Silicon Integrated Circuits", IEEE Transactions on Components, Hybrids, and Manufacturing Technology, Vol. 13, No. 2, pp. 284-293, June 1990.

209   R. Padmanabhan, "Corrosion Failure Modes in a TAB200 Test Vehicle," IEEE Transactions on Components, Hybrids, and Manufacturing Technology, Vol. CHMT-8, No. 4, pp. 435-439, December 1985.

210   R. Padmanabhan, "Metal Corrosion in An Experimental Tape Automated Bonded Die," Proceedings of IEEE Electronic Components Conferences, pp. 309-313, May 1985.

211   N.T. Panousis and P.M. Hall, "Thermocompression Bonding of Copper Leads Plated with Thin Gold," Proceedings of IEEE Electronic Components Conference, pp. 220-225, 1977.

212   N.T. Panousis and P.M. Hall, "Reduced Gold-Plating on Copper Leads for Thermocompression Bonding-Part I: Initial Characterization," IEEE Transactions on Parts, Hybrids, and Packaging, Vol. PHP-13, No. 3, pp. 305-309, September 1977.

213   N.T. Panousis and P.M. Hall, "Reduced Gold-Plating on Copper Leads for Thermocompression Bonding-Part II: Long Term Reliability," IEEE Transactions on Parts, Hybrids, and Packaging, Vol. PHP-13, No. 3, pp. 309-313, 1977.

214   N.T. Panousis and P.M. Hall, "The Effects of Gold and Nickel Plating Thicknesses on the Strength and Reliability of Thermocompression-Bonded External Leads," IEEE Transactions on Parts, Hybrids, and Packaging, Vol. PHP-12, No. 4, pp. 282-287, December 1976.

215   N.T. Panousis and R.C. Kershner, "Thermocompression Bondability of Thick-Film Gold- A Comparison to Thin-Film Gold," IEEE Transactions on Components, Hybrids, and Manufacturing Technology, Vol. CHMT-3, No. 4, pp. 617-623, December 1980.

216   N.T. Panousis, "Copper and Silver Contamination of Thick film Gold Surfaces," Thin Solid Films, Vol. 64, pp. 41-45, 1979.

217   N.T. Panousis and P.M. Hall, "Application of grain boundary diffusion studies to soldering and Thermocompression Bonding," Thin Solid Films, Vol. 53, pp. 183-191, 1978.

218   A. Parthasarathi, "Bumped Tape Bonding Mechanisms", Proceedings of the 2nd International TAB Symposium, pp. 25-40, 1990.

219   W. Patstone, "Beam-Tape Technology Brings Benefits (Mostly Hidden) to IC Users," EDN, pp. 39-46, August 1977.

220   W. Patstone, "Tape-Carrier Packaging Boasts Almost Unlimited Potential," EDN, pp. 30-37, October 1974.

221   M. Patterson and A. Millerick, "Laser Dambar and Flash Removal for Plastic Molded Surface Mount Components With Lead Specing Under 20 Mils," Proceedings of IEEE International Electronic Manufacturing Technology Symposium, pp. 18-21, October 1988.

222   N.K. Pearne and M.G. Sage, "An Interconnection Overview for Microelectronics," Proceedings of the Printed Circuit World Convention II, Vol. 1, pp. 26-31, 1981.

223   W.E. Pence and J.P. Krusius, "The Fundamental Limits for Electronic Packaging and Systems," IEEE Transactions on Components, Hybrids, and Manufacturing Technology, Vol. CHMT-10, No. 2, pp. 176-183, June 1987.

224   R. Pennisi, "Characterization of Epoxy Encapsulants Formulated for TAB Applications," Proceedings of NEPCON West, pp. 790-796, February 1988.

225   R. Pennisi, "Simulation and Analysis of Thermal and Microwave Curing of TAB Encapsulants," Proceedings of IEEE International Electronic Manufacturing Technology Symposium, pp. 40-45, October 1987.

226   R. Pennisi, F. Nounou and S. Machuga, "Cure Optimization of TAB Encapsulants," Proceedings of NEPCON West Conferences, pp. 225-232, March 1989.

227   E. Philofsky, R. Bowman and W. Miller, "Aluminum Ultrasonic Joining in Spider and Wire Connections," Proceedings of IEEE Electronic Components Conference, pp. 289-294, 1971.

228   D.E. Pitkanen, J.P. Cummings J.A. and Sartell, "Compatibility of Copper/Dielectric Thick Film Materials," Proceedings of ISHM International Symposium on Microelectronics, pp. 148-156, 1979.

229   D.E. Pitkanen, J.P. Cummings and C.J. Speerschneider, "Status of Copper Thick Film Hybrids," Solid State Technology, pp. 141-146, October 1980.

230   T. Pitkanen, T. Salo, H. Sundquist, S. Lalu and P. Collander, "Improved IC Reliability with TAB Packaging", Proceedings of the 39th IEEE Electronic Components and Technology Conference, pp. 190-193, 1989.

231   D.J. Quinn and R.H. Bond, "New Process for Automated IC Assembly Manufacturing," Proceedings of IEEE International Electronic Manufacturing Technology Symposium, pp. 137-141, 1986.

232   B. Reding, M. Hagen and A. Russo, "A Development of TAB for Large Gate Array Devices," International Journal of Hybrid Microelectron, Vol. 6, No. 1, pp. 574-579, 1983.

233   A. Reubin, B. Bohon and R. Smith, "Transfer Molded TAB Package", Proceedings of NEPCON West, pp. 894-905, 1989.

234   P.W. Rima, "TAB Technology: An Historical Overview of Lessons Learned, Relating to Current Technology and Material Issues," Proceedings of NEPCON West, pp. 587-590, February 1988.

235   P.W. Rima, "Recent Developments in Tape Automated Bonding Equipments," Proceedings of NEPCON WEST, pp. 130-131, 1987.

236   P.W. Rima, "The Basics of Tape Automated Bonding," Hybrid Circuit Technology, pp. 15-21, November 1984.

237   L.A. Robinson, "Carrier System for Testing and Conditioning of Beam Lead Devices," Proceedings of IEEE Electronic Components Conference, pp. 1-9, 1977.

238   W.R. Rodrigues de Miranda, R.G. Oswald and D. Brown, "Lead Forming and Outer Lead Bond Pattern Design for Tape-Bonded Hybrids," IEEE Transactions on Components, Hybrids, and Manufacturing Technology, Vol. CHMT-1, No. 4, pp. 377-383, December 1978.

239   W.R. Rodrigues de Miranda, "Hybrids with TAB-At the Threshold of Production," Solid State Technology, pp. 115-122, October 1980.

240   W.R. Rodrigues de Miranda and R.G. Oswald, "Advances in TAB for Hybrids -TC Outer Lead Bonding," Proceedings of ISHM International Symposium on Microelectronics, pp. 105-114, 1977.

241   W.R. Rodrigues de Miranda, "Tape Bonding of Large Chips," Proceedings of International Microelectronic Conference, pp. 110-116, 1980.

242   W.R. Rodrigues de Miranda and E.R. Smith, "Tape Automated Bonding Test Board," United States Patent #4195195, March 25, 1980.

243   A.S Rose, F.E. Scheline and T.V. Sikina, "Metallurgical Considerations for Beam Tape Assembly," Solid State Technology, pp. 49-68, March 1978.

244   P.K. Rouhan and D. Sorrells, "Reliability of Tape Automated Bonded Integrated Circuits," Proceedings of the Surface Mount '88 Conference, pp. 117-128, 1988.

245   W.D. Ryden and E.F. Labuda, "A Metallization Providing Two Levels of Interconnect for Beam-Leaded Silicon Integrated Circuits," IEEE Journal of Solid-State Circuits, Vol. SC-12, No. 4, pp. 376-382, August 1977.

246   M.G. Sage, "I/C Packaging and Interconnection Developments," Circuit World, Vol. 7, No. 1, pp. 59-68, 1980.

247   K. Saito, "Fine Lead Pitch Technology in Tape Carriers and Double Metal Layer Tape Carrier", Proceedings of the 2nd International TAB Symposium, pp. 199-213, 1990.

248   V. Sajja, "Thermocompression Joining of Copper To Copper For TAB," Proceedings of IEEE International Electronic Manufacturing Technology Symposium, pp. 40-43, October 1988.

249   J.S. Sallo, "Bumped Beam Tape for Automatic Gang Bonding," Proceedings of ISHM International Symposium on Microelectronics, pp. 153-156, 1978.

250   J.S. Sallo, "An Overview of TAB Applications and Techniques," Proceedings of NEPCON EAST, pp. 139-145, 1985.

251   S. Sato and H. Tsunemitsu, "A New Approach to Reliability Built-In Connections for LSI Terminals," Proceedings of IEEE 29th Electronic Components Conference, pp. 94-98, 1979.

252   A.G. Saunders, "Trends in Packaging and Interconnection of Integrated Circuits," Microelectronics Journal, Vol. 12, No. 1, pp. 23-29, 1981.

253   T.A. Scharr and M.D. Nagarkar, "Outer Lead Die Bond Reliability in High Density TAB", Proceedings of the 39th IEEE Electronic Components and Technology Conference, pp. 177-183, 1989.

254   T/A/ Scharr, "TAB Bonding A 200 Lead Die," Proceedings of ISHM International Microelectronic Symposium, pp. 561-565, 1983.

255   G.L. Schnable, "State of the Art in Semiconductor Materials and Processing for Microcircuit Reliability," Solid State Technology, pp. 69-73, October 1978.

256   W.H. Schroen, "Plastic Packaging, All There Is To Know," IEEE International Reliability Physics Symposium, Tutorial Notes, pp. 4.1-4.18, 1985.

257   S.E. Scrupski, "ICs on Film Strip Lend Themselves to Automatic Handing By Manufacturer and User, Too," Electronics, pp. 44-48, February 1971.

258   K. Sekimoto, H. Nakata and M. Miura, "Analysis of Trapped Energy Resonators with TABs," Proceedings of IEEE 39th Annual Frequency Symposium, pp. 386-391, 1985.

259   B. Sharenow, "Package Design Considerations Using Tape Automated Bonding Materials," Proceedings of NEPCON WEST, pp. 123-126, 1987.

260   B. Sharenow, "Tape Automated Bonding High Lead Count Cost Analysis," Proceedings of NEPCON West, February 1988.

261   N.K. Sharma, D.H. Klockow, N.T. Panousis and D. Jaffe, "A New Generation of Hybrid Packages for LSI Devices," Proceedings of ISHM International Symposium on Microelectronics, pp. 172-176, 1979.

262   I. Shibata, S. Okada and K. Tokura, Y. Arao, I. Abiko and K. Nihei, "400 Dots/in. LED Printing Head With High Accuracy Die Bonding Technology," IEEE International Electronic Manufacturing Technology Symposium, pp. 158-163, 1986.

263   R.W. Shreeve, "Automated On-Chip Burn-In System for TAB Reels", Proceedings of the 39th IEEE Electronic Components and Technology Conference, pp. 187-189, 1989.

264   J. Simon and H. Reichl, "Single Chip Bumping for TAB", Proceedings of the 2nd International TAB Symposium, pp. 49-53, 1990.

265   J. Simon, E. Zakel and H. Reichl, "Electroless Deposition of Bumps for TAB Technology", Proceedings of the 40th IEEE Electronic Components and Technology Conference, pp. 412-417, 1990.

266   D.J. Small, "Tape Automated Bonding and its Impact on the PWB," Circuit World, Vol. 10, No. 3, pp. 26-29, 1984.

267   D.J. Small and A.A. Blain, "Tape Automated Bonding," British Telecom Research Journal, Vol. 3, No. 3, pp. 86-92, July 1985.

268   J.M. Smith and S.M. Stuhlbarg, "Hybrid Microcircuit Tape Chip Carrier Materials/Processing Trade-Offs," IEEE Transactions on Parts, Hybrids, and Packaging, Vol. PHP-13, No. 3, pp. 257-268, September 1977.

269   C.J. Speerschneider and J.M. Lee, "Solder Bump Reflow Tape Automated Bonding", Proceedings of ASM International Electronic Materials and Processing Congress, pp. 7-12, 1989.

270   P.J. Spletter and R.T. Crowley, "A Laser Based System for Tape Automated Bonding to Integrated Circuits", Proceedings of the 40th IEEE Electronic Components and Technology Conference, pp. 757-761, 1990.

271   V. Solberg, "Design Considerations for Solder Mask and Solder Paste Application on Fine Pitch SMT", Proceedings of NEPCON West, pp. 593-601, 1989.

272   R.P. Stapleton, "Surface Elongation and Interfacial Sliding During Gold-Gold Thermocompression Bonding," Master Thesis, Lehigh University, 1974.

273   D.W. Still, "A 4ns Laser-Customized PLA with Pre-Program Test Capability," Proceedings of IEEE International Solid-State Circuits Conference, pp. 154-155,

1983.

274   W. Su and S.M. Riad, "Wideband Characterization and Modeling of TAB Packages Using Time Domain Techniques", Proceedings of the 40th IEEE Electronic Components and Technology Conference, pp. 990-994, 1990.

275   S.M. Sze, "VLSI Technology," McGraw-Hill Book Company, New York, NY, 1983.

276   N. Tajima, Y. Chikawa, T. Tsuda and T. Maeda, "TAB Design for ESD Protection", Proceedings of the 2nd International TAB Symposium, pp. 77-87, 1990.

277   H.P. Takiar, "TAB Technology- A Review," Electro and Mini/Micro Northeast Conferences Record, pp. 6/2-1-3, 1986.

278   H.P. Takiar and B.J. Shanker, "Enhancing System Performance Using Tape Automated Bonding," Proceedings of the Surface Mount '88 Conference, pp. 109-115, 1988.

279   H.P. Takiar and J. Belani, "Stresses in TAB Bonded Packages," Proceedings of NEPCON WEST, pp. 127-129, 1987.

280   H.P. Takiar and B.J. Shanker, "Thermal Characterization of TAB Packages," Professional Program Session Record 23, Wescon, November 1987.

281   R.S. Tarter, "Apple Computer's Approach to Instituting the TAB Process," Proceedings of the Surface Mount '88 Conference, pp. 103-107, 1988.

282   H. Teoh, M. McGeary, and J. Ling, "Laser/Infrared Evaluation of TAB Innerlead Bond Integrity", Proceedings of the 40th IEEE Electronic Components and Technology Conference, pp. 442-449, 1990.

283   T. Tsuda and M. Hayakawa, "New Inner Lead Bonder for Multistation Tape Automated Bonding," Solid State Technology, pp. 167-169, August 1985.

284   J. Tuck, "Chip-On-Board Technology," Circuits Manufacturing, pp. 78-83, March 1984.

285   C.W. Umbaugh, "New Packaging Technology for Honeywell Large Scale Computer System," Proceedings of IEEE 14th Computer Society International Conference, pp. 263-266, 1977.

286   C.W. Umbaugh, "Tape Automated Mass Bonding," Electronic Packaging and Production, pp. 49-53, October 1976.

287   R.F. Unger and J.W. Kanz, "BTAB's Future-An Optimistic Prognosis," Solid State Technology, pp. 77-83, March 1980.

288   R.F. Unger and J.W. Kanz, "Bumped Tape Automated Bonding (BTAB) Past, Present and Future," Proceedings of ISHM International Microelectronics Symposium, pp. 161-168, 1978.

289   R.F. Unger, E.P. Kelley and C. Burns, and N. Laybhen, "Hermetic Tape Carrier- A New High Density Leaded Ceramic Chip Carrier," International Journal of Hybrid Microelectronic, Vol. 5, No. 2, pp. 373-380, November 1982.

290   R.F. Unger, "Overview of Ongoing Navy Manufacturing Technology (MT) programs," Proceedings of International Microelectronics Conference, pp. 64-70, 1981.

291   R.F. Unger, C. Burns and J. Kanz, "Bumped Tape Automated Bonding (BTAB) Applications." Proceedings of International Microelectronic Conference, pp. 71-77, 1979.

292   A. Van Der Drift, W.G. Gelling and S. Rademakers, "Integrated Circuits with Leads on Flexible Tape," Philips Technical Review, Vol. 34, No. 4, pp. 85-95, 1974. Also, Solid State Technology, pp. 27-35, 1976.

293   E.J. Vardaman, "New TAB Developments in The United States and Japan: A Market/Technology Comparison", Proceedings of NEPCON West, pp. 863-876, 1989.

294   O. Vaz, G. Iverson and J. Lynch, "Effect of Bond Process Parameters on Au/Au Thermal Compression Tape Automated Bonding (TAB)," Proceedings of IEEE International Electronic Manufacturing Technology Symposium, pp. 22-32, October 1987.

295   J. Walker, "THETAj1: A New Definition of Package Thermal Measurement," Professional Program Session Record 23, Wescon, November 17-19, 1987.

296   J. Walker, "Molded Versus Direct Mount TAB: A Comparison," Proceedings of NEPCON West, pp. 775-784, February 1988.

297   D. Walshak, "The Effects of Bonder Parameters on Gold/Gold TAB Inner Lead Bonding", Proceedings of NEPCON West, pp. 906-928, 1989.

298   S.B. Weinstein, "Smart Credit Cards: The Answer to Cashless Shopping," IEEE Spectrum, pp. 43-49, February 1984.

299   J.D. Welterlen, "Trends in Interconnect Technology," Proceedings of ISHM International Microelectron Symposium," pp. 111-113, 1980.

300   E.A. Wilson, "An Analytical Investigation into The Strain Distribution Within IC Bumps," International Journal of Hybrid Microelectron, Vol. 4, No. 2, pp. 233-239, 1981.

301   E.R. Winkler, "High Performance Packaging," Semiconductor International, pp. 350-355, May 1985.

302   E.R. Winkler, "Current and Future High Performance Packaging," Proceedings of NEPCON West, pp. 437-457, February 1985.

303   G. Wolfe, "Tape Carrier Technology, Will It Automate Microcircuit Production?,"

Circuits Manufacturing, pp. 74-79, September 1977.

304   M. Wong, "A High Yield, High Volume, and High Reliability TAB Outerlead Bonding Method", Proceedings of the NEPCON East, pp. 772-787, 1989.

305   M. Wong, "TAB Outerlead Bonding and SMT," Proceedings of NEPCON West, pp. 785-789, February 1988.

306   S. Wong, "TAB Inner Lead Bonding for High-Lead-Count, High-Volume Applications," Proceedings of NEPCON West, March 1989.

307   F-J Wu, and S.P. Sun, "Rework and Repair of TAB Devices", Proceedings of the NEPCON East, pp. 359-366, 1990.

308   F-J Wu, and D. Mackersie, "Inspection of Tape Automated Bonding Devices", Proceedings of the NEPCON East, pp. 788-798, 1989.

309   F.J. Wu, "Inspection of Tape Automated Bonding Devices," Proceedings of NEPCON West, March 1989.

310   J. Xie, and Y. Wang, "The Technique of The Multichip Tape Carrier," Proceedings of the 37th IEEE Electronic Components Conference, pp. 70-73, May 1987.

311   E. Zakel, and H. Reichl, "Investigations of Failure Mechanisms of TAB-Bonded Chips During Thermal Aging", Proceedings of the 40th IEEE Electronic Components and Technology Conference, pp. 450-459, 1990.

312   G. Zimmer, "Using Advanced Pulsed Hotbar Solder Technology for Reliable Positioning and Mounting of High Lead Count Flat Packs and TAB Devices", Proceedings of the 2nd International TAB Symposium, pp. 230-249, 1990.

STRESS ANALYSIS FOR

COMPONENT-POPULATED

CIRCUIT CARDS

PETER A. ENGEL

2.1. Introduction

In the mechanical design of second level packaging structures, circuit cards are considered to be load carrying elements. Their mechanical tasks include the support of chip carriers which, generating heat during operation, contribute the functional loading to be withstood by the leads and solder joint interconnections. In addition, however, vibrational loads are almost always present during operation. The processes of manufacturing and handling may introduce crucial stresses; besides posing a "time zero" concern, they may critically influence the longevity (fatigue life) of the joints. Mechanical stresses arising in circuit cards may be greatly influenced by their attachment and support on planars, larger boards, frames, or through connectors; these are usually referred to as "boundary conditions."

      The engineering properties of circuit cards play a paramount role in packaging design. For multi-layered construction, it is desirable to reduce the thickness and dielectric constant of the polymer sandwiching together copper conductors. Good heat conduction is desirable. Thermal mismatch between copper and polymer must be reduced in the plane of the card; since the stiffness of the polymer is smaller in the transverse direction, orthotropic behavior results. For the traditional materials of construction, Table 2.1 shows data, based on those given by Tummala and Rymaszewski [1].

      By the method of attachment of chip carriers we distinguish those of leadless interconnections, pin-grid arrays and peripheral leaded arrangements. The interconnections devised are also strongly related to the particular chip carrier used, such as ceramic (CCC) or plastic (flatpack), etc. Typical representative schemes of these methods and their analysis will constitute the bulk of this chapter.

      Quality assurance requires an understanding of the solder joint failure mechanisms arising due to thermal, vibrational and handling loads. The critical failure parameter (e.g. stress or strain) can often be determined by analysis, be it performed by classical or finite element procedures. The failure parameter must naturally reflect repeated loading (fatigue) conditions.Thus mechanical evaluation methods, coupled with testing, can provide the desired reliability for the structure.

2.2. Leadless Chip Carriers (LCC)

For LCC's, thermal stress analysis for the mismatch u between CTE (coefficients of thermal expansion) of the ceramic substrate (α_1) and the circuit card (α_2) was developed by Hall [2]. The mismatch can be written,

Table 2.1 Engineering properties for circuit card and module materials.

	Dielec Constant	TCE ppm/$^\circ$C	Thermal Conduct'y w/m.$^\circ$C	Elastic Modulus GPa
Glass-ceramics	4-8	3-5	5.0	300
Cu clad Invar (10% Cu) /glass coated		3	100	200
Epoxy-Kevlar, x-y (60%)	3.6	6	0.2	
FR-4 (x-y plane)	4.7	16	0.2	14
Polyimide	3.5	50	0.2	2.7
Teflon	2.2	20	0.1	0.5
Copper		17	393	124
Gold		14	297	40
Aluminum		23	240	70
Pb-5%Sn		29	63	18
Kovar		5.3	17	138

assuming uniform heating:

$$u = \Delta T.(\alpha_2 - \alpha_1).L \qquad\qquad [2.1]$$

The solder posts receive a shear and a moment; Hall applied strain gages on top and bottom of both substrate and card to measure the state of stress (or strain) obtained. Figure 2.1(a) shows the schematic including the positions of the four strain gages, e_1, e_2, e_3, e_4.

In deriving an equation for the shear force F experienced by a solder post, Hall assumed that an equivalent polar symmetrical model can replace the square chip carrier. This simplification has the advantage of yielding radial shears F_r which can be related to the moments M_r; for F_r see Equation [2.2]:

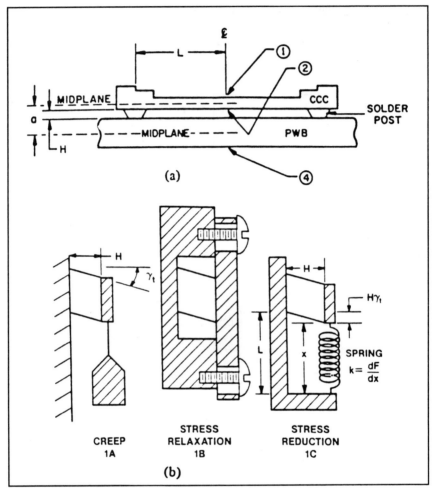

Figure 2.1 Schematic for leadless chip carrier system.

$$F_r = \frac{s}{12\left(H+\dfrac{h_1+h_2}{2}\right)}\left[\frac{E_1 h_1^2(e_2-e_1)}{1-v_1}+\frac{2E_2 h_2^2(e_4-e_3)}{1-v_2^2}\right] \qquad [2.2]$$

Here $E_{1,2}$ and $v_{1,2}$ are the elastic constants and $h_{1,2}$ the thicknesses of the ceramic and circuit card, while s is the solder post spacing.

The above system was generalized by Hall [3], showing that the constitutive equation of solder can be obtained by measuring the strains as the temperature T of the assembly was raised and lowered. Figure 2.1(b) shows the scheme of "stress reduction" composed of both creep and relaxation. The total behavior of the structure is a combination of the spring element k which represents the elasticity of the substrate and circuit card on one hand, and the shearing γ of solder posts on the other. When the solder is solid, the k factor can, rather accurately, be obtained by using Timoshenko's bimetallic thermostat formula [4] for the curvature, Equation 2.3, which governs the behavior of the sandwich in the colder region, up to about 40°C.

$$\frac{1}{R} = \frac{6}{h_1+h_2}\left[\frac{(\alpha_2-\alpha_1)(1+h_1/h_2)^2\,\Delta T}{3\left[1+\frac{h_1}{h_2}\right]^2\left[1+(\frac{h_1}{h_2})(\frac{E_1}{E_2})\right]\left[(\frac{h_1}{h_2})^2+\frac{1}{(h_1/h_2)(E_1/E_2)}\right]}\right] \qquad [2.3]$$

This analysis explains the hysteresis property of solder (Figure 2.2) during thermal chamber cycling. At low temperatures the assembly of Figure 2.1(a) is convex from above, while above 40°C the curvature reverses to concave.

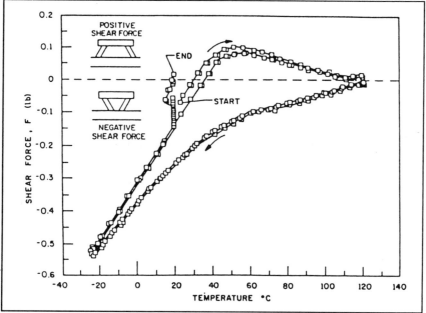

Figure 2.2 Shear force loop obtained from thermal cycling [3].

The large strains induced in solder by leadless chip carrier systems require application of a low thermal expansion circuit card material [5] such as copper-Invar-copper systems can give. Figure 2.3 shows the equivalent CTE in terms of relative Invar content in the sandwich. Simple formulae for the equivalent modulus and CTE respectively, of a multi-material sandwich, allowed to expand in one dimension only, are as follows:

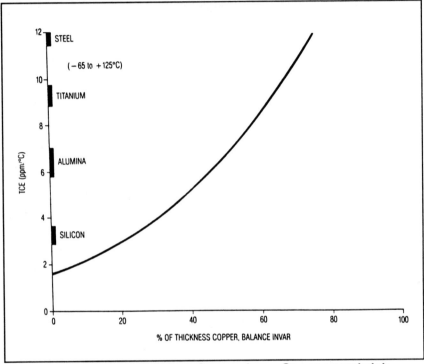

Figure 2.3 Equivalent CTE in a Copper-Invar-Copper sandwich.

$$E = \frac{\sum h_i E_i}{\sum h_i} \qquad [2.4]$$

$$\alpha = \frac{\sum h_i E_i \alpha_i}{\sum h_i E_i} \qquad [2.5]$$

It is noted that even a complete matching of CTE's between chip carrier and circuit card may produce substantial transient thermal stresses in the case of uneven heating of the two components. As a counter-measure, flexible leads can be introduced for mechanical and electrical continuity.

2.3. Pin Grid Array

The vast majority of pin-grid array structures used today is of the pin-in-hole type. A schematic, including thermal deformations, is shown in Figure 2.4. The pin-connection to the ceramic substrate is brazed or swaged, this joint acting as a fixity. The bottom is soldered in a card hole, and this can be modelled as an elastic foundation. Engel et al [6] described thermal analysis of such systems. Thermal mismatch is now taken up by the rather flexible leads, Amzirc and Kovar being some of the popular materials. Table 2.2 shows mechanical properties.

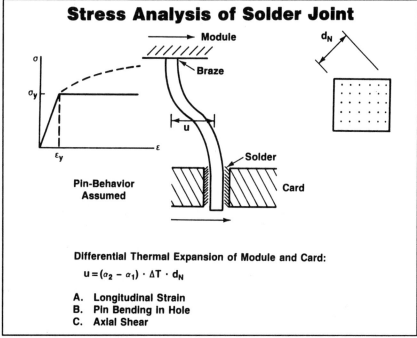

Stress Analysis of Solder Joint

Differential Thermal Expansion of Module and Card:

$$u = (\alpha_2 - \alpha_1) \cdot \Delta T \cdot d_N$$

A. Longitudinal Strain
B. Pin Bending in Hole
C. Axial Shear

Figure 2.4 Schematic of pin grid array structure and thermal deformations.

Large numbers of specimens subjected to accelerated thermal

Table 2.2 Mechanical properties of pin materials [6].

	Kovar	Alloy 42	Amzirc
	Fe/Ni/Co alloy	Fe/Ni alloy	Zr/Cu alloy
E Gpa (k-psi)	138 (20,000)	148 (21,500)	127 (18,500)
α (ppm/$^\circ$C)	5.9	5.4	16
ANNEALED:			
σ_y (k-psi)	345 (50)	276 (40)	89.7 (13)
σ_u (k-psi)	517 (75)	517 (75)	255 (37)
ϵ_u	0.30	0.30	
UNANNEALED:			
σ_y (k-psi)	724 (105)	731 (106)	-
σ_u (k-psi)	724 (105)	731 (106)	-
ϵ_u	-	-	-

tests (ATC) proved that corner pin-joints failed most often. The mode of failure was barrel cracking of the solder. Limited barrel cracking of the solder is allowed by most specifications, recognizing that by cracking, the effective pin-length is increased, and so a partial strain relief is achieved.

The thermal mismatch, Equation 2.1, between chip carrier (module) and circuit card induces bending in the pin. An approximate fixed-fixed beam model of the structure ignores deformation of the solder and, consequently, rotation and translation of the pin in the soldered hole. Such a simplification would lead to an overestimate of pin stress; meanwhile, it disallows calculation of solder pressures caused by pin anchorage. It is thus both realistic and useful to consider the role of solder as an elastic foundation.

The essentials of elastic foundation analysis [7] for the pin embedment include a shear force V and moment M acting at the entry, $x = 0$, while V and M vanish at the opposite end of the card, $x = a$ (Figure 2.5). The differential equation for the embedded beam is

$$w^{IV} + 4\lambda^4 w = 0 \tag{2.6}$$

where

$$\lambda = \left[\frac{k}{4EI}\right]^{1/4} \tag{2.7}$$

The general solution is

$$w = \cosh\lambda x(C_1\cos\lambda x + C_2\sin\lambda x) + \sinh\lambda x(C_3\cos\lambda x + C_4\sin\lambda x) \tag{2.8}$$

the coefficients C of which must be determined by four boundary conditions.

The analysis determining V_o and M_o on the solder joint due to a thermal mismatch displacement u can utilize the stiffness coefficients [k] on a beam embedded in solder ($0 \leq x \leq a$); the 2x2 coefficient matrix is obtained for the shear V_o and bending moment M_o required for causing unit displacements w_o and w_o' at the entry, $x = 0$. The stiffness matrix reduces to a particularly simple form if the non-dimensional quantity $\beta = \lambda a$ is larger than 3, a condition satisfied for most designs. We then get

$$\begin{bmatrix} V_o \\ M_o \end{bmatrix} = 2EI\lambda \begin{bmatrix} 2\lambda^2 & \lambda \\ \lambda & 1 \end{bmatrix} \begin{bmatrix} w_o \\ w_o' \end{bmatrix} \tag{2.9}$$

The displacements at $x=0$ can now be determined from the algebraic system of equations:

$$V_o = \frac{12EI(u-w_o)}{L^3} - \frac{2EIw_o'}{L^2}$$

$$M_o = \frac{6EI(u-w_o)}{L^2} - \frac{4EI}{L}$$

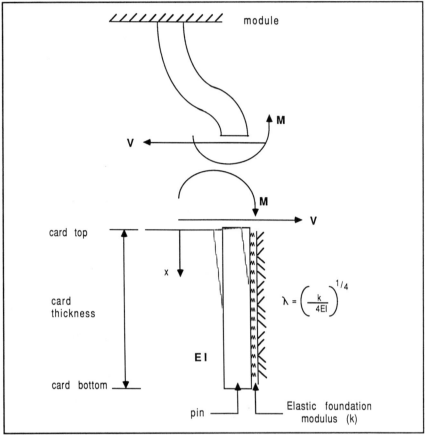

Figure 2.5 Schematic of elastic foundation treatment of pin-through-hole solder joint.

Figures 2.6 show the effects of both the card thickness "a" (i.e. soldered length) and the standoff length L on the maximum non-dimensional solder pressure p* and on the maximum pin strain (i.e. curvature x half-thickness).

The pin may plastically deform under thermal excursions encountered in practice, necessitating plastic analysis for the pin subjected to known end-displacement u.

Another failure mechanism of the solder joint is the axial force P that is developed by the lengthening ("S-shape") of the pin during flexural deformation. P induces a shear stress on the solder, while it adds a slight amount of bending moment on the pin as well. Computation of the axial force P is based on elongating the pin by an extra length ΔL.

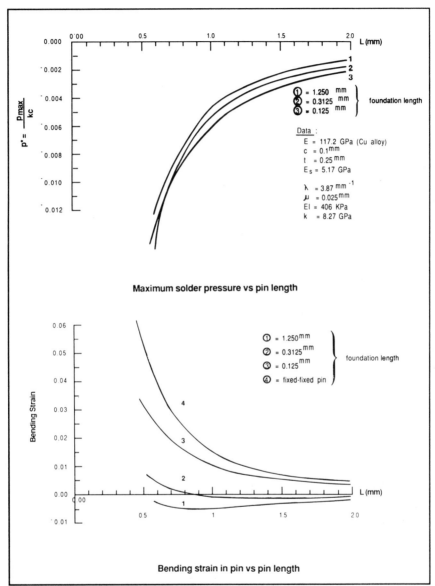

Figure 2.6 Effect of pin length and solder embedment on stresses in a pin-through-hole system. (a) Solder pressure. (b) Pin strain.

The formula resulting from elastic pin deformation is

$$P = \frac{3\pi E c^2 \Delta L^2}{5L^2}$$ [2.10]

Analysis shows that the pin's ability to plastically deform actually reduces P. Such an elastic-plastic analysis was made by Engel et al. [6], for both bending and stretching of a pin-through-hole arrangement.

Thermal stress analysis may be refined by including the finite flexibility, hitherto neglected, of both chip carrier and circuit card; this tends to reduce the pin stresses by an amount proportional to the size of the chip carrier. The "system reduction factor" f for various structures was obtained by Engel et al [8], by use of finite element methods. Defining it as the ratio of lead bending stress that includes carrier- and card flexure to the lead bending stress in a single pin, it yielded values ranging between 0.63 and 1.0; it is proportional to pin spacing, and inversely proportional to module size and standoff.

Figure 2.7 shows the system reduction factor for various pin-grid array systems. Figure 2.8 shows the dependence of the allowable (square-) module size on the pin length and diameter.

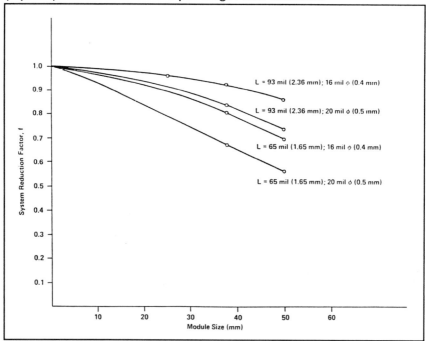

Figure 2.7 System reduction factor.

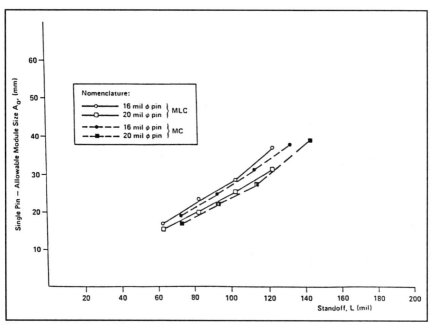

Figure 2.8 Allowable module size versus pin length.

It is remarked that pin-grid arrays can be produced not only by use of pin-through-hole technology, but by surface soldering of the pin to the card as well [9].

2.4. Flatpack Module

Plastic leaded chip carriers (PLCC) have flexible leads (typically J-lead or gullwing constructions) which are surface soldered to cards. For lead spacing, 1-2 mm has been popular, but much smaller dimensions are being introduced in the industry. These interconnections allow a great deal of thermal mismatch. However, the use of larger modules may still be critical for the maximum lead stress; vibration loads must also be checked for each particular design, especially above the 28 mm square size. Vibration analysis of leaded modules idealized as rigid frames was given by Steinberg [10]. In what follows, we shall review the loading of a module-lead-card assembly through static deformation, along the lines of a recent study by Engel [11].

In flexing, the cause of lead stress is the curvature of the card that is transmitted to the module by leads of appreciable stiffness. Figure 2.9 shows an assembly, with the card and module having distinct plate constants D (or "rigidities"). A flexible card coupled to a stiff module will tend to induce large lead stresses; however, the leads must be sufficiently stiff to transmit adequate structural interaction from card to module. This interaction produces a participation of the card and module in the bending moment M, in proportion to their rigidities. While this participation

$$\frac{M_m}{M} = \frac{D_m}{D_m+D_c} \; ; \; \frac{M_c}{M} = \frac{D_c}{D_m+D_c} \qquad [2.11]$$

must be developed gradually from the corner lead toward the inner leads, the approximate total rigidity of an area of module-populated card is somewhat less than the total rigidity $D_T = D_m + D_c$.

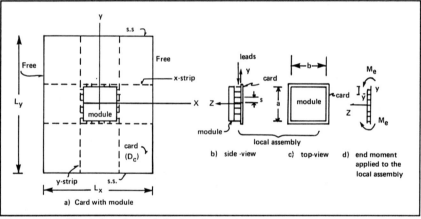

Figure 2.9 Models of module-lead-card system.

Apart from determining the total rigidity, say for vibration analysis purposes [12], the analyst needs the lead forces for stress analysis. These lead forces, whether from bending or twisting (the combined action is called "flexing") are difficult to determine from experiments, because of the small size of the leads. Closed-form analytical determination is out of the question, because of the great complexity of the structure. Numerical (finite element) results can be used to advantage. The engineering theory quoted above has been worked out for rapid calculation purposes; it was corroborated with finite element computations. All but the axial lead force components were neglected, since the latter were the most substantial; inclusion of other components (shears and moments) would tend to yield non-conservative results.

In analyzing a card-and module assembly (Figure 2.9(b)), a strip in the (y,z) plane, including the module in the center was studied. The row of leads (each having an individual spring constant K) on the side parallel to the y-axis was idealized as an elastic foundation of foundation modulus $k = 2K/s$; the factor of 2 includes leads on both sides of the module. Recognizing that the module develops structural interaction with the card gradually as we pass from the corner toward the center of a row of leads, the "moment development function" $g(y)$ was derived from D_m, D_c, a, s and k. This non-dimensional function, which stands for lead stiffness, has a maximum value of 1, meaning immediate load transmission from card to the module, as later used in Equation 2.14. The "continuous" $g(y)$ function derived for a beam of continuous elastic foundation is not as realistic as the "discrete" one that considers large additional end-springs representing the leads marching in the perpendicular direction for square modules. Figure 2.10 shows a family of such discrete $g(y)$ functions; the non-dimensional parameter ρ is proportional to lead stiffness, since

$$\rho = \frac{\lambda a}{2} , \qquad\qquad [2.12]$$

the elastic foundation constant of the row of leads being

$$\lambda = \sqrt[4]{\frac{K(1/D_c+1/D_m)}{4a}} \qquad\qquad [2.13]$$

For a perpendicular x-strip, a similar analysis was performed. Superposing, at last, the lead force contributions from both y- and x-strip, there resulted

$$P = \frac{D_m}{D_m+D_c}\left[g(y)M_y'+g(x)M_x'\right] \qquad\qquad [2.14]$$

where the bending moments M_x', M_y' are taken per unit width of strip.
    Note that the rigidity of the module, that of the card and that of the leads is represented in the above equation, but the corner force is, above all, proportional to the local bending moment(s). This means that the curvature of the module locality enters decisively into the calculation

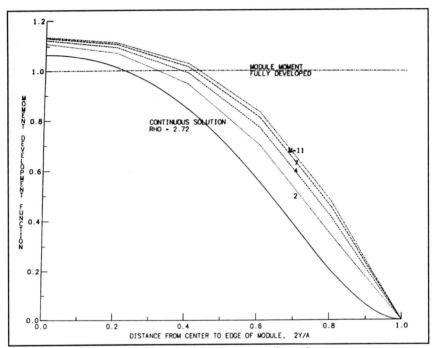

Figure 2.10 Discrete moment development functions.

of the maximum lead force.

A simpler, more conservative formula was also derived,

$$P = \frac{kMa}{4D_c\lambda^2 m}$$ [2.15]

The parameter m stands for the number of leads perpendicular to the direction of bending.

Finite element answers were found to be straddled by the two equations [2.14] and [2.15]. For different boundary conditions, the local plate moments M_x' and M_y' could be inserted in the above formulae.

This engineering theory can be used to investigate the effect of design changes of the following: the material (stiffness) of the card, module, and the leads; the lead spacing; the size of module and card; and use of various module clusters. Simple calculations extended its use to double-sided module arrangements, and stacked module arrangements as well, Figure 2.11.

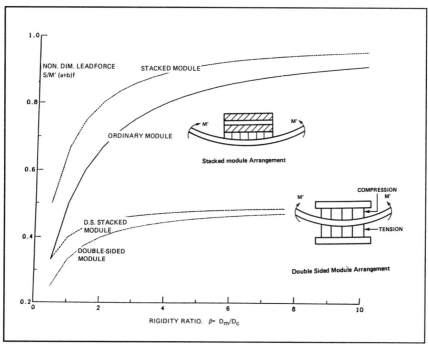

Figure 2.11 Lead force versus rigidity ratio of module to card, for double sided and stacked configurations.

Often module-populated circuit cards are handling-tested in torsion [13] instead of bending. It is again the corner lead forces which prove to be the largest. An engineering analysis by Engel and Vogelmann [14] calculated maximum corner lead forces Q in various configurations, utilizing the equivalence of the external card torque to that taken up by the leads. Simulation of bending and twisting of surface mounted assemblies have been performed and analyzed by Wong et al [15] using a hybrid experimental/analytical approach.

2.5. References

1   R.R.Tummala and E.J.Rymaszewski, Microelectronics Packaging Handbook, (Van Nostrand, New York, 1989).

2   P.M.Hall, "Forces, Moments, and Displacements During Thermal Chamber Cycling of Leadless Ceramic Chip Carriers Soldered to Printed Boards," IEEE Trans. CHMT-7, 314-327 (1984).

3   P.M.Hall, "Creep and Stress Relaxation in Solder Joints in Surface Mounted Chip

Carriers," Proc. 1987 IEEE ECC Conference, CHMT, pp 579-588.

4   S.P.Timoshenko, Collected Papers, (McGraw Hill, New York, 1953).

5   R.J.Hammond, "Digital Electronic Engine Controls and Leadless Chip Carriers,"
        Proc. IEEE/AIAA 7th Digital Avionics Systems Conference (1986), Ft. Worth TX, pp
        625-632.

6   P.A.Engel, C.K.Lim, M.D.Toda and R.Gjone, "Thermal Stress Analysis of Soldered
        Pin Connectors for Complex Electronics Modules," Comput. Mech. Eng., 2, 59-69
        (1984).

7   M.Hetenyi, "Beams on Elastic Foundation," Chapter 31 in Handbook of
        Engineering Mechanics, edited by W.Flugge (McGraw-Hill, New York, 1962).

8   P.A.Engel, M.D.Toda and A.K.Trivedi, "Design Guide for Solder Cracking in
        Module-Card Systems," IBM Technical Report 01.2678, Endicott (1983).

9   P.A.Engel and L-C.Lee, "Surface Solder Stress Design for Thermal Loading of
        Modules," Proc. 1986 NEPCON East Conference, Boston MA, pp 263-270.

10   D.S.Steinberg, Vibration Analysis for Electronic Equipment, (Wiley, New York,
        1973).

11   P.A.Engel, "Structural Analysis for Circuit Card Systems Subjected to Bending," J.
        Electronic Packaging, Trans. ASME, 112, 2-10 (1990).

12   J.M.Pitarresi, "Modeling of Printed Circuit Cards Subjected to Vibration," Proc.
        1990 IEEE ISCAS Conference, New Orleans LA, pp 2104-2107.

13   P.A.Engel, "Torque Stress Analysis for Printed Circuit Boards Carrying Peripherally
        Leaded Modules," 1986 ASME Winter Annual Conference, Anaheim CA, Paper No.
        86-WA-EEP-2.

14   P.A.Engel and J.T.Vogelmann, "Structural Analysis for Circuit Card Systems
        Subjected to Torsion," (to be published).

15   T-L.Wong et al, "Strength Analysis of Surface Mounted Assemblies Under Bending
        and Twisting Loads," J. Electronic Packaging, Trans. ASME 112, 168-174 (1990).

MODELING CONCEPTS FOR

THE VIBRATION ANALYSIS

OF CIRCUIT CARDS

JAMES M. PITARRESI

3.1 Introduction

In today's electronics packaging, a great variety of surface mount technology (SMT) as well as pin-in-hole (PIH) configurations are being employed. These module configurations vary widely with respect to size, shape, arrangement, and material. Additionally, these modules are then mounted on a printed circuit card (PCC), which is typically composed of laminations of conducting and non-conducting materials, through the use of soldered leads or pins. Since circuit cards in a given electronic package can vary widely with respect to the module type, configuration, size, and boundary conditions, it is easy to see that a typical PCC populated with modules is a structure of impressive complexity.

Usually, one of the most important mechanical design considerations is that of thermally induced stresses. However, in addition to thermal management, other sources of loading must be considered. Dynamic effects arising in an electronic package, whether owing to environmental, manufacturing, shipping, or other causes, may result in vibration amplitudes and/or acceleration levels which must be sustained by the package [1,2]. These vibrations are transmitted throughout the PCC, inducing stress in the connectors, modules, and perhaps most importantly, the leads and solder joints attaching the modules to the PCC. Over stressing of any component of a populated PCC can lead to performance degradation and ultimately to the complete failure of the system. Consequently, it is the task of the electronics packaging engineer to provide a reliable, cost-effective structure to meet these demands. Toward this end, it is the purpose of this chapter to present the following topics:

1. A brief review of the theory of vibrations.
2. Basic concepts of the finite element method.
3. Fundamentals of experimental modal analysis.
4. Validation techniques for the correlation of finite element and experimental data.

Additionally, a case study employing the above techniques will be presented.

3.2 Vibration Theory Fundamentals

One question that often occurs in the design of a structure is whether or not to include dynamic effects in the analysis. The basic rule-of-thumb in vibration analysis is that if inertial forces contribute significantly to the response, then a dynamic investigation is in order. How to include these inertial effects is the subject of this section. As this section is intended as a brief overview, only the key concepts are included. Details may be found in a number of texts on vibrations, including Craig [3].

The mathematical modeling of a structural system can be divided into two broad categories: continuous modeling and discrete modeling. Continuous models are those that are described by functions of spatial variables and time. Discrete models reduce the continuous representation to that of a collection of discrete components, such as mass or stiffness components. When assembled, these discrete components typically represent an approximation to the continuous system. Discrete models are further divided into those which are single-degree-of-freedom (SDOF) models, and those which are multi-degree-of-freedom (MDOF) models. It is the description of the SDOF and MDOF systems that is of current interest in this chapter.

The components of a discrete parameter model are those that are related to the displacement, velocity, and acceleration. Each, in turn, is related to a force. By invoking Newton's Law, the dynamic equilibrium equations are then generated.

The component that relates a displacement to force is termed a spring. In linear elastic systems, the spring force F_s, is directly proportional to the elongation e, experienced by the spring, and is expressed by

$$F_s = K.e \tag{3.1}$$

where K is the spring rate, given in dimensions of force/length. Note that the elongation, which we can accept as acting either positively or negatively, causes energy to be stored in the spring. In addition to energy storage, a component is also available for the dissipation of energy, namely the viscous dashpot. The damping force, which acts in opposition to the motion, is assumed to be a viscous damping mechanism, with the force proportional to the relative velocity, given in equation form by

$$F_d = C.\frac{de}{dt} \tag{3.2}$$

where we see that the damping force F_d, is directly proportional to a damping coefficient C, times the time rate of change of the elongation of the dashpot. The dimensions of C are force-time/length. Lastly, through the Principle of D'Alembert, the inertial force F_i, is proportional to the mass of the discrete body M, times its acceleration a, relative to an inertial reference frame, given by

$$F_i = -ma \qquad [3.3]$$

Combining the components of a SDOF system, we can readily use Newton's Law to determine the equation of equilibrium of the body. Figure 3.1(a) shows a typical SDOF system under the influence of the effects we have just described, with the addition of a time varying forcing function P(t). Applying Newton's Second Law to the freebody diagram of Figure 3.1(b), we have

$$p(t) - F_s - F_d = m . \frac{d^2x}{dt^2} \qquad [3.4]$$

where the overdots indicate differentiation with respect to time. Upon substitution of Equations [3.1-3.3] into the above expression, the following differential equation of equilibrium is obtained.

$$m\frac{d^2x}{dt^2} + c\frac{dx}{dt} + kx = p(t) \qquad [3.5]$$

Equation 3.5 often represents a good approximation for many dynamic systems. However, it is often necessary to include may springs, dashpots, and masses in order to effectively model a complex system. Consequently, MDOF models, based upon the principles previously derived, must be used. As can be expected, such models drastically increase the difficulty of the modeling effort. In the next section, a technique for handling the modeling of complex systems is presented.

3.3 The Finite Element Method

3.3.1 Basic concepts

The finite element method of analysis has been extensively developed over the past three and one-half decades to become a powerful and

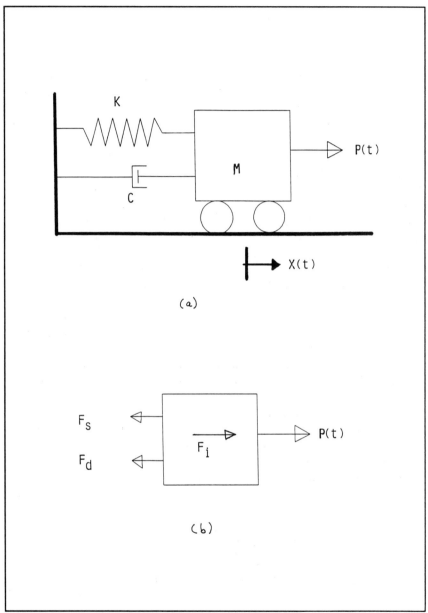

Figure 3.1 Idealized SDOF system. (a) Spring-mass-dashpot. (b) Free body diagram.

popular tool of both researchers and practicing engineers. Essentially, the finite element method (FEM) is an approximation technique that transforms the governing differential equations of a system to a less complex form that is more amenable to solution on the computer. Currently, the finite element method has been successfully used in virtually all areas of engineering science, including structural and solid mechanics, heat transfer, fluid flow, and electrical and magnetic theory, to name only a few. The broad scope of these applications is due, in no small part, to the basic steps of the FEM that remain essentially unchanged, regardless of the problem. These steps include: 1) the discretization of the domain of the problem into smaller subdomains, called elements, 2) making a reasonable guess as to the behavior of the field variable (and perhaps its derivatives) within the domain, and 3) the formulation of the problem in either a variational or method of weighted residuals form. Upon completion of these steps, a collection of matrices, representing the behavior of the element, is obtained. By assembling (connecting) the elements together over the entire domain, an algebraic approximation to the problem results. The boundary conditions are imposed, and the remaining equations are solved for the unknown field variables. Back-substitution of these results is often performed to obtain additional information such as stress, heat flux, etc..

The formulation of the element stiffness matrix is perhaps best illustrated by a simple example. Consider a simply supported beam of total length 2L, with a concentrated load acting in the center (Figure 3.2(a)). We will discretize this structure into two equal sized elements as shown in Figure 3.2(b). A typical beam element is shown in Figure 3.2(c) where we have indicated both transverse displacement and rotational degrees-of-freedom at each of the two nodes of the element, denoted by a_1 to a_4. It is at these nodes that we shall connect our element with other elements, and/or enforce boundary conditions. The differential equation for our beam element in the domain between the nodes, with no transverse loads, is given by [4]

$$\frac{d^4w(x)}{dx^4} = 0 \qquad\qquad [3.6]$$

The general solution of this equation is given as

$$w(x) = \alpha_1 + \alpha_2 x + \alpha_3 x^2 + \alpha_4 x^3 \qquad\qquad [3.7a]$$

or in matrix form

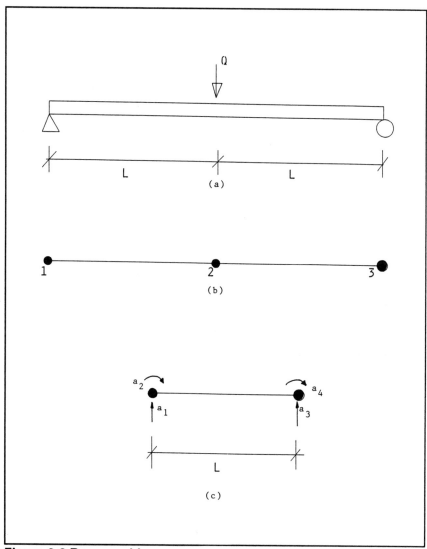

Figure 3.2 Beam problem. (a) Beam with loading. (b) Discretized beam. (c) Typical beam element.

$$w(x) = \begin{bmatrix} 1 & x & x^2 & x^3 \end{bmatrix} \begin{Bmatrix} \alpha_1 \\ \alpha_2 \\ \alpha_3 \\ \alpha_4 \end{Bmatrix} \qquad [3.7b]$$

where the α's are constants. We can see that we have four DOFs for our element and four constants in Equation 3.7. Therefore, by using Equation 3.7 as our "guess" of the element's behavior, we can relate the nodal constants to the nodal DOFs. The first step is to relate the nodal DOFs to the displacement components of $w(x)$ as follows:

$$w(0) = a_1 \quad w'(0) = a_2 \qquad\qquad [3.8a]$$

$$w(L) = a_3 \quad w'(L) = a_4 \qquad\qquad [3.8b]$$

Upon substitution of Equation 3.7, we have in matrix form

$$\begin{Bmatrix} a_1 \\ a_2 \\ a_3 \\ a_4 \end{Bmatrix} = \begin{bmatrix} 1 & 0 & 0 & 0 \\ 0 & 1 & 0 & 0 \\ 1 & L & L^2 & L^3 \\ 0 & 1 & 2L & 3L^2 \end{bmatrix} \begin{Bmatrix} \alpha_1 \\ \alpha_2 \\ \alpha_3 \\ \alpha_4 \end{Bmatrix} \qquad\qquad [3.9]$$

or symbolically

$$\{a\} = [A]\{\alpha\} \qquad\qquad [3.10]$$

However, we desire to describe the element's behavior in terms of its nodal DOFs (i.e. $\{a\}$), not in terms of the polynomial coefficients of Equation 3.7. That is, we require

$$\{\alpha\} = [A]^{-1}\{a\} \qquad\qquad [3.11]$$

so that now we can express the displacement of the element as

$$w(x) = \begin{bmatrix} 1 & x & x^2 & x^3 \end{bmatrix} [A]^{-1} \{a\} \qquad\qquad [3.12]$$

It is customary to express Equation 3.12 as

$$w(x) = [N] \{a\} \qquad [3.13]$$

where the shape function matrix [N] is a matrix of interpolation functions which interpolate through the domain of the element the values obtained at the nodal DOFs. Their role in finite element analysis is critical, as their selection influences the accuracy and convergence characteristics of the solution [4-6]. Carrying out the operations indicated in Equation 3.13, we see that the shape functions for our element are given by

$$[N]^T = \begin{bmatrix} 1-3\dfrac{x^2}{L^2}+2\dfrac{x^3}{L^3} \\[2mm] x-2\dfrac{x^2}{L}+\dfrac{x^3}{L^2} \\[2mm] 3\dfrac{x^2}{L^2}-2\dfrac{x^3}{L^3} \\[2mm] -\dfrac{x^2}{L}+\dfrac{x^3}{L^2} \end{bmatrix} \qquad [3.14]$$

For the beam element, a suitable variational statement is total potential energy functional π, given as

$$\pi = \int_0^L \frac{1}{2}EI\left(\frac{d^2w}{dx^2}\right)^2 dx - \{a\}^T\{q\} \qquad [3.15]$$

where E is the modulus of elasticity, I is the second moment of area of the beam cross section and {q} is a vector of externally applied, statically equivalent loads. Defining the generalized strain ϵ as

$$\{\epsilon\} = \left[-\frac{d^2w(x)}{dx^2}\right] \qquad [3.16]$$

and the generalized stress as

$$\{\sigma\} = \left[-EI\frac{d^2w}{dx^2} \right]$$

[3.17]

we have, on invoking the stationary of Equation 3.15, the following

$$\delta\pi = \int_0^L \{\delta\varepsilon\}^T\{\sigma\}dx - \{\delta a\}^T\{q\} = 0$$

[3.18]

From Equations 3.16 and 3.17 we can see that the displacement function w, is operated upon by the following linear differential operator

$$[L] = \left[-\frac{d^2}{dx^2} \right]$$

[3.19]

Since the displacement function is expressed by Equation 3.13, we see that the differential operator of Equation 3.19 will operate only on the shape functions [N]. Defining the [B] matrix as follows

$$[B] = [L]\,[N]$$

[3.20a]

or

$$[B] = \left[\frac{6}{L^2}-12\frac{x}{L^3} \;,\; \frac{4}{L}-\frac{6x}{L^2} \;,\; -\frac{6}{L^2}+\frac{12x}{L^3} \;,\; \frac{2}{L}-\frac{6x}{L^2} \right]$$

[3.20b]

we can then express Equation 3.18 as

$$\int_0^L ([B]^T[D][B])dx\{a\} = \{q\}$$

[3.21]

The constitutive matrix [D] simply relates stress to strain. For the beam element, it may be readily verified that [D] = EI. We can now define the beam element stiffness matrix as

$$[K] = \int_0^L [B]^T [D][B] dx \qquad [3.22a]$$

carrying out the indicated operations yields

$$[K] = \frac{EI}{L^3} \begin{bmatrix} 12 & 6L & -12L & 6L \\ 6L & 4L^2 & -6L & 2L^2 \\ -12 & -6L & 12 & -6L \\ 6L & 2L^2 & -6L & 4L^2 \end{bmatrix} \qquad [3.22b]$$

so that Equation 3.21 is now given in the familiar form

$$[K]\{a\} = \{q\} \qquad [3.23]$$

3.3.2 Finite element method in vibration analysis

For the problem of determining the natural frequencies and mode shapes, the following generalized eigenproblem must be solved

$$([K] - \omega^2 [M])\{\phi\} = \{0\} \qquad [3.24]$$

where [K] and [M] represent the FE stiffness and mass matrices, respectively, $\{\phi\}_i$ is the i-th mode shape vector, and ω_i is its corresponding natural frequency of vibration [7]. The mass matrix is a function of the mass density and the element shape functions and is expressed in the following form, where ρ is the mass density [5].

$$[M] = \int\int\int [M]^T \rho [M] dt \qquad [3.25]$$

A typical FE model may contain thousands of DOFs. Consequently, Equation 3.24 may become time consuming to solve. This is highlighted by the fact that in vibration problems, usually only the first few (i.e., lower) frequencies are desired. Therefore, reduction techniques are often used to reduce the size of the problem represented by Equation 3.24. The most common of these is the Guyan reduction method [6,8]. This method reduces the size of the matrix equations in Equation 3.24 by selectively picking key DOFs as "master" DOFs (MDOFs). The remaining DOFs, often referred to as "slave" DOFs, are eliminated from appearing explicitly in the reduced equation set. Their influence on the reduced stiffness matrix is correctly accounted for, however the redistribution of the mass is approximate. In selecting MDOFs, the frequency and mode shape accuracy is enhanced if their locations are somewhat evenly distributed throughout the structure. Additionally, MDOFs should be chosen at nodes for which the inertial forces, compared with the elastic forces, are significant (i.e, at locations of high mass to stiffness ratio).

3.3.3 Application to electronics packaging

The finite element method has been shown to be an effective and versatile computational tool for determining the stresses [9-11] and the dynamic characteristics [12-14] of a populated PCC. However, a detailed finite element (FE) model can be time consuming to generate and expensive to solve. This is heightened by the fact that a FE modal analysis often involves an iterative technique, thereby making the solution cost many times that of a static analysis of the same model [5]. Moreover, in the initial design stage, the engineer is often assessing the trade-offs between potential configurations, dictating a new model and analysis for each new configuration. An efficient (i.e., rapid and accurate) analysis technique is therefore required.

With a reduced complexity model, the designer can now rapidly estimate the natural frequencies and mode shapes of competing configurations. From this information, trade-offs and parameter studies may then be undertaken concerning such factors as lead forces, fatigue life, and dynamic response. Additionally, the optimal layout of components to reduce the effects of the shock and vibration environment becomes more practical with a reduced complexity model [15]. The net result is that the designer will be in a position to choose the best available configuration in the least possible time, thereby reducing the design time while increasing the reliability of the package.

One of the objectives of this chapter is to investigate the accuracy of the "smearing" or homogenizing technique for approximating the material and structural properties of a FE model of a populated PCC. The effects of both SMT and PIH components are included. Specifically, the natural frequencies and their corresponding mode shapes, obtained from experiment and from a smeared FE analysis, are compared. Conclusions as to the degree of correlation, frequency distribution, and the effects on the dynamic response are addressed. To define the validation of the FE model, the modal assurance criterion (MAC) is used as a measure of the correlation between the FE and experimental mode shapes [16]. Direct comparisons of correlated mode frequencies are used to indicate the departure of the FE natural frequencies from those obtained by means of testing. Additionally, other considerations are included in this study. These considerations include the sensitivity of the correlation to such influences as the number of elements used (i.e., mesh density), the location of the master degrees-of-freedom (i.e., matrix reduction techniques), and the procedures used for determining the smeared composite module/lead/card properties, must be addressed.

3.3.4 Smeared properties determination

The fundamental motivation for smearing the material and structural properties is to generate a structure which is not as complex to model and analyze as was the original. However, before one can construct a simplified model, one must comprehend the behavior of the original structure. To this end, a populated PCC can be envisioned as an assemblage of many plate and beam structural elements; the modules and the PCC behave as plates while the leads attaching the modules to the card behave as beams (see Figure 3.3(a)). In a smeared representation, it is the effective stiffness and mass density of the assemblage that is desired. In this case, the assemblage of plates and beams representing the module/lead/card configuration can be modeled as a single plate but with the effective stiffness and mass density of the original assemblage. Once the effective properties are known, the original configuration may be replaced by the smeared model (see Figure 3.3(b)). The net result is a model of greatly reduced complexity. Note that this technique may be applied to each module on a card, or to any group of similar and/or dissimilar modules, including the entire card itself.

For our purposes, it is desired to smear the original module/lead/card assemblage to that of an orthotopic plate with an equivalent stiffness and mass. The thickness of the smeared plate is taken as that of the original PCC so as to facilitate surface strain calculations, which are of critical importance in calculating the lead stresses [17].

One way to determine the smeared properties is to perform a bending test to the region of interest. For the cards considered in this paper, the regions were cut-out of the existing cards and subjected to a three point bend test along both of the principle directions of the region. The effective elastic moduli E_x and E_y are determined from Equations 3.26(a) and 3.26(b) as follows

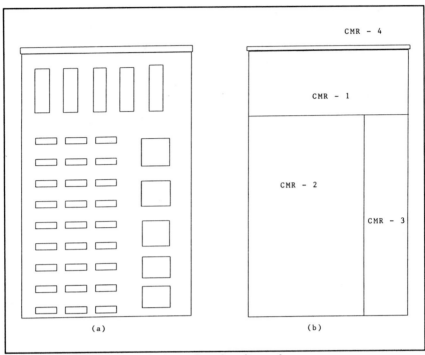

Figure 3.3 Smearing technique for circuit cards. (a) Actual card. (b) Smeared representation.

$$E_x = K_x \frac{a^3}{4bt^3} \qquad\qquad [3.26a]$$

$$E_y = K_y \frac{b^3}{4at^3} \qquad\qquad [3.26b]$$

where K_x and K_y are the slopes of the load-deflection curves obtained from bending about the y and x axes, respectively, and a and b are the unsupported dimensions of the sample (see Figure 3.3). The thickness t is the desired thickness for the smeared model which, for reasons cited earlier, is taken to be the thickness of the PCC. The smeared mass density is determined by

$$\rho = \frac{M}{abt} \qquad\qquad [3.27]$$

where M is the actual mass of the sample.

By repeating this procedure for various module types, a library of smeared material properties may be compiled. If samples are not available or if the design is at the conceptual stage, it is possible to simulate the bending test by using the finite element method. In this case, a model of the sample would be developed and the smeared properties are readily determined by simulating the loading and using Equations 3.26 and 3.27.

For the case studies considered in this chapter, the finite element mesh was chosen to be as simple as possible. Consequently, wherever a group of similar modules were located, the region was modeled as a single smeared composite region. Repeating this procedure of grouping similar regions into smeared property regions produces a "patch work" finite element mesh where the original populated PCC is replaced by an equivalent smeared card composed of several composite modeling regions (CMRs). The number and distribution of the CMRs is included as a parameter for study regarding their effect on the accuracy of the proposed smearing method.

3.4 Experimental Procedures

Experimental modal analysis is the process whereby the natural frequencies, modes shapes, and damping values are estimated from measurements made on an existing structure [16]. These modal parameters are determined via data curve-fitting algorithms from a set of frequency response functions (FRFs) obtained at various points throughout the structure. The FRFs are generated by exciting the structure, typically through a random shaker or an instrumented impact hammer, and measured by means of an acceleration transducer. In order to obtain the modal characteristics, the following assumptions are presumed of the structure [18]:

1.   The structural vibrations can be represented by a set of linear second order differential equations.
2.   The structure obeys Maxwell's Reciprocity Law.
3.   There are no repeated frequencies.
4.   The modal properties are globally, not locally, defined.

A distinct difference between testing and FE analysis is that testing yields, in addition to mode shapes and frequencies, estimates of the damping parameters of the system. In the FE analysis, damping is neglected when computing the natural frequencies and mode shapes. This suggests that the damping estimates from the experiment can be used in a forced response FE model for predicting amplitudes and stresses in the structure.

Typically, the dynamic equation of equilibrium is expressed in the Laplace domain as

$$s^2[M]\{X(s)\} + s[C]\{X(s)\} + [K]\{X(s)\} = \{F(s)\} \qquad [3.28]$$

where $\{X(s)\}$ and $F\{(s)\}$ are the Laplace transforms of the displacement and force vectors, respectively, and s is the Laplace variable. In systems form

$$[B(s)]\{X(s)\} = \{F(s)\} \qquad [3.29]$$

where

$$[B(s)] = s^2[M] + s[C] + [K] \qquad [3.30]$$

If no forcing function is included in Equation 3.29, it then reduces to a complex eigenproblem. The complex eigenvalues (also known as poles) are the modal frequencies and damping values.

From a testing point of view, it is necessary to impart some form of energy into the system so that its response, and hence its dynamic characteristics, may be determined. Defining the transfer matrix as

$$[H(s)] = [B(s)]^{-1} \qquad [3.31]$$

we can then measure the ratio of response to input, thereby building [H(s)]. As long as the roots of the determinant of [B(s)] are all distinct, [H(s)] can be written in partial fraction form, and hence, curve fitting techniques can be used on the data [18].

For the case study considered in this paper, the following experimental procedure was used. The populated PCC was suspended, with an elastic cord, so that it approximated free-free boundary conditions. An impact testing philosophy was assumed in which the structure was repeatedly impacted with an instrumented hammer at a number of predetermined points. These points where chosen so as to yield a good spatial representation of the lower modes of interest. Care was taken to avoid impacting the modules directly whenever possible. The response to the impacts were recorded at a fixed point on the PCC. The corresponding FRFs were then analyzed using standard modal analysis curve fitting techniques [18].

3.5 Correlation of the Modal Data

It is important when comparing results from the experiment and FE analysis that both the natural frequencies and the mode shapes should be investigated. It is possible to have the frequencies match in magnitude but have their corresponding mode shapes represent two different displaced configurations. Since the dynamic response of the populated PCC is a function of both the mode shapes and the frequencies (among other factors), it is logical that good correlation of both modes and frequencies is vital to the reliable prediction of the response. To this end, the modal assurance criterion (MAC) is used as a measure of the correlation of the mode shapes between the experimental and FE data sets. Basically, MAC is goodness-of-fit, in a least squares sense, between each mode shape from the two data sets. In matrix form it is expressed as [16]

$$[MAC(\{\phi_e\}_j , \{\phi_f\}_k)] = \frac{(\{\phi_e\}_j^T \{\phi_f\}_k)^2}{\{\phi_e\}_j^T \{\phi_e\}_j \{\phi_f\}_k^T \{\phi_f\}_k} \qquad [3.32]$$

where

$\{\phi_e\}_j$ = Experimental mode shape vector j = 1, n_e
$\{\phi_f\}_k$ = FE mode shape vector k = 1, n_f
n_e = number of experimental mode shape vectors
n_f = number of FE mode shape vectors

A MAC value of unity indicates that the two modes correlate perfectly; a value of zero indicates no correlation. In practice, it is difficult for even an experienced analyst to achieve MAC values for all correlated modes higher than 0.90 (i.e., 90% correlation) [19]. The required MAC for good modeling representation, however, is problem dependent. Research is currently under way to define such limits.

3.6 Case Study

In this section, a description of the case studies is presented. The results from both the experimental modal analysis, as well as the smeared FE analysis, are included. In all of the experimental test setups, the populated PCCs were supported in such a way as to simulate free-free boundary conditions. The reasoning for this is twofold. First, by approximating the free-free state, any bias towards a particular card support condition is eliminated. Additionally, the modeling of the "true" boundary conditions for actual circuit cards increases the complexity of the FE model due to the presence of edge connectors, guide rails, damping pads, and so forth. For the present study, this increase in modeling complexity would lead to ambiguity in the results, thereby obscuring the effectiveness of the smearing technique. The second reason for using free-free boundary conditions is that typically, more modes will be excited at lower frequencies. These additional modes, such as bending about both the x and y axes, and torsional modes, will give the researchers an opportunity to see how the proposed smearing method works for a wide variety of mode shapes. Since the FE model must also reflect the free-free condition, it is important to use a FE package that can permit the extraction of rigid body modes. For this research, the ANSYS FE software package was used [20].

The PCC shown in Figure 3.4 incorporates both SMT and PIH modules. Specifically, a tee shaped pattern of ceramic pin grid array (PGA) PIH modules is flanked by two identical arrays of 26 lead small outline J-lead (SOJ) SMT modules. Card edge connectors were soldered at opposite ends of the PCC.

For the modal test, the 35 impact points and a single response measurement point where chosen as shown in Figure 3.5. The resulting frequencies, and damping values are listed below in Table 3.1, and their corresponding mode shapes are shown in Figure 3.6. In this study, natural frequencies above 200 Hz were not of interest. Consequently, the modal data reported for this card, as well as for the other cards in this study, reflect this view point. The damping levels for all the reported modes are less than 1% of critical damping, indicating a very lightly damped structure.

For the smeared FE model, a logical approach to identifying regions of similar properties is to use the 'natural' divisions already present in the card, that is, the two ceramic PGA, and SOJ areas, as well as the two edge connectors, are treated as separate CMRs thereby creating a 6 CMR FE model, as shown in Figure 3.7. The smeared orthotopic plate properties for the modules are given in Table 3.2. The

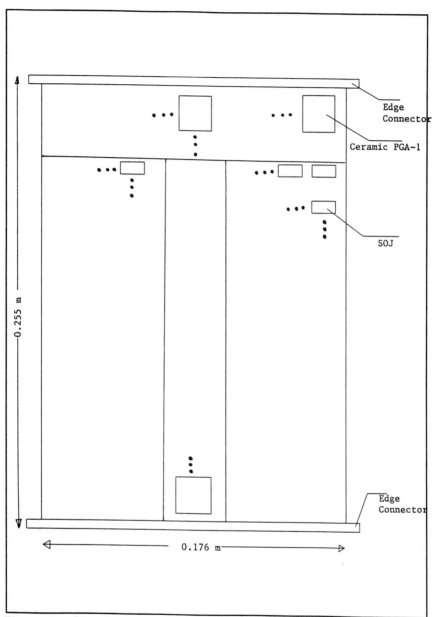

Figure 3.4 Circuit card for case study.

edge connectors are modeled as massless beams using an effective
bending rigidity determined from a three point bending test. Note that
the actual mass of the connectors was accounted for by smearing it over

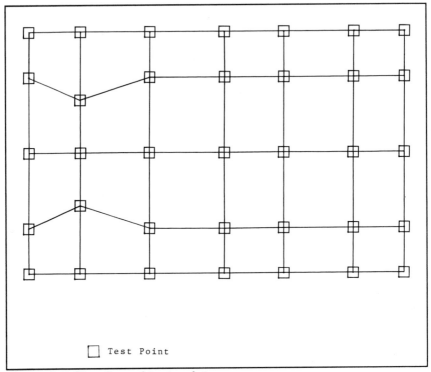

Figure 3.5 Circuit card test points.

the entire PCC. This approximation may not be valid if the mass of the connector is significant with respect to the mass of the card.

One difficulty that was encountered in determining the smeared properties via the three point bend test is that only information concerning the elastic bending moduli (i.e., E_x and E_y) can be directly obtained. In orthotopic plate theory, four material constants are necessary; E_x, E_y, ν_{xy}, and G_{xy}, which represent the two elastic moduli for the x and y direction, Poisson's ratio, and the in plane shear modulus, respectively. Typically, a reasonable value of ν_{xy} is assumed leaving the shear modulus to be determined. However, G_{xy} may be computed from the relationship [21]:

$$G_{xy} = \frac{E_x E_y}{E_y + E_x(1 + \nu_{xy})} \qquad [3.33]$$

The shear modulus computed from Equation 3.33 does not fully account

Table 3.1 Experimental modal data: Card A.

Frequency	Damping
(Hz)	(% critical)
33.8	0.785
62.1	0.759
93.1	0.499
126.0	0.798
152.0	0.509
182.0	0.450
201.0	0.594

for the increased torsional rigidity of the PCC afforded by the modules. That is, the module/lead/card assemblage, when viewed in cross-section, resembles a collection of closed tubes. Therefore, one would expect that its torsional rigidity would be greater than the torsional rigidity of a flat plate with its shear modulus obtained from Equation 3.33 [17]. This is indeed the case. Table 3.3 summarizes the shear modulus as computed by Equation 3.33 and as obtained via torsion testing. The average shear modulus from testing, G'_{xy}, is approximately twice that obtained from Equation 3.33. It is interesting to note that the increases in torsional rigidity for both the SOJs and the ceramic PGA CMRs are roughly equal.

To answer the question of which value of the shear modulus (i.e., Equation 3.33 or the average value from testing) should be used in the smeared FE model, two comparison runs were made. The FE model was composed of 99 orthotopic thin plate quadrilateral elements (4 nodes/element, 3 DOFs/node) and 18 three-dimensional beam elements (2 nodes/element, 6 DOFs/node), which are ANSYS elements Stiff 63 (with the plate option), and Stiff 4, respectively [20]. Table 3.4 compares the frequencies and MAC values for the two FE runs with experiment.

In general, the frequencies are slightly more accurate using the average shear modulus obtained from torsional testing; however, for the frequencies corresponding to the torsional modes (see Figure 3.6), substantial improvement is observed. Note that the MAC values are virtually identical for the two runs. Since the average shear modulus was increased by roughly the same amount for the SOJ and ceramic PGA CMRs, it is not surprising that the mode shapes were not significantly influenced. In the following analyses, the average shear modulus

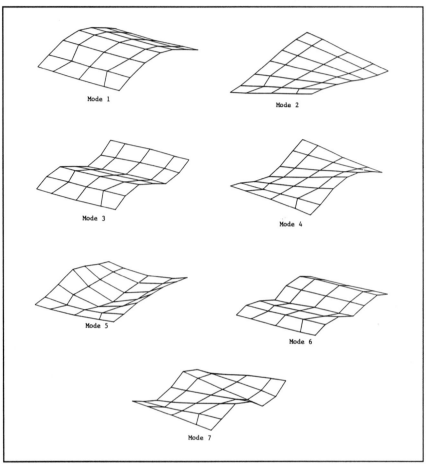

Figure 3.6 The first seven mode shapes of the circuit card.

obtained from torsion testing was used.

From Table 3.4 we can start to get an appreciation for the degree of correlation of the frequencies and mode shapes for a given number of finite elements. By keeping the same number of CMRs and changing the number of elements, the sensitivity of the correlation may be appraised. Tables 3.5 and 3.6 summarize the number and type of elements versus the frequency and mode shape correlation. As can be observed, the frequency and mode shape correlation of the smeared FE model is relatively insensitive to the number of elements used.

In the preceding analyses, the Guyan reduction technique [8] was used with the MDOFs incorporating the transverse displacements only (i.e., perpendicular to the plane of the card); all rotational DOFs were

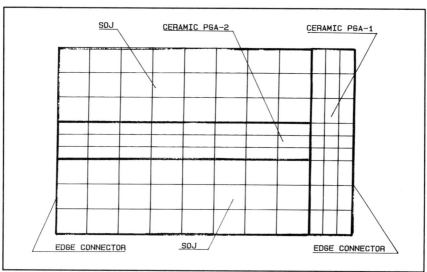

Figure 3.7 Circuit card FE model.

Table 3.2 CMR material properties: Card A.

CMR	E_y $\times 10^{10} Pa$	E_x $\times 10^{10} Pa$	ρ $\times 10^3 kg/m^3$
PGA - 1	6.11	6.11	7.60
PGA - 2	6.11	6.11	9.00
SOJs	2.53	3.67	6.65
Connectors	0.764	-	0.00

slave DOFs. To investigate the validity of this assumption and to assess the impact of the number of MDOFs on the mode shape and frequency correlation, the previously used FE model with 117 total elements and 6 CMRs was re-analyzed. By keeping the FE model unchanged except for the number of translational MDOFs, and maintaining a uniform distribution of the MDOFs, the frequency and mode shape sensitivities to the number of MODFs is obtained. Frequencies from the model are tabulated with the experimental frequencies (Table 3.7). The mode shapes are correlated to the full set of 360 DOFs (i.e., both translation and rotation DOFs) from the smeared FE model (Table 3.8).

Table 3.3 Shear modulus data from torsion test. (All values x10^{-10} Pa.)

CMR	G_{xy} Eqn 3.33	G'_{xy} about y	G'_{xy} about x	G'_{xy} average
PGA-1	2.35	5.40	-	5.40
PGA-2	2.35	-	5.40	5.40
SOJ	1.20	3.45	1.76	2.61

Table 3.4 Comparison of frequency and MAC using smeared analytical and experimental shear moduli.

Expt (Hz)	FE Run 1 with G_{xy} (Hz)	FE Run 2 with G'_{xy} (Hz)	FE Run 1 MAC values	FE Run 2 MAC values
33	34	34	0.90	0.90
62	42	56	0.93	0.94
93	94	97	0.71	0.71
126	99	126	0.82	0.82
152	141	164	0.77	0.76
182	180	193	0.44	0.44
200	187	216	0.50	0.50

The first observation is that, for the FE mesh used, the number of MDOFs required for good correlation need only be a fraction of the total DOFs in the model. It is not until as few as 10 MDOFs are used that the mode shapes and frequencies display a conspicuous reduction in accuracy. Furthermore, the effect of the rotational DOFs is seen to be negligible in this case.

Next, the effect of the number of CMRs on the accuracy was investigated. In this case, the ceramic PGA and SOJ regions of the card were smeared into a single orthotopic plate region. The smeared properties were obtained by performing the bend test to the entire card, without the edge connectors. The edge connectors were modeled separately, as before, with beam elements. Ninety-nine plate and eighteen beam elements were used with only translational DOFs retained as MDOFs. The comparison of the frequency and mode shape

Table 3.5 Frequency comparison for various FE mesh densities.

Number of Elements: Plates/ Beams/ Total	Frequency (Hz) - Mode Number -						
	1	2	3	4	5	6	7
270/30/300	34	56	96	126	164	191	215
240/30/270	34	56	96	126	164	192	215
168/24/192	34	56	96	126	164	192	215
99/18/117	34	56	97	126	164	193	216
81/18/99	34	56	97	126	163	197	215
48/12/60	34	57	97	127	164	196	217
15/6/21	34	57	100	130	167	208	226
Experiment	34	62	93	126	152	182	200

correlation of the 2 CMR and 6 CMR models is summarized in Table 3.9. As can be seen, the mode shape correlation between the two models shows no significant differences. However, the frequencies from the 2 CMR model did show more significant error, most noticeably at the important lower frequencies, than the 6 CMR model.

Finally, the influence of the frequency and mode shape correlation on the forced dynamic response is addressed. The procedure used is as follows. The card was modeled as shown in Figure 3.8. Along the two opposite ends of the card, i.e., at the locations of the edge connectors, simply supported boundary conditions were assumed. A sine sweep base motion was then applied to the supports, given by Equation 3.34, with the frequency range of 10 - 500 Hz.

$$x(t) = \frac{A}{\Omega^2}\sin(\Omega t) \qquad [3.34]$$

where A is a constant and was taken to be one-half the acceleration of gravity and Ω is the sweep frequency. The net effect of Equation 3.34 is that a constant acceleration is imposed on the card from the supports. To simulate the effects of different levels of correlation on the forced

Table 3.6 Correlation of FE and experimental mode shapes for various FE mesh densities.

Number of Elements: Plates/ Beams/ Total	Modal Assurance Criterion (MAC) - Mode Number -							
	1	2	3	4	5	6	7	Avg
270/30/300	.90	.94	.70	.81	.81	.38	.53	.724
240/30/270	.92	.95	.76	.87	.82	.48	.62	.774
168/24/192	.88	.93	.65	.79	.79	.34	.48	.694
99/18/117	.90	.93	.71	.82	.77	.44	.50	.724
81/18/99	.88	.91	.71	.78	.77	.37	.52	.706
48/12/60	.89	.94	.73	.81	.82	.37	.53	.727
15/6/21	.89	.92	.71	.79	.80	.40	.56	.724

response, the locations of the MDOFs for the FE mesh were varied. Three meshes were considered: Mesh 1, for which all translational DOFs were selected as MDOFs, and Meshes 2 and 3 for which the numbers and locations of the MDOFs were varied so that their correlation with Mesh 1 would represent different values.

Table 3.10 lists the first ten natural frequencies from each of the three meshes. As is observed, both Mesh 2 and 3 exhibit frequency shifting from the values obtained for Mesh 1. Furthermore, only the first few frequencies show reasonable correlation with Mesh 1. The correlation of the mode shapes for Mesh 2 and Mesh 3 are listed in Table 3.11. The benchmark for the three meshes was Mesh 1. The MAC values shown are for those modes whose corresponding frequencies were less than 500 Hz, which is the cut-off frequency of the sine-sweep forcing function to be used later. The average MAC value is used as a rough indicator of the overall mode shape correlation. Therefore, we can say that, with respect to Mesh 1 (the benchmark mesh), Mesh 2 correlates better than Mesh 3.

The response of the three meshes to the base excitation of Equation 3.34 is shown in Figures 3.9, 3.10, 3.11. Figure 3.9 shows the displacement of the center of the card (point A on Figure 3.8) in the vicinity of the first natural frequency as predicted by all three meshes, that is, from 15-17 Hz. The peak response from all three meshes is essential the same. Note, however, that the location of the peak is

Table 3.7 Sensitivity of smeared FE frequencies to the number of MDOFs.

Number of Trans-lational MDOFs	Frequency (Hz) - Mode Number -						
	1	2	3	4	5	6	7
10	36	62	105	164	210	293	448
20	34	57	100	131	177	207	242
40	34	57	97	127	166	197	219
80	34	57	97	126	164	195	217
100	34	57	97	126	164	195	217
120	34	56	97	126	164	193	216
360	34	56	97	126	164	193	216
Experiment	33	62	93	126	152	182	200

frequency shifted in each case. The second range of interest, from 60 - 90 Hz, is shown in Figure 3.10. Here we start to see major differences in the dynamic response. Mesh 1, the benchmark mesh, has only a slight peak in this region. Both meshes 2 and 3, however, display relatively large peaks, at frequencies quite different form each other and Mesh 1. These differences are further in evidence in the last region, from 140-200 Hz, shown in Figure 3.11. In this region, Mesh 3 predicts relatively no dynamic response, while both meshes 1 and 2 display large peaks in the response.

From the preceding forced response analysis, we have observed the effects of both frequency and mode shape correlation. The frequency shifting of the peak responses is directly influenced by the predicted natural frequencies. Clearly, in a given FE model, the correlation of the frequencies is of critical importance, not only for resonance checking, but also, as demonstrated in Figures 3.9 through 3.11, for predicting peak response due to external forcing functions. In addition, the mode shape correlation indirectly plays a role in the peak response. Since the response at any given location can be thought of as the summation of the contributions of all the modes, a FE model that accurately predicts the mode shapes will also accurately predict the peak response. This can be seen in the degradation of the forced response

Table 3.8 Sensitivity of modal correlation to the number of MDOFs: full 360 MDOF model used as benchmark.

Number of Trans- lational MDOFs	Modal Assurance Criterion (MAC) - Mode Number -						
	1	2	3	4	5	6	7
10	.875	.790	.979	.863	.934	-	-
20	.996	.997	.923	.968	.891	.743	.831
40	1.00	1.00	.993	.998	.998	.928	.987
80	1.00	1.00	.998	.999	1.00	.960	.994
100	1.00	1.00	.998	.999	1.00	.961	.994
120	1.00	1.00	1.00	1.00	1.00	.998	1.00

Table 3.9 Frequency and MAC comparisons of two and six CMR models.

Mode No	Frequency (Hz)			MAC	
	Expt	2 CMRs	6 CMRs	2 CMRs	6 CMRs
1	33	43	34	.91	.90
2	62	50	56	.94	.93
3	93	119	126	.72	.71
4	126	117	126	.83	.82
5	152	162	164	.75	.77
6	182	234	193	.46	.44
7	200	210	216	.51	.50

results for meshes 2 and 3 as the forcing frequency is increased. This is particularly true in Figure 3.11, where Mesh 3 predicts almost no dynamic response, while meshes 1 and 2 shown pronounced peaks.

In the preceding analysis, the effects of the correlation on the dynamic response were tentatively investigated. Clearly, more extensive

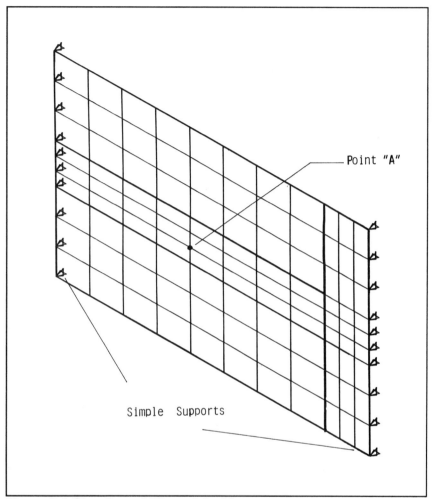

Point "A"

Simple Supports

Figure 3.8 Circuit card FE model with supports.

and rigorous analysis of these effects needs to be performed.

3.7 Summary and Conclusions

In this chapter, the basic ingredients of vibration analysis of circuit cards were covered. Particular emphasis was directed at the technique of smearing the material and structural properties of a printed circuit card populated with modules.   The motivation for using the smeared properties was to produce a less complex finite element model for

Table 3.10 Comparison of frequencies for forced response evaluation.

Mode #	FEM Frequencies (Hz)		
	Mesh 1	Mesh 2	Mesh 3
1	15.4	15.5	15.5
2	62.9	69.4	76.7
3	70.3	81.8	85.0
4	144	170	285
5	150	220	446
6	165	308	647
7	247	540	1340
8	262	640	2453
9	270	929	3607
10	343	2390	4732

Table 3.11 MAC values for the three test meshes.

Mode #	Mesh 2	Mesh 3
1	0.999	0.999
2	0.956	0.843
3	0.867	0.804
4	0.727	0.407
5	0.585	0.350
6	0.558	n.a.
Average	0.782	0.681

subsequent analysis. To verify the use of the smearing technique, a case study was presented. This case study involved both experimental and finite element analysis of a circuit card populated with modules. The reduced modeling technique was verified by direct comparison of the

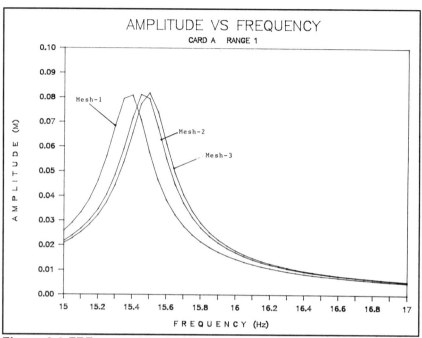

Figure 3.9 FRF range 15 - 17 Hz.

frequencies with those obtained experimentally, and by correlating the
mode shapes between experiment and the reduced finite element model.
Additionally, the forced harmonic excitation of the card was investigated
in order to better understand the influence of the degree of correlation on
the dynamic response. In general, the following is concluded:

1.   The technique of smearing the material and structural properties
       provides the designer with a methodology for reducing the
       complexity of finite element modeling of printed circuit cards
       populated with modules. The reduced model can then be used
       in a static or dynamic analysis.

2.   For the card considered, the number of finite elements used did
       not significantly affect the results.   Consequently, it is
       recommended that the number of elements used be driven by
       the geometric complexity of the model.

3.   For the card considered, it was found that only the translational
       degrees-of-freedom need be retained in the reduced finite
       element model. Furthermore, the number of master degrees-of-

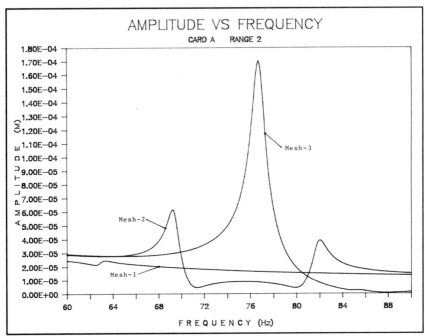

Figure 3.10 FRF range 60 - 90 Hz.

freedom need only represent a fraction of the total. Care must be taken to ensure that the master degrees-of-freedom are adequately distributed in the reduced model.

4. For the card considered, the number of modeling regions (i.e., groupings of modules smeared as a single region) did not significantly affect the mode shape correlation as indicated by the MAC coefficient. However, the accuracy of the FE natural frequencies increased with the number of modeling regions used.

5. The use of the modal assurance criterion proved useful in determining the accuracy of the mode shapes. However, the degree of correlation necessary to ensure accurate estimation of the forced dynamic response was not addressed in this paper but is the subject of current research.

6. If torsional modes of the card are of interest, then an effective shear modulus obtained from testing should be used.

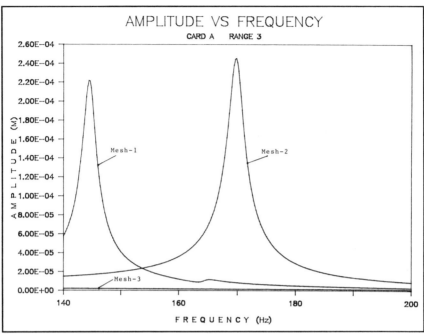

Figure 3.11 FRF range 140 - 200 Hz.

7.   The dynamic forced response is strongly influenced by the degree of correlation of both the frequencies and mode shapes. Further study is necessary to define guidelines as to the degree of correlation required for acceptable dynamic response prediction.

3.8 Acknowledgments

Support from the National Science Foundation, grant number DMC 8909533, under the direction of Dr. Marvin DeVries, and IBM Corporation, grant number 9003, under the direction of Mr. Mark Toda and Dr. Bahgat Sammakia, is greatly appreciated. Additionally, the assistance of Mr. Jerry Lavine is gratefully acknowledged.

3.9 References

1   D.S.Steinberg, <u>Vibration Analysis for Electronic Equipment</u>, (John Wiley & Sons, New York, 1988).

2   H.W.Markstein, "Designing Electronics for High Vibration and Shock," Electronic

Packaging & Production 42, 40-43 (April, 1987).

3   R.R.Craig, Structural Dynamics: An Introduction to Computer Methods, (John Wiley & Sons, New York, 1981).

4   I.H.Shames and C.L.Dym, Energy and Finite Element Methods in Structural Mechanics, (Hemisphere, New York, 1984).

5   K.J.Bathe, Finite Element Procedures in Engineering Analysis, (Prentice-Hall, New Jersey, 1982).

6   R.D.Cook, D.S.Malkus and M.E.Plesha, Concepts and Applications of Finite Element Analysis, (John Wiley & Sons, New York, 1989).

7   C.R.Wylie and L.C.Barrett, Advanced Engineering Mathematics, (McGraw-Hill, New York, 1982).

8   R.Guyan, "Reduction of Stiffness and Mass Matrices," J.Amer. Inst. Aeronautics & Astronautics, 3 (2), 380 (1964).

9   P.A.Engel and C.K.Lim, "Finding the Stresses with Finite Elements," Mech. Eng. 108 (10), 46-50 (1986).

10   P.A.Engel and C.K.Lim, "Stress Analysis in Electronics Packaging," Finite Elements in Analysis and Design 4, 9-18 (1988).

11   P.A.Engel, "Structural Analysis for Circuit Card Systems Subject to Bending," Trans. ASME 112, 2-10 (March 1990).

12   T.E.Renner, "Mechanical Design of Circuit Boards Using Finite Element Analysis," 1983 ANSYS Confer. Proc., Pittsburgh, PA, April 17-20, pp 1.2-1.22.

13   C.A.Zager, "Vibration Studies of Freestanding Electronic Enclosures," Design Automation Confer., Cincinnati, OH, Sept 10-13, 1985 pp 1-7.

14   J.M.Pitarresi, "Modeling of Circuit Cards Subject to Vibration," Proc. 1990 Internat. Sympos. Circuits & Systems, New Orleans, LA, May 1-3, 1990 pp 2104-2107.

15   R.J.Kunz and J.M.Pitarresi, "Analytical and Experimental Optimization of Support Locations for Vibrating Printed Circuit Boards," Proc. 9th IEPS Confer., San Diego, CA, Sept 11-13, 1989.

16   D.Ewins, Modal Testing: Theory and Practice, (Research Studies Press, Ltd and John Wiley and Sons, New York, 1984).

17   P.A.Engel (Private Communication), 1989.

18   Star Theory and Applications, (Structural Measurement Systems, Inc., San Jose, CA 95134, 1990).

19   "A Case Study in the Correlation of Analytical and Experimental Analysis," (KIT

Corporation, St. Paul, MN, 1989).

20   ANSYS Users Manual, (Swanson Analysis Systems Inc., Houston, PA, 1989).

21   C.C.Chamis, Structural Design and Analysis, Part I, (Academic Press, New York, 1975).

POWER TECHNOLOGY

PACKAGING FOR THE 90s

CRAIG D. SMITH

4.1 Introduction

There have been few major advances in power technology during the computer age. Solid-state linear regulators appeared in the mid 60s. Switching regulators came into common usage in the mid 70s. We are now at the beginning of the next significant technology - high frequency distributed architectures. This technology is highly dependent upon packaging solutions, and will have a large impact on overall system-level packaging of Data Processing and Telecommunications products developed in the 90s.

We will first review the traditional linear and switching regulator technologies from the viewpoints of system packaging and development. The new distributed technology will then be described, along with the industry needs that are accelerating its introduction. Architectures, converter topologies, components, magnetics, thermal management, and power silicon devices will also be briefly discussed. We will conclude with predictions of future power packaging trends.

Much of the content will be rather generic in nature. This is due to both the limited space available, and the proprietary nature of some of the technologies involved. For those wishing to pursue these topics in more detail, a list of references [1-43] has been included in Section 4.7, organized by subject matter. Finally, many of the opinions and predictions herein are very subjective, are strictly those of the author, and do not necessarily reflect those of the IBM Corporation, SUNY, the editor or publisher.

4.2 Power Regulator Technology - History

Throughout the history of electronic systems, there has always been a need for power regulators. Almost all electronic devices operate from a DC source voltage rather than from the generally available AC powerline. Consequently there is a need for a device to convert the AC power into a source of one or more DC voltages. In many cases, these DC voltages need to be tightly regulated, be sequenced on and off in a certain way, and contain minimal ripple and noise. Thus, the power supply industry was born.

In spite of this universal need for power supplies, there have been surprisingly few fundamental technology advances within the computer age. Until the 60s, vacuum tube regulators were used if tight regulation requirements existed. By today's standards, these regulators were extremely large, inefficient, and unreliable. The first major technology change was discrete transistorized linear regulators in the mid

60s. These were in common usage until the mid 70 s, when switching mode regulators started to appear in commercial and industrial products. To put the benefits of the high-frequency distributed power architectures of the 1990s into proper perspective, it is important to review and compare the previous two technologies. As a basis for comparison, a classic unregulated power supply will first be considered.

4.2.1 Unregulated power supply

This is the classical, minimal element power supply model often seen in introductory electronics texts. It is the fundamentally simplest topology that can provide the required functions of isolation, rectification, and filtering. It does not provide the function of regulation.

In order to provide a common basis for comparison between the topologies to be discussed, and to give a meaningful and practical example of each, the approach will be to use an actual design objective for the comparison. This objective is shown in Table 4.1. This would be a very typical requirement for an AC input power supply that was to power a load consisting of TTL logic.

Table 4.1 Power supply design objective.

INPUT VOLTAGE:	115 VAC - 1ϕ - 60 Hz
INPUT LINE VARIATION:	±20%
OUTPUT VOLTAGE:	5 Volts
OUTPUT REGULATION:	±5%
OUTPUT CURRENT:	60 Amps
OUTPUT POWER:	300 Watts
OUTPUT RIPPLE:	100 mV peak to peak

Figure 4.1 shows an attempt to design to the above requirement using an unregulated power supply. The transformer turns ratio is selected to provide 5V output with the input line at a nominal 115VAC. One problem is evident immediately. The 20% input line voltage variation is reflected through the transformer, so that the best case output voltage regulation can be no better than 20%. Allowing for an additional 3% deterioration due to resistive losses in the transformer windings, rectifiers, and interconnections gives an expected output voltage regulation of 23%

- not even close to the 5% design requirement.

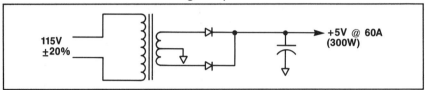

Figure 4.1 Unregulated power supply.

For purposes of estimating the power losses in the circuit, the following assumptions can be used:

Average Diode Current: 30A each
Diode Forward Voltage Drop: 0.75V
Transformer Efficiency: 95%

Applying the above assumptions:

Diode Losses =	(2)(30A)(0.75V) =	45 Watts
Transformer Losses =	(345W)(0.05) =	17 Watts

TOTAL LOSSES		62 Watts

$$\text{EFFICIENCY} = \frac{\text{Output Power}}{\text{Input Power}} = \frac{300W}{362W} = 83\%$$

The ripple frequency will be twice the input line frequency due to the full-wave rectification (120 Hz). The amplitude of the ripple will be a function of the complexity and size of the output filter elements. For practical implementations, the output ripple will be in the range of several hundred millivolts - more than the design objective. To summarize, this approach does not meet the required regulation and ripple specifications. It is physically large due to the low frequency transformer and the size of the output filter elements required to reduce the output ripple. The efficiency, however, is very good.

4.2.2 Solid state linear regulator

The general availability of power transistors in the 1960s allowed the development of solid state linear regulators. The concepts and topologies were not new - similar approaches had previously been used with vacuum tubes as the series control element. However the transistor (first germanium, then silicon) made these circuits much more practical. Fortunately, this development occurred in the same time frame as the need for tightly regulated voltage sources for use with integrated circuits. The previous discrete logic and analog circuits tended to be much less sensitive to variations in the power supply voltage(s).

An implementation of the specification of Table 4.1 using a typical linear regulator topology is shown in Figure 4.2. The front end of the circuit is identical to the classical power supply discussed in Section 4.2.1, although the output voltage at this point, labeled $V_{secondary}$, is designed for a DC voltage higher than the 5V power supply output voltage. The selection of this intermediate voltage will be discussed later. The main regulating element in this circuit is the transistor Q1, through which all of the 60 Amp output current passes. The topology is essentially an emitter follower circuit with the load as the emitter resistor. The voltage drop across Q1 changes as required to regulate the output voltage at 5 Volts. This is done by changing the drive current available to Q1 by means of the feedback loop consisting of the resistive divider, the operational amplifier, and the drive transistor Q2. $V_{reference}$ is a DC voltage reference typically generated by a zener diode or, more recently, by a reference voltage chip.

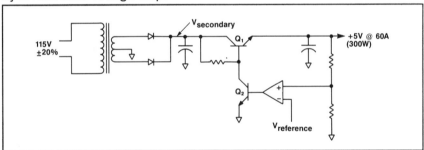

Figure 4.2 Solid state linear regulator.

The active feedback loop controls the output voltage very accurately. Regulation within a millivolt is easily achieved. Because the bandwidth of the regulation loop is much higher than the ripple frequency (120 Hz in this case), this circuit also regulates out almost all of the ripple component on $V_{secondary}$. Consequently, the output ripple is very low - often a millivolt or less. This circuit gives very high performance, easily surpassing all of the required specifications.

But this performance comes at a price. The price is efficiency. In order to regulate properly, a minimum collector-emitter voltage must be maintained across Q1, typically on the order of 3 volts. With the emitter voltage fixed at 5 volts, the collector voltage must be at least 8 volts under all conditions, including variations in the input AC line voltage. To accomplish this, the transformer turns ratio must be designed to provide 8 volts at $V_{secondary}$ at the lower extreme of the line voltage (-20% from nominal). Thus, at nominal line voltage, which is the typical condition, the voltage $V_{secondary}$ will be 20% higher than 8 volts, or 9.6 volts. Using the same assumptions for diode and transformer characteristics as were used for the unregulated power supply example, and assuming that the losses in the feedback loop elements are negligible, the losses in this circuit can be estimated as follows:

Diode Losses =	(2)(30A)(0.75V) =	45 Watts
Q1 Losses =	(9.6V - 5V)(60A) =	276 Watts
Transformer Losses =		
	(300W + 45W + 276W)(0.05) =	31 Watts

TOTAL LOSSES		352 Watts

$$EFFICIENCY = \frac{Output\ Power}{Input\ Power} = \frac{300W}{652W} = 46\%$$

The resulting efficiency is very poor, and is the biggest disadvantage of the linear regulator approach. The dissipation in Q1 necessitates several transistors to be used in parallel, a large heatsink assembly, and lots of cooling air. This results in a very large power supply (very low power density). The efficiency varies with the required output voltage, with bigger efficiency penalties for lower output voltages. Generalizing the above analysis as a function of output voltage, V_{out}, gives the following expression:

$$Maximum\ Efficiency = \frac{V_{out}}{1.26\ (V_{out} + 3.63)}$$

The above relationship is plotted in Figure 4.3. Since most of today's circuitry requires voltages of 5V or less, the usage of linear regulator technology results in unacceptably low efficiencies that more than offset the other advantages of this approach. The need for higher efficiency was the motivation for the development of the switching regulator.

4.2.3 Solid state switching regulator

The switching regulator is a fundamentally different device from the linear

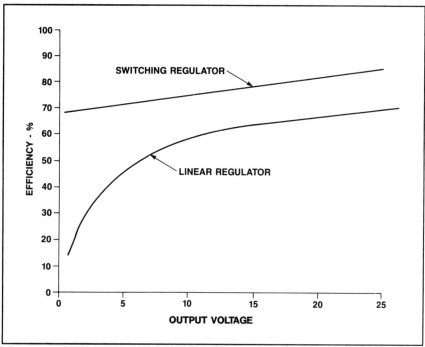

Figure 4.3 Efficiency vs. output voltage.

regulator. It was developed in the late 60s for military use, where low weight and high density were of paramount importance. Its additional complexity and higher cost prevented its usage in commercial systems for several years, until the late 70s when it became commonplace in data processing and telecommunications applications.

The linear regulator, as its name implies, operates with the power regulating element in a linear region - a region that dissipates significant amounts of wasted power. The switching regulator tries to avoid this situation by operating the semiconductors in a switched mode - either saturated or completely turned off. An "ideal" device would dissipate zero power in either of these two states. Actual devices naturally do not achieve this. They have finite, but small,dissipations both in the conducting region (due to the saturated collector to emitter voltage) and in the non-conducting region (due to the collector to emitter leakage current). In practical implementations, the transitory dissipations during the turn-on and turn-off time intervals are also significant. In spite of this, the total switching element losses in a switching regulator are substantially less than those in the linear regulator series pass element - perhaps only 2% as much.

But this is not the only major advantage of the switching regulator approach. The linear regulator operated at the frequency of the AC power line. The transformer operates at 60 Hz. The energy storage elements (filter capacitors and inductors) are operating with 120 Hz AC components. The operating frequency of a switching regulator is not tied to that of the power line. Conceptually, it could be anywhere from 1 Hz to GigaHertz. Practical considerations, some of which will be discussed later, have resulted in typical operating frequencies in the range of 20 KHz to 200 KHz for most regulators. This increase in frequency allows a very significant reduction in the size of the energy storage elements and the transformers, since, for a fixed power level, the amount of energy stored or transferred per cycle is inversely proportional to the frequency. With this approach, transformers, inductors, and filter capacitors become a fraction of their former size. This, in conjunction with the lower power dissipation in the semiconductors discussed above, results in a much smaller regulator - an order of magnitude smaller. The advent of switching regulators is the most fundamental and significant development in the history of electronic power systems.

There are dozens of different topologies that fall within the classification of switching regulators. We will select only one for purposes of illustration, the topology known as the "half bridge". This is one of the more commonly applied topologies as well as one whose operation is relatively easy to understand. The switching regulator is a complex device. It requires a large amount of "support and overhead" circuitry, such as bias voltage generators, reference generators, clock generators, drive circuits, voltage and current sensors, and protection and diagnostic circuits. These auxiliary support circuits are both digital and analog, but mostly operate at a low power level and are consequently candidates for integration. Due to space constraints only the main elements that are required to understand the operation of the regulator will be addressed. For the half bridge topology, these are shown in Figure 4.4.

The high frequency transformer T1 divides the circuit into two sections. The primary side of T1 is shown on the left hand side of Figure 4.4, and is associated with components that are connected and referenced to the input AC power line. The secondary side of T1, shown on the right, is associated with components that are connected to the output of the regulator. T1, in addition to its voltage transformation function, serves the very important function of providing DC isolation between the high-voltage circuits on the primary and the low voltage, user accessible circuits on the secondary.

The half bridge is a DC to DC converter. The source of input DC is developed by directly rectifying the AC line voltage, most typically with

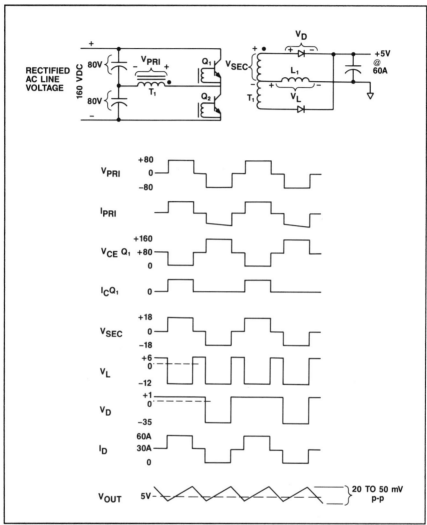

Figure 4.4 Solid state switching regulator.

a 4 element full bridge diode configuration. This results in a no-load DC voltage of about 160 volts. This is shown as the DC input voltage in Figure 4.4. The two capacitors minimize the ripple on the input voltage and also act as a voltage divider, which is required for this topology. These capacitors are typically fairly large aluminum electrolytic devices. Their imposed ripple is at twice the line frequency. All other components shown operate at the regulator switching frequency or above.

First, consider the operation of the primary side of the circuit.

Transistor Q1 is first turned on with Q2 off. This imposes an 80V step function across the transformer primary. Q1 is kept on only for a few microseconds, so that the transformer does not saturate, and is then turned off. After a short delay, Q2 is turned on. This reverses the polarity of the voltage imposed on the primary of T1. After another delay, the cycle repeats with Q1 being turned on. The result is an approximation of a "high frequency AC" waveform across the T1 primary, shown as V_{pri} in the waveforms section of Figure 4.4. The delay between on times is required to prevent simultaneous conduction of Q1 and Q2 which would result in a direct current path across the 160V input and destruction of the transistors.

Now, consider the secondary side of the circuit. The T1 center tapped secondary along with the diodes configure a full wave rectified output operating at twice the switching frequency of the primary. Energy is transferred from primary to secondary during both the Q1 and Q2 on times, but not during the delay time when both Q1 and Q2 are off. During this off time, current must still be provided to the output through the diodes. The source of this current is the energy stored in inductor L1. During the off time, current flows from the inductor through both diodes in parallel. During each on time, only one of the diodes is conducting, resulting in the diode current waveform labeled I_d in Figure 4.4. Other waveforms are also shown to indicate the voltage and current stresses imposed on the circuit elements. Note, for example, that this 5V output regulator requires a transformer secondary voltage of 18V, and that the output diodes must be rated at least 60 amps at 35 volts.

Further analysis of the waveforms of Figure 4.4 will show that:

Average diode current = 30 Amps

Average Transistor Vce = 80 Volts

If we assume that the transistor loss is dominated by the on state saturation voltage, and that it is 1 volt, we can estimate the losses in the converter as follows:

Diode Losses =	(2)(30A)(0.75V)	= 45 Watts
Transformer Losses =	(300W + 45W)(0.05)	= 17 Watts
Transistor Losses =	(362W)(1.0V)/(80V)	= 5 Watts

TOTAL LOSSES   67 Watts

$$\text{EFFICIENCY} = \frac{\text{Output Power}}{\text{Input Power}} = \frac{300W}{367W} = 82\%$$

The analysis above is very rough, and does not include many of the smaller losses in other circuit elements. But it illustrates the major improvement in efficiency over the linear regulator. As with the linear regulator, the efficiency is somewhat dependent upon the output voltage. This dependence is much less for the switching regulator, however. Typical efficiencies range from perhaps 70% at 3 volts output to 85% at 25 volts output. Most of this dependence is due to the predominance of the output diode losses, and the higher diode currents for a fixed power level as the output voltage decreases. The efficiencies of switching converters can be compared to those of linear regulators by referring to Figure 4.3.

The performance of switching converters falls midway between the unregulated power supply and the linear regulator. The regulation is good (typically several millivolts), but not as tight as the linear regulator. Ripple is also higher than for the linear regulator, but still within the specification limits. In short, it meets all the requirements defined in Table 4.1, and has the added advantage of being significantly smaller than either of the other approaches we have addressed. This power density advantage (Watts delivered per cubic cm of power supply) is what has made the switching converter the technology of choice over the past decade. Table 4.2 shows a high level comparison of the three approaches.

Table 4.2 Power supply comparison.

	UNREGULATED	LINEAR	SWITCHING
REGULATION	Bad	Excellent	Good
EFFICIENCY	Excellent	Bad	Excellent
RIPPLE	Bad	Excellent	Moderate
DENSITY	Bad	Very Bad	Excellent

As was noted previously, this discussion of switching converters has been a much simplified one. In actuality, they are extremely complex electronic sub-systems requiring very high skill levels to design. Hundreds of volts and hundreds of amps are being switched in a few nanoseconds. The transitory stresses to components are very high and very difficult to predict analytically. These stresses are, to a large extent, dependent upon high frequency parasitic component characteristics that are not documented in the component data sheets. The potential for both conducted and radiated electro-magnetic interference must be recognized and dealt with. The design, component selection, layout, packaging, and manufacture of switching converters is a very time consuming and demanding skill. As will be seen later, this complexity has created a problem for development of products using these converters.

4.3 Conventional Power System Architecture

In order to address the packaging aspects of power regulators, we must turn our attention to power system design. The converter fundamentals established in Section 4.2 allow an understanding of how regulators and converters operate, but any actual application involves interfaces (electrical, mechanical, environmental) to the power supply. Also, many systems utilize more than one power regulator. Power system design encompasses the tasks of defining the power architecture, specifying power regulators, defining all interfaces to the power regulators, power distribution, control and diagnostics support, mechanical packaging, and cooling analysis.

Most of today's systems incorporate one or more power supplies. These power supplies typically are custom designs that are developed to meet the unique voltage and current demands of the system. They tend to be physically located in one area of the system package, and the outputs distributed to the load circuits by means of wires or bus bars. The supplies are typically solid-state switching regulators using the AC line voltage as an input, and operating at a frequency of 50 to 200 KHz. This concept is referred to as a custom centralized power system, an example of which is shown in Figure 4.5.

The system shown in Figure 4.5 is not a real one, but rather an arbitrary example developed for tutorial purposes. It does, however, reflect the type of architecture and packaging that is in widespread use today for mid-range systems. Typical products that would use a power system similar to this would be mini-computers or telecommunications controllers.

AC power is distributed throughout the box to provide input power to the switching regulators and to drive fans for cooling. DC is distributed from the power supplies to the load. The DC distribution can be a challenging and costly design, since much of today's logic requires very tight regulation. Even with remote sensing for the regulators, the distribution drop must be kept small in order to keep overall losses and efficiency to reasonable levels. The machine package does not allow the power supply to be located in close proximity to the load. The DC distribution runs are typically a meter or two. This can be very expensive,

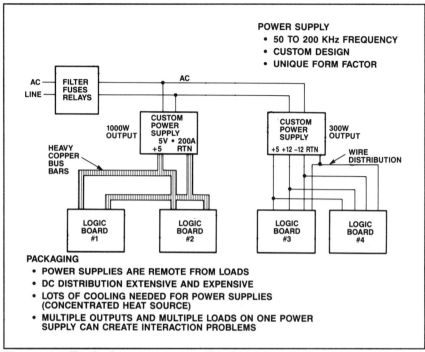

Figure 4.5 Typical custom centralized power system.

especially for high current feeds such as the 200 Amp case shown in the example. The bus bars for such systems are often solid copper, over 10 mm thick, and 50 mm wide. This adds considerable cost, weight, and labor content to the system.

Another problem often encountered with centralized systems is load interaction. Consider logic boards 1 and 2 in Figure 4.5. Even with remote sensing, the voltage can be set accurately at only one of the boards, so there must be some DC voltage differential between the boards. Similarly, there will be some ground shift. These effects will get worse as the load current varies in a dynamic fashion on one or both boards. A different exposure exists on the 300W multiple output regulator. Not only is there the potential for interaction due to the two logic boards, but there can also be interaction within the power converter between the 3 voltage levels. Converters such as this one typically have a single main control loop on the highest power output (probably 5V in this case), and use additional secondaries on the high frequency transformer to develop the secondary outputs such as the 12 volt outputs shown here. This can create cross regulation problems on these secondary outputs as the load on the main output changes.

In addition to the possible technical problems with centralized power architectures, they present development and manufacturing problems that can be quite severe from a program management point of view. We must explore some of these issues in order to gain an appreciation for the exposures and limitations of centralized systems. We will do this by following the development and manufacturing start-up process for a mythical centralized power system, and observing some of the pitfalls along the way.

Before design can begin, requirements and specifications must be developed. The most important specifications are the voltage, current, and regulation requirements for the load circuits within the machine. A partitioning of the load circuitry into cards, boards, frames, etc. must also be known or assumed. This is rarely a smooth process. Typically, the designers of the machine have very little knowledge about the power requirements during the design process. Many times, the machine is being built with circuit technology that has not been used previously, so that other machines cannot be used as a model for the estimate. Also, the development time for the power supply is often longer than that for the other circuitry, and power converter design must actually begin before design of the load circuits. This is especially true for processor design, where automated design systems allow for very rapid design and production of logic circuits. No such systems exist for the design of custom centralized power supplies. The most common outcome of this situation is that a power supply design is started based on the best possible estimate of the system requirements. Later, as the system design progresses, many changes are made, and the estimate no longer represents the actual system. The power system design must then be modified or, many times, completely redone. The design process is iterative, with two or more passes typically required. This has the effect of lengthening what is already a very long and involved (at least two years) development process.

The challenges are not all electrical. Each custom power supply has a unique mechanical package, some of which needs to be very precisely designed. Connector locations, for example, must often need to line up with mating connectors within fractions of a millimeter. Centralized power supplies represent a centralized, concentrated source of heat. The thermal density is large enough so that forced air or liquid cooling is commonly used to cool the power supply. The cooling interface often is the source of problems. For forced air cooling, airflow direction, volume, and temperature must be estimated during the power supply design prior to having a real machine to use for testing. When the resulting power supply is integrated with the actual machine, the cooling environment is often different enough to create thermal problems within the supply.

Components are also a problem. Since the power converters are custom designs, an attempt is made to optimize the component selection for that particular application. There are hundreds of components in a switching regulator. There are dozens of different types - logic circuits, analog ICs, discrete low level resistors and capacitors, fuses, relays, connectors, power semiconductors, complex high frequency magnetics, electrolytic capacitors, etc. They are produced by many different vendors utilizing different processes. The regulator operation may be dependent upon some second-order parasitic characteristic of several of these components - characteristics not documented and guaranteed by the component vendor. Many parts have more than one source. The regulator may work with one vendor's part, but be marginal with the same part number from a different vendor. These substitutions are usually made by the manufacturer of the power supply months or years after the design is completed. The exposure for problems is severe. Compare this environment to that of the logic circuitry in the equipment. All the logic usually is from one circuit family manufactured by one vendor using one tightly controlled process. Even though there may be hundreds of logic chips, they are, in actuality, just replications of one basic component. Consequently, component problems within power supplies are much more common.

As the custom power supply moves from design into manufacturing, other disadvantages of its uniqueness become apparent. Because each such supply has different dimensions, circuit board sizes, layouts, and test requirements, many of the manufacturing operations must also be unique. Customized handling equipment, process flows, tooling for covers, and test programs typically need to be developed. These items create significant non-recurring manufacturing charges for capital and programming support. These front-end charges are particularly bothersome because many production runs are relatively low in volume (less than 10,000), and the non-recurring costs per power supply are a large percentage of the total manufacturing cost.

After the first power supplies come out of the manufacturing process, the power supply usually needs to be qualified. The qualification is done to verify that the design is sound and to gain confidence that the needed reliability levels will be achieved. The qualification process is lengthy and costly due to the high reliability levels specified for today's power supplies. The test program will need either large sample sizes (high hardware expense) or long test durations (long test times and support expense). Each of these options is very expensive, and the results are not always available soon enough to mesh with the remainder of the program schedule. In addition to this internal qualification, power supplies must be submitted to Underwriter's Laboratory and various international agencies for safety approval. This process can also be lengthy. Any problems discovered during these qualifications must be addressed. The problem could be related to design, component quality, manufacturing processes or just be a random failure. In any event, the solution to the problem, especially design and component problems, can be very complex and time consuming. Countless product programs have been delayed for months by a problem discovered during custom power supply qualification.

Usually at about the same time that the qualification testing is occurring, the final integration of the power supplies with the system is taking place. During this process more subtle problems with the custom power supplies or incompatibilities between the power supply and the system begin to surface. A common class of problems is Electromagnetic Compatibility. Switching converters generate a broad spectrum of electrical noise due to the fast high current transients associated with the switching activity. The main switch transistors and the output diodes are the two most common sources of such noise. Without very careful electrical and layout design of the power supply, this noise can be coupled into the system either by means of conduction or radiation. This is perhaps the most difficult aspect of regulator design, and even the most skilled and experienced designer sometimes encounters problems during system integration. Again, solving these problems is not easy. It requires re-design or additional shielding, either of which requires re-qualification and time delays.

For cost reasons, much of the actual manufacturing of custom power supplies is done at assembly vendors, either domestic or off-shore. The vendor often is given responsibility for procuring the components. This presents additional problems and exposures in terms of assuring component quality, understanding the assembly processes, and providing general quality control for the power supply. These problems are potentially severe due to the length of the "pipeline" between the vendor and the final product. Power supplies from Far East vendors are shipped ocean freight, and can take 2 or 3 months to reach the product. By the time a problem is discovered and corrective action taken, there can be 3 or 4 months worth of defective production in transit and in storage, adding to the expense and delay in fixing the problem.

The custom power supply approach is also limited in terms of its ability to adapt to changes in the product requirements. This is important, as most products incur design changes to allow for additional features to be added after product introduction. If the power system has to be re-designed every time there is a system level change, the resulting costs and schedule impacts can be enormous. This lack of flexibility is one of the biggest limitations of centralized power.

In order to adequately design and support switching power converters, an engineering team with high skill levels and years of experience in many fields is required. Developing and maintaining a team such as this is not easy or inexpensive. This, in conjunction with the long development cycle described above, results in very high non-recurring engineering expenses for custom power supply designs. This is one of the main reasons why alternatives to the custom approach are receiving attention.

The design environment described above has another disadvantage. Due to the already high expense associated with development, and the shortage of people with these design skills, very little of this resource is assigned to technology development. Consequently, almost all of these custom power supplies are designed with standard topologies and off-the-shelf components. Very little genuine innovation is seen in the areas of developing new circuit topologies, new power semiconductor devices, new components, new packaging techniques and new assembly methods. The result is custom power supplies that are all very similar to each other in technology, with relatively little differentiation between manufacturers. This has contributed to the power supply being thought of as a "commodity" item.

4.4 Power Requirements for the 90s - Industry Trends

As the discussion in Section 4.3 indicates, there is growing dissatisfaction with centralized custom power within the information processing and telecommunication industries. This dissatisfaction stems from the pure economics of high non-recurring expense and lengthy development times. In addition to this, there are also several technical and environmental trends which are now starting to form. These trends represent additional motivation for abandoning the traditional centralized custom approach and adapting other techniques for power regulation and distribution. Some of these trends are discussed below.

4.4.1 Lower voltages

The average DC voltage level for digital circuitry (and analog) is going down. For years, the 5 volt TTL level was a standard voltage level for most equipment. In order to maximize the performance (speed) of digital logic circuitry, the operating voltage has been reduced, so that voltages like 1.7 volts and 3.3 volts are becoming more and more common. Analog circuitry used to be predominately discrete and use plus and minus 12 or 15 volts as a power source. Now that analog integrated circuits are replacing most of the discrete components, operating voltages are moving downward.

The reduction in voltages has four effects on the power supply requirements - all negative. The power required by the load is remaining approximately constant, so that the current requirements are increasing. The higher current levels (approximately double) require heavier and more expensive DC distribution cables and bus bars. Since the cost of switching regulators is highly dependent on output current (the output diodes and filter network represent a big percentage of the cost), the lower voltage also increases the cost of the power supply. As described in Figure 4.3, the lower voltage output reduces the efficiency of the converter, adding to heat loss and cost while reducing reliability. Finally, the lower voltage levels still require very tight regulation - up to 1 or 2percent. Two percent of 1.7 volts is only 34 millivolts. This is a very tight requirement to meet considering the hundreds of amps of current present in many of the DC distribution systems. Even with remote sensing, it requires extra cost to be put into the distribution system.

As the voltage levels go down, centralized power makes less and less sense. Big penalties are being paid for distributing the high current DC levels over large distances. It's a classic example of optimizing one area (logic density) at the expense of another (power system complexity and cost).

4.4.2 Powerline interface concerns

Several changes have taken place in electrical and electronic equipment over the past several years. Just a few years ago, most computing equipment was located in environmentally protected rooms and supplied with power from motor generator sets and back-up generators so that it was isolated from the effects of the powerline. Now, computers are proliferating in the office and home, and operating directly off the powerline. The powerline is not a pure sinewave source. It contains high frequency noise, fluctuations in voltage, voltage spikes, brownouts and dropouts. Most office and household loads were immune to this kind of thing. They used AC motors, resistive heaters, and relay-based control logic. Today, even inexpensive household appliances such as VCRs, microwave ovens, and clock radios contain computers (microprocessors) and memory that are very sensitive to changes in the power supply voltage. Suddenly, they are no longer immune to the less-than-perfect powerline. Transients on the AC line can erase memories or completely destroy some of this circuitry.

Classical switching converters, such as the one shown in Figure

4.4 do not present a resistive load to the powerline. The input rectifier and capacitive filter present a non-linear load, with a low conduction angle. Current tends to flow only near the peak of the applied AC voltage rather than during the complete cycle, as shown in Figure 4.6. Because the powerline source does not have zero impedance, this non-linear current flow creates a voltage distortion on the powerline, which can affect other equipment.

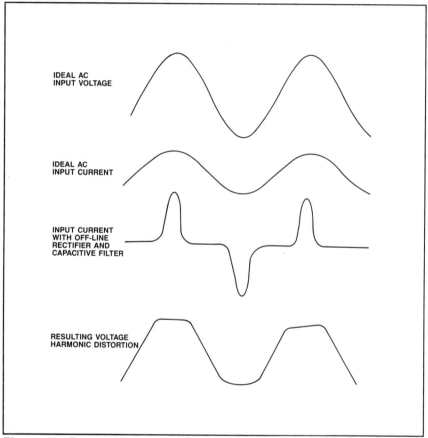

Figure 4.6 Power line distortion.

Since switching regulators were previously confined only to "high end" applications, this condition used to be common only in data processing centers where it could be easily managed. Today switching converters are used in all kinds of equipment such as low-end personal computers, printers, and FAX machines. In large office buildings, it is not uncommon to have hundreds or thousands of these units all connected to one AC power system. The combined effect of all these regulator front-ends is to severely distort the powerline. In many cases this has created problems with the useability of the AC power.

The classical switching regulator input rectifier and filter also result in a poor power factor - typically around 0.6. This wastes much of the available powerline capability. It limits the DC power available from a single phase 115V outlet to about 700 watts rather than the over 1000 watts that would be available with a unity power factor. It also increases the user's cost of powering the equipment, since the utility metering measures the apparent power rather than the real power.

To complicate things further, the quality of the AC supplied by the utilities is getting worse rather than better. Very little new investment is being done in new and better generation and distribution facilities. More power is being used, increasing the loading on existing lines. More and more equipment is using the switching converter, adding additional noise and harmonic distortion as described above.

Because of the environment just described, there is increasing demand for solutions such as power factor correction, battery back-up systems, and uninterruptible power supplies. These trends are also resulting in changes in the specifications and characteristics of power supplies. As will be seen later, it is also increasing the acceptance of distributed power systems.

4.4.3 Shorter development cycles

The electronics industry is coming under increasing competitive pressure both from other domestic sources and from abroad. The rate of technology improvements continues to increase, and system offerings, especially in the low-end and mid-range information processing fields, have a competitive lifetime that is less than it was in years past. These pressures are resulting in the desire for faster product introductions and shorter development cycles for products of all types. More emphasis is also being placed on development productivity - getting designs completed with less total resource and non-recurring expense. This environment exists in all areas of the system design, both hardware and software. But, as indicated in Section 4.3, the development requirements for custom centralized power systems are not compatible with this environment -they traditionally require long development lead times and a significant amount of highly skilled development resource. Consequently, when projects are attempted using the new, very aggressive development schedules, the power system often is the element that is hardest to integrate into the schedule. This has led to the search for alternative methodologies for power development. As will be seen, the distributed power architecture results in advantages that resolve many of these conflicts.

4.4.4 Increased focus on reliability

In many of today's market segments, reliability is becoming extremely important, even to the point of being the key competitive parameter. Some such markets are mainframe processors, DASDs, and telecommunications hardware. Advances in logic and memory technology have resulted in continual increases in the scale of integration and packaging with resultant benefits in simplicity and reliability. Conventional centralized power systems have seen no such improvement. The operating frequency has been going up, and the power supplies are getting smaller as a result, but the centralized approach has not had a fundamental advancement in the area of integrated packaging. The supplies, however, are becoming more reliable as the experience level increases in such areas as component stresses prediction, knowledge of the standard topologies, and assembly processes.

As the product becomes more reliable, the "unreliability" of the powerline becomes a factor. In fact, for most systems running from an unbuffered powerline, the system will go down much more often due to faults on the utility service than for faults within the equipment itself. Selzer [43] reports that for a typical service in the United States there is some kind of powerline disturbance about twice a day. This could be an undervoltage, overvoltage, outage, or spike. Most systems will operate through these types of disturbances except for the outage, although they may contribute additional electrical stress and reduce the long-term reliability. The outages are the types of faults that will bring the equipment down, and these are estimated to occur about once every 51 days on the average. We thus have the scenario where equipment is designed to operate reliably for years, yet is unavailable about once every month or two due to powerline faults. This is another reason why solutions such as battery back-up and uninterruptible power systems are becoming more popular.

4.4.5 Continued cost pressures

In addition to reliability, cost is the other major competitive battleground. This is especially true now that a significant amount of manufacturing and assembly is being done in the Far East and other areas with much lower labor rates than are available domestically. But conventional centralized power supplies are difficult to reduce in cost. They utilize hundreds of unique and specialized components, each of which has to be procured, inventoried, assembled and tested. Unlike the situation for digital circuitry, much of the assembly of conventional switching converters must be done by hand rather than by automated techniques. This is because many of the components, such as large electrolytic capacitors and magnetics are physically too large for automated equipment. Also, the large wire sizes used and the special requirements for lead dressing preclude using any of the readily available automated assembly equipment, and the production runs for customized power regulators tend to be too small to justify development of specialized assembly machinery.

Since the last major improvement, in terms of topology, was the switching regulator over 10 years ago, the production of custom switching converters has become almost a "commodity" market. Most of the cost leverage achievable from clever design has been utilized by most all of the better power supply vendors, and the cost battle is being fought in terms of manufacturing value add, which gives a big advantage to vendors with off-shore assembly facilities.

4.4.6 Desire for North American manufacturing

The emphasis on reducing costs described above has increased the percentage of power supplies procured from vendors in the Far East. This is a partial solution to the cost situation, but has created a more serious problem that has gotten the attention of many executives in the electronics industry. When so much of the power system content is sourced from overseas, there is very little product differentiation in terms of the power system. There is also very little motivation for any fundamental development work in power technology or in power packaging. This can have a long-term negative impact on our competitive position. Also, the shear magnitude of the dollars flowing overseas to procure power supplies is helping to contribute to the national balance of trade problem.

What is desired is a new technology or new process that will revitalize the power industry, and allow power systems to be configured with domestically produced hardware while still maintaining a competitive edge. The solutions described in Section 4.5 meet this need and also represent fundamental improvements to the power system architecture and regulator topology. High frequency converters along with distributed architectures represent the first significant advancement in power systems since the introduction of the switching regulator.

4.5 Power and Power Packaging Design - Industry Trends

In Section 4.4 we have reviewed many of the needs of the electronics industry for power system technology, and observed some developing trends in terms of technical requirements, cost and reliability objectives, and the economic environment of the development cycle. We will now discuss some of the recent developments in power system technology and design that address these needs. It will be seen that the solutions described below work synergistically to provide for not just new technology, but for a new design and development environment that allows significant competitive leverage.

4.5.1 Higher frequency

As was briefly discussed in Section 4.2.3, the size of a switching converter can be drastically reduced by increasing the operating frequency. Within a limited range (up to approximately 200 KHz), this technique has been applied to conventional switching regulators. At about this frequency, however, the transient switching losses in the silicon start to dominate the total losses, and increasing the frequency further increases the losses, lowers the efficiency, and negates the size advantage that would otherwise be realized. So the limitation with conventional technology is really a component limitation. The first barrier is the high frequency performance of the silicon (power MOSFETS in most of today's designs). Much of this limitation is due to the package used for the silicon. The wirebonds internal to the package in conjunction with the length of the package pins add significant inductance in series with the device. This packaging inductance, at the high transient current levels used in a switching converter, create unacceptable losses as the operating frequency exceeds a few hundred KHz.

The next barrier is the magnetic structures. Conventional materials, winding methods, and assembly techniques also run into trouble as the frequency goes up due to problems with coupling and lead inductances.

Electrolytic capacitors also are not suited for very high frequencies, due to their internal impedance characteristics. They are efficient only from DC up to about 200 KHz.

Consequently, even though power converter designers have long realized the advantages of higher frequencies, there have been so many difficult component problems to overcome that not much progress was made. Most converter design organizations do not possess the resource, funding, or time to do advanced component development. The power component industry is segmented. One vendor makes MOSFETs. Another makes magnetics. Others make capacitors. None of these vendors is motivated to unilaterally develop a component suited for very high frequency, knowing that the other needed components do not exist and that there is no guarantee that it will be a profitable item.

One answer for this dilemma is to take a quantum leap in frequency rather than slowly working up from 200 KHz to 300 KHz to 400 KHz, etc. What if you are determined to build a 2 MHz converter? What could it look like topologically? What kinds of components could you use? What kind of package would be required? What assembly techniques could you use? This approach "leap-frogs" over the limitations of today's components and packaging and forces the developer to think about designing the required solutions with a "clean sheet of paper". It appears that those organizations who are successfully participating in the new world of high-frequency power systems have used this type of reasoning.

What follows below is a summary of the types of solutions that have surfaced once the psychological high frequency barrier was crossed. Some of these areas are just starting to be explored, and consequently are incomplete. Other areas are very controversial within the power community, with vocal advocates for several different approaches still contending for consensus. Some are well accepted by most power system designers but are awaiting acceptance from product designers. The very newness of many of these concepts along with the degree of innovation represented in the solutions makes this a very exciting time to be involved with power electronics.

4.5.2 Components and magnetics improvements

To allow efficient converter operation in the Megahertz frequency range, traditional components, especially energy storage and power handling elements, must be abandoned in favor of new approaches. In this section, we will address some of the component-related issues, and discuss how new component solutions and packaging approaches have changed the role of components from an inhibitor to a facilitator, allowing high frequency distributed power to become a practical reality.

Perhaps the biggest benefit of this approach is the elimination of electrolytic capacitors. Designers of these devices were struggling to produce parts that operated efficiently at the 100 to 200 KHz frequencies dominant in modern switching regulator designs. The principal limitations were the equivalent series resistance and equivalent series inductance inherent within the device. These parasitic elements cause internal heating due to the AC ripple current imposed upon the device in typical switching converter topologies. The internal heating accelerates the depletion of the chemical electrolyte sealed within the capacitor, making these devices limited lifetime circuit elements, with a known end-of-life wearout mechanism. Contrast this with other circuit elements, such as resistors, inductors, and semiconductors that have no known wearout mechanism, and theoretically should operate forever. With careful design, attention to understanding the voltage and ripple current stresses, and conservative derating, designs can be done using electrolytic capacitors that meet most of the present reliability specifications and equipment lifetime requirements. However, the capacitors still end up being the highest failure rate item in the converter. Elimination of the electrolytic capacitors would dramatically improve the reliability. The trend for tighter and tighter reliability requirements provides additional incentive for finding a way to eliminate electrolytic capacitors from power regulators.

Another disadvantage of electrolytic capacitors is their size. They are, with the exception of the magnetics, the largest devices in a switching converter. They are typically cylindrical in form factor (which does not allow optimal packing density), with a height that does not allow for a low profile package. To make matters worse, the larger sizes cannot be automatically inserted by automated board assembly equipment, requiring costly manual assembly to be used. These devices are also polarized, but without any mechanical keying. This allows them to be inserted into the board backwards. If this happens, they become at best ineffective, and in many applications can violently self-destruct. Consequently, much care must be taken during assembly to insure that they are correctly polarized. This often takes the form of time-consuming and expensive inspection processes.

Since the energy handled per cycle is inversely proportional to operating frequency, capacitors need to store only one tenth as much energy at 1 MHz than at 100 KHz. This allows alternative capacitor implementations to be used instead of electrolytic devices. The most useful capacitor for high frequency operation is the multilayer ceramic (MLC). These devices are fabricated by stacking several layers of metallized conductors applied to a ceramic dielectric. They have several characteristics that make them attractive in high frequency power applications. They are physically small and very inexpensive relative to electrolytic devices. They are not sensitive to polarity, and their small size allows for convenient assembly with automated equipment. They are readily available in surface mount (SMT) packages, making them very suitable for the types of packaging used for high frequency power regulators. Unlike the electrolytic, they have no chemical electrolyte, and consequently no wearout mechanism. They allow significantly higher reliability levels to be achieved. The MLC capacitor thus provides advantages in cost, size, reliability, and ease of assembly - a very impressive list of attributes.

As the operating frequencies of conventional switching converters has increased up to the 200 KHz range, magnetic devices have gotten somewhat smaller. They still are predominantly conventional in terms of materials and structure. Nickel-Zinc or Manganese-Zinc ferrite are the most commonly used core materials. The core structures most widely used are E-I cores and toroids. The winding structure is especially critical in determining the operating characteristics of the device, and as the operating frequency increases, the design and consistent manufacture of the windings becomes an ever more challenging problem. Wire sizes, physical relationships between windings, twisting of windings, location of windings on core, lead lengths and lead dress all must be repeatedly controlled to achieve a successful design. This is primarily a manual process. It depends upon the skill and consistency of the operators at specialized magnetics vendors.

At frequencies up to one or two MHz, these conventional approaches can still be used, but success depends upon extremely close attention to all the details mentioned above. The completed device becomes smaller. At converter output power levels up to 100 watts, conventional magnetic structures are being fabricated that are less than 12 millimeters high - so called "low profile" magnetics. When constructed in a suitable package or carrier, these devices become suited for automated surface mount assembly into the converter.

Although most present high frequency transformers use conventional wire windings, in some applications windings can be fabricated using circuit board traces or flexible printed circuits. These approaches tend to introduce limitations in terms of the electrical design flexibility, but offer the advantages of cost (no manual winding operations), repeatability (winding structures controlled by artwork), and lower profile (planar winding structures).

At frequencies above one or two MHz, more innovative magnetic solutions will be needed. These solutions will involve materials, structures, and assembly process techniques. Goldberg et al [27] have addressed the design and analysis of magnetic structures up to 10 MHz. This paper discusses materials, structures, and performance, and is recommended reading for those wishing to further explore the world of high frequency magnetics.

There are other approaches being discussed for magnetics that need to operate at frequencies above a few MHz. One approach is to use a thick film magnetic material that is screened onto the circuit board or component carrier. This would be used in conjunction with windings implemented with circuit board traces or flexible circuitry. The use of thick film magnetic materials is presently very rare, other than for experimental purposes. The ultimate extension of this concept would be to integrate magnetic structures within the internal layers of the circuit board.

Another approach that is currently receiving attention is the so-called "matrix" transformer. This concept uses a flat array of very small toroidal cores to implement an equivalent single large magnetic structure. This approach provides for very low profiles and a reasonable degree of design flexibility while utilizing readily available conventional magnetic cores. The paper by Sum [29] describes this concept further.

Power semiconductor devices must also change if high frequency operation is to be practical. Conventional bipolar and MOS power devices are packaged in leaded plastic or ceramic packages intended for insertion into circuit boards. At high frequencies, the lead inductance becomes an inhibitor to efficient circuit operation. Going to a surface mount package helps somewhat, but SMT packages are limited in terms of available power dissipation and in heat sink hardware. Several high frequency designs are instead using direct attachment of the silicon chip to the circuit card or carrier and wiring to the chip by means of wire bonds or tape automated bonding (TAB). This reduces the effects of the device leads and also offers opportunities for alternative thermal treatments as will be described in Section 4.5.4. Another aspect of silicon packaging that has allowed extension of converters into the high frequency range is "integrated power" or "intelligent power". These devices combine power handling elements with low level control and logic elements on a single silicon chip, and will be discussed in more detail in Section 4.5.3.

Other types of components, such as connectors, resistors, fuses, and integrated circuits do not derive as much benefit to increased operating frequency as the components discussed above. They stay about the same size they would be at lower frequencies. Some savings in circuit board area are obtained, however, by using SMT versions of these components rather than the pin-in-hole offering.

4.5.3 Silicon integration

For several years, switching converters were constructed using mostly discrete semiconductors. Standard digital and analog integrated circuits such as operational amplifiers, comparators and flip-flops were used whenever possible, but the rest of the silicon was discrete. Consequently, such functions as voltage references, sensors, unique diagnostic circuitry, bias voltage generators, and drivers were implemented with discrete semiconductors along with all the other discrete components required to form these circuits. The first relief from this situation became readily available in about 1980, when specialized integrated circuits for switching regulator control appeared on the market. This allowed for a savings of dozens of low power components and improved the cost and reliability of switching converters. The functionality of these control chips has been steadily improving, and some of them now include drivers so that they can directly interface with the main high powered switching devices.

The next level of sophistication is "integrated power" silicon. This approach combines two or more circuit technologies on the same silicon chip. One of these technologies supports power handling devices. DMOS or bipolar is the most common choice. The other technologies are used to support low-level signals, high accuracy applications, or logic. These technologies include small-signal bipolar and CMOS. A typical, but sophisticated, integrated power chip might include CMOS, bipolar, and DMOS devices. The CMOS would be used to implement high density logic for diagnostics and controls. The bipolar would be used for high accuracy analog applications such as comparators or for intermediate power levels, such as drivers. The power DMOS could be used for the main switching devices in a switching converter.

Several semiconductor vendors offer "standard" versions of such chips that are designed for applications such as switching regulator controllers, DC motor drivers, solenoid drivers, and printer hammer drivers. Other vendors offer "custom" versions. With the custom approach, the user can define a unique chip that is configured to satisfy the requirements of one particular application. The chip can then be designed, typically using some kind of silicon "macro" approach, whereby standard functional cells are implemented in silicon by customized masks and interconnected with customized metallization. This approach yields very high levels of integration, often replacing literally hundreds of discrete components covering several square inches of circuit board area with a single silicon chip.

Whether standard or custom, the integrated power chip offers a significant set of advantages over more discrete approaches. Of course, it reduces the component count and improves cost and reliability. It also improves the electrical performance of high speed power circuits by reducing the interconnection lead length between stages of a circuit. The driver and power switch can now be located a millimeter from each other on the same piece of silicon with low inductance interconnections between them. With conventional approaches, they would typically be several cm apart and also separated by the internal package inductance of each device, at least two solder joints, and some circuit board trace inductance. Consequently, the performance of the integrated device is much better. Higher frequencies can be achieved without sacrificing efficiency. Another advantage is the ability to achieve better thermal tracking between power devices and their control and drive circuits. Since all these circuits are on the same piece of silicon, they tend to track each other thermally. This allows for improved implementations of such functions as thermal protection of drivers and current and power sensing.

4.5.4 Enhanced thermal management

We have seen how innovative approaches to components and packaging have allowed operating frequencies of switching converters to reach the Megahertz range. We have also seen that this, in conjunction with surface mount packaging, has significantly reduced the size of regulators. Figure 4.7 gives a "before and after" comparison of regulator size, efficiency, and power density. It is based on measurements from hardware that is actually available today, and indicates that there is another problem to be solved before high frequency power conversion becomes practical - a thermal management problem. The thermal density - the power per unit area dissipated within the converter - has increased by approximately a factor of ten. Without attention to cooling, the high frequency regulator will not be practical due to excessive internal temperatures.

The first thought might be to solve this problem using conventional heatsinks. Several disadvantages to this approach become apparent after some thought. Heatsinks are bulky. They, by their nature, take up lots of air space, so that the ideal low-profile form factor for the converter will not be possible. They are also not compatible with automated assembly, typically requiring at least one manual process step for installation. They also add to the cost of the product, both hardware cost and assembly cost.

One technique that is beginning to see usage is to construct the converter on some sort of "thermal carrier" substrate. This carrier could be as simple as a fairly normal glass-epoxy circuit board for lower power applications or could be a specialized high thermal conductivity carrier for higher power applications. In either case, the concept is simple - conduct heat from the bottom of the components and through the carrier. The heat is then removed by cooling the bottom of the carrier by means of convection, forced air, or conduction. An example of this technique is shown in Figure 4.8. In this example, the semiconductor devices are contained in surface mount packages and a special thermal carrier is used with a thermally conductive dielectric, heavy metal backer plate, and forced air cooling.

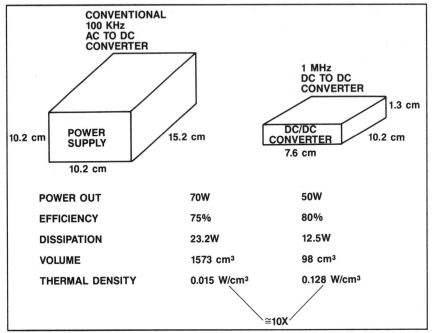

Figure 4.7 Thermal management problem.

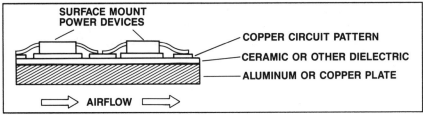

Figure 4.8 Thermal carrier solution.

Thermal carrier material is presently available from several vendors. The dielectric is typically thermally enhanced epoxy, although ceramics can also be successfully applied in this application. The backer plates are usually aluminum or copper. The top layer of metallization is a single layer of copper, very much like that of a single sided printed circuit board.

An extension of this approach, that provides for even better thermal performance, is to remove the SMT package from the silicon and mount the chip directly to the carrier (direct chip attach). The chip can then be connected to the top-side circuit pattern by means of wirebonding. This results in a very low profile package with excellent thermal and electrical performance, since the thermal and electrical impedances inherent in the SMT package are no longer present.

The thermal carrier approach appears to be adequate to address the thermal management tasks for high frequency converters for the next few years. Longer range, more sophisticated techniques may be required if power densities are to increase even further. Some of these techniques, which today tend to be either inefficient or too costly, include cold plates, liquid cooling, vapor phase cooling, heat pipes, cryogenic cooling, and thermoelectric coolers.

4.5.5 New converter topologies

All the switching converters we have discussed have been pulse width modulated. They operated with their switching device(s) either in a fully conducting or fully off condition, and with rectangular drive waveforms. The transition periods between on and off were kept as short as possible to minimize the time spent in this region. These pulse width modulated (PWM) approaches work well up to the lower MHz range (perhaps 2 to 5 MHz).

At frequencies higher than this, the picture changes. At these high frequencies, the transition time begins to be an appreciable fraction of the total cycle time, so that the efficiency starts to degrade. Also, the parasitic circuit elements associated with interconnection inductances begin to be especially bothersome. If the designer accepts the fact that these parasitic elements are not going to go away, and instead uses them to advantage in the circuit, a new class of converters appears - resonant converters. Resonant converters are fundamentally different from PWM converters. PWM converters operate at a constant frequency and use pulse width modulation to provide regulation.   Resonant converters operate at a constant "pulse" width and achieve regulation by varying the frequency. PWM converters ideally have rectangular voltage and current waveforms. Resonant converters have sinusoidal current waveforms.

There are dozens of resonant topologies and various classification schemes. The reader is referred to the references for additional detail in this area, as it can become a very complex study. It is, in fact, one of the "hottest" areas for research at universities specializing in power electronics. For our purposes, it is sufficient to list the major classification schemes without going into operational details on each one. The most widely used types of classification are:

* Zero voltage switching vs. zero current switching
* Series resonant vs. parallel resonant
* Fully resonant vs. quasi-resonant vs. transition-resonant

As an example of one topology, Figure 4.9 shows a series resonant fully resonant converter. This topology is series resonant because the transformer primary is in series with the resonant circuit. The resonant circuit is composed of C1 and L1. In actuality, L1 is typically composed mostly of the transformer primary leakage inductance. There may or may not be a separate inductive component. In this circuit, line and load regulation is achieved by changing the operating frequency, and consequently the average energy transferred to the secondary. The frequency of the resonance remains unchanged, but at high converter operating frequencies, the full amplitude of the resonant tank circuit is transferred. At lower converter operating frequencies, the energy being transferred to the secondary lowers the average resonant current - the primary current begins to look like a damped sine wave.

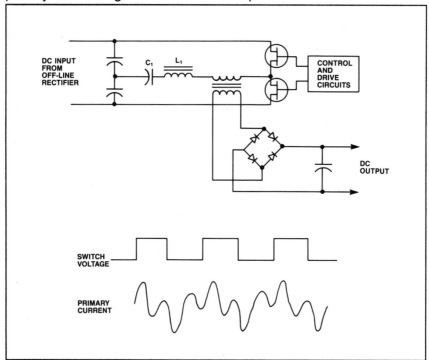

Figure 4.9 Resonant converter example.

Resonant topologies have some interesting advantages

compared to PWM designs. The zero current switching operation removes one of the major sources of noise generated in PWM topologies. The voltage waveforms still have sharp transitions, but these do not generate as much noise as the fast current transitions do. The lower silicon stresses at turn-on and turn-off in the resonant converters can result in greater reliability. The sinusoidal output current waveforms are also compatible with synchronous rectifiers, which may offer advantages at very low output voltages. At frequencies of 5 MHz and above, the resonant approaches should also offer advantages in efficiency.

But resonant converters also have disadvantages. Even though the switching stresses on the semiconductors are less than for PWM topologies, the semiconductor utilization is not as efficient. This is because the peak to average ratios of voltages and currents encountered with the sinusoidal waveforms are higher than those for the rectangular PWM waveforms. Consequently, the silicon chips must be "oversized" in terms of voltage rating. This typically results in devices with higher saturation resistances and lower efficiencies. The variable operating frequency creates a very challenging electromagnetic compatibility problem. The magnitude of the noise generated is less because of the "softer" switching, but its fundamental frequency changes with the load current, so that the effects and interaction with the system must be understood over a very wide frequency range. Another disadvantage is the "newness" of the resonant topologies, their operational dependence on parasitic circuit elements, and the lack of a substantial component base to support these topologies. These factors combine to make the design job a much more difficult one than for the well understood PWM approaches.

In spite of the initial enthusiasm that occurred when the resonant approaches were first being promoted, the general consensus today is that at frequencies up to 2 or 3 MHz the PWM converter generally offers advantages over a resonant one except in very specialized applications. As a result, the vast majority of the converter designs for the immediate future will continue to utilize PWM techniques. However as operating frequencies increase over the next few years, resonant approaches will become more and more attractive. This long-range potential for efficient operation at very high frequencies warrants the continued attention to development of these topologies.

4.5.6 Distributed architectures

When the technological developments discussed in the preceding sections are combined, the power system designer has some very powerful tools with which to construct electronic power systems that were not previously realizable or practical. Perhaps the best example of this is the class of power systems referred to as distributed power architectures. Distributed power systems are just now becoming very prominent in several types of products, and represent the most significant development in power system design since the advent of the switching converter.

Distributed power architecture is such an important development that we should attempt to define it. The following definition captures the spirit of distributed systems:

A distributed architecture is a functional and physical partitioning of a power system that utilizes advantages of two or more power topologies or technologies. Distributed power architectures frequently contain the following elements:

* Centralized AC to DC Converter to interface with the AC powerline

* The AC to DC Converter provides the functions of isolation, DC conversion, noise suppression and power factor correction

* An isolated intermediate DC voltage (12 to 60 V) is distributed within the product

* Individual load converters (DC to DC) are used for each load function or load package

* DC to DC converters are physically located at, or very close to, the load

* The DC to DC converters are small, dense, and may or may not contain isolation

* Provision is made for easy product upgrades or featurization

* Provision for redundancy

* Provision for battery back-up or other technique to provide immunity from powerline faults

A system does not need to contain all the above attributes in order to be considered a distributed power system, but most such systems contain several items on the list. For purposes of discussion, a typical distributed power system is shown in Figure 4.10.

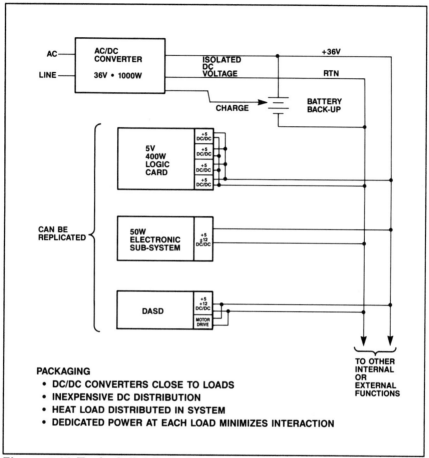

Figure 4.10 Typical distributed power system.

The system shown in Figure 4.10 is not an actual system, but does depict what would be a reasonable system design for a mini computer or communications controller type of product. There is an AC to DC converter that provides isolation from the powerline and converts the incoming AC into a DC voltage (36V in this example) that is then distributed throughout the product. Three major loads are shown. The 400 W logic load is powered by 4 paralleled DC/DC converters that are located at the logic board. These could be 100 W converters, or they could be 135W converters in a N+1 redundant configuration. This would allow the load to continue to operate if one of the converters experienced a failure.

The 50W load is powered with a triple output DC/DC converter, and the DASD by a double output device. All these DC/DC converters would typically be high frequency pulse width modulated units, with an operating frequency somewhere between 500 KHz and 3 MHz. The motor drive electronics for the DASD can also use the distributed 36V as a power source as shown. Many of the same technologies and packaging approaches used for power converters can be successfully applied to motor drive applications. This tends to maximize the commonality in the technology and packages used in the product.

If desirable, the 36 V intermediate voltage can also be distributed external to the product to provide a source of power for external expansion units. This approach has the advantage of minimizing cost, since the AC/DC converter does not need to be duplicated in the other unit. Also shown is a battery back-up function to allow operation of this product to continue in the event of disruptions on the powerline.

When the distributed system of Figure 4.10 is compared with the centralized system of Figure 4.5, many of the advantages of distributed power become apparent. Some of these advantages are discussed below. They are divided into technical advantages and development advantages.

TECHNICAL ADVANTAGES

* Better DC regulation is achieved at the loads. The elimination of much of the distribution conductors reduces losses and allows the correct voltage setting to be obtained at each load.

* Much lower cost for DC distribution. The heavy bus bars needed for distributing low voltage DC are no longer required.

* The heat load is distributed throughout the system. The centralized source of heat dissipation represented by the power supplies in a centralized system is removed, and replaced by several smaller sources of heat.

* The interaction problem between loads has been eliminated. Each load function has its own local DC converter so that dynamic current changes in one load will not affect the voltage going to another load.

* For availability enhancement, redundancy is easy to implement by using such techniques as N+1 configurations of DC to DC converters.

* Battery back-up is very easy and inexpensive to implement. The intermediate voltage DC bus is selected such that it can also be

supported by a battery. Battery monitoring and charging can be accomplished easily by adding very little additional circuitry to the AC to DC converter.

* The intermediate DC voltage is a convenient source voltage for motor controls and other power electronics applications. Here it is shown as a source for the motor drives for the DASD unit.

* Product upgrades and featurization are very easy. The "entry level" product needs only the DC to DC converters to support its function, so that the cost is minimized. As other function is added, its power hardware is also added so that the power system cost grows with the system functionality. This is not the case for centralized systems, where the power supply must be big enough to support the total load of a completely configured system.

* The distributed system offers easier fault isolation. With centralized approaches, a short in one of the loads pulls down the centralized power supply, and diagnosing the location of the fault is very difficult. With distributed systems, the local regulators and their respective loads form natural and easily identifiable fault groups.

DEVELOPMENT ADVANTAGES

* Many power systems can be put together with "off-the-shelf" distributed power modules. This allows for the creation of truly custom power systems without the disadvantages of custom power supply development.

* Development time for the power system is substantially improved. Savings of up to a year are not unreasonable.

* There is significantly less exposure to design and qualification problems. Most of the unique design is no longer required, and many of the modules needed will have already been qualified.

* Much quicker response to system design changes is possible. Since the load interdependence typical of centralized systems is no longer present, many changes can be accommodated by just changing the part number of a standard module or by increasing the usage of a module already present.

* A much lower level of design skills is required. Detailed knowledge of the switching regulator design process is not needed. The power system

designer needs only skills in how to select and interconnect the modules needed to configure the desired power system.

* The massive exposure to problems from component characteristics and component vendor changes that is a part of custom power supply designs is largely eliminated. The distributed power modules are mostly replications of a standard design and all use the same set of components, which are very well understood and qualified.

* By the same reasoning, exposure to assembly process changes is drastically reduced. The distributed modules are all assembled by the same vendors using common processes and techniques.

* The agency approval process is substantially easier. Many of the modules would already be approved. The approval of new modules would be very streamlined due to the similarity of design to existing units.

* The net result of the above is a very large savings in non-recurring development cost and in manufacturing start-up costs.

As can be seen, there is a very impressive list of reasons why the distributed architecture approach should be seriously considered for any new product designs. There is, in fact, considerable interest at the present time in distributed approaches. The concept is not new. What is new is the availability of hardware that can implement such systems in a cost competitive way. This is due largely to the advent of the small high frequency DC to DC converters we have previously discussed.

As with most technology decisions, the centralized vs. distributed architectural decision is many times made on the basis of cost. At the present time, the cost of distributed solutions is steadily decreasing. For many systems it is already less than the cost for a centralized solution. Examples of such systems are mini computers and telecommunications products. In many very low end applications, such as personal computers, the cost situation may favor a centralized solution. Even so, the many other advantages that the distributed architecture offers has many system designers considering its use in spite of some additional expense.

4.5.7 Improved powerline interface

As was discussed in paragraphs 4.4.2 and 4.4.4, there is increasing awareness on the part of equipment users of the effects of the AC powerline on the reliability and availability of electronic products. The power market is beginning to respond to this awareness by providing cost-effective solutions that satisfy these needs.

Power factor correction is becoming much more common, especially in information processing and telecommunications equipment. With discrete circuit solutions, it is presently economically feasible to include power factor correction in equipment that draws AC power levels of 2 KW or more. Very recently, integrated circuit manufacturers have made available integrated chips that provide on one chip all the low level circuitry required for power factor correction. Typically, external power devices are still required. This technology has reduced considerably the cost and size of the power factor correction function, and has made its incorporation feasible in products at lower power levels - in the 300 W to 2 KW level.

There is also a trend towards decreasing the dependence upon the AC powerline. Products such as home computers have made most users more aware that there are interruptions in the utility service - interruptions that can impair the useability and availability of electronic equipment. One solution to this problem is to provide battery backup. This solution is becoming more popular due to the improved battery technologies now available. Many of these battery improvements have come about from the need for light weight, high power density, highly reliable batteries for portable and laptop computers. As shown in Figure 4.10, battery backup is especially attractive in distributed power systems. This is due to the low cost of adding the battery and charger/monitor functions to the intermediate voltage DC bus that is already in place. In many such systems, battery backup capability can be added for only a nominal increase in price. In combination with the other advantages of distributed power, the addition of battery backup creates a power system with unsurpassed performance and flexibility. It allows uninterrupted operation of the system during transients or temporary outages in the AC input voltage, resulting in significant improvements in user availability.

4.6 Predictions for the 90s

As we have seen, power system architectures and power packaging technologies have recently entered a period of rapid improvement in size, cost, reliability, and functionality. It is expected that these improvements will continue well into the 1990s as the high frequency converter technologies become easily and inexpensively manufacturable. As with any technology prediction, it is very difficult to estimate the growth rate of new developments in the field of power system electronics. None the less, below are the author's predictions of the kinds of technological advances that could occur within the next decade.

* By 1995, power factor correction will become common at all power
levels

* Distributed power architectures will dominate information processing
and telecommunications applications by the end of the decade

* Secondary regulation will move closer to the load circuits
    - Regulator on board common by 1991
    - Regulator on card common by 1993
    - Regulator on module common by 1996
    - Regulator on chip common by 2000

* Switching frequencies will continue to increase
    - 1 MHz common by 1992
    - 5 MHz common by 1996
    - 10 MHz common by 1998

* PWM topologies will dominate throughout the first half of the decade,
but resonant topologies will be in widespread usage for high frequency
designs by 1996

* By the middle of the decade, most power converter development will
be done by power module vendors rather than by the equipment
developers. The very sophisticated high frequency designs then being
used will require unique materials and processing capabilities as well as
very specialized design talent. The environment for power converter
design will be not unlike that of today's integrated circuit industry, with a
few specialized companies supplying the vast majority of the industry's
needs.   Along with this will come increasing standardization and
economies of scale in production.

4.7 References

For more detailed information on the topics discussed in this chapter, the
following references will be useful:

SWITCHING REGULATOR TECHNOLOGY

1   M.H. Rashid, Power Electronics, (Prentice Hall, Englewood Cliffs,New Jersey,
         1988).

2   A.I. Pressman, Switching and Linear Power Supply, Power Converter Design, (Hayden, Rochelle Park, New Jersey, 1977).

3   P. Wood, Switching Power Converters, (Van Nostrand Reinhold, New York, 1981).

4   E.R. Hnatek, Design of Solid-State Power Supplies, 2nd Edition, (Van Nostrand Reinhold, New York, 1981).

5   R.D. Middlebrook and S. Cuk, Advances in Switched Mode Power Conversion, Vol 1, (TESLAco, Pasadena, California, 1981).

6   R.D. Middlebrook and S. Cuk, Advances in Switched Mode Power Conversion, Vol 2, (TESLAco, Pasadena, California, 1981).

RESONANT CONVERTERS

7   Recent Developments in Resonant Power Conversion, Edited by K.K. Sum, (Intertec Communications, Ventura, California, 1988).

8   R. Vinsant, "Basics of series-resonant converters", Powertechniques Magazine, Sept 1985.

9   D.C. Hopkins, M.M. Jovanovic, F.C. Lee and F.W. Stephenson, "Two Megahertz off-line hybridized quasi-resonant converter", Proc. 2nd Annual IEEE Applied Power Electronics Confer, San Diego, CA, 1987.

10   J.G. Kassakian, M.F. Schlecht, "High-frequency high-density converters for distributed power supply systems", Proc. IEEE, 76 (4), Apr 1988.

11   P.C. Todd, "Classifying resonant-mode converters", Powertechniques Magazine, Aug 1988.

12   P.C. Todd, "Resonant converters: to use or not to use? That is the question", Powertechniques Magazine, Oct 1988.

13   F.E. Sykes, "Resonant-mode power supplies: a primer", IEEE Spectrum, May 1989.

14   W.A. Tabisz and F.C. Lee, "5MHz, 50W, Zero-Voltage Switched Multi-Resonant Converter", PCIM, Aug 1988.

15   G. Mulcahy, "PWMs versus resonant converters", Electronic Engineering Times, March 26, 1990.

DISTRIBUTED POWER ARCHITECTURE

16   L. Thorsell, "Mini DC/DC supplies simplify redundancy in parallel systems", EDN, April 28, 1988.

17   F.F. Kunzinger and M.E. Jacobs, "Powering the next generation of electronics equipment with distributed architectures", Powertechniques Magazine, Dec 1988.

18   S.D. Cogger, "Simplifying system power distribution with high power-density DC/DC converters", Powertechniques Magazine, Sept 1988.

19   A. Bindra, "Toward distributing DC/DC power", Electronic Engineering Times, March 26, 1990.

20   C. Lazarovici, "Fault-tolerant power systems", Powertechniques Magazine, Jan 1989.

HIGH FREQUENCY DC/DC CONVERTERS

21   P. O'Farrell, "Market trends in power supplies", Electronics Industry, May 1985.

22   D. Haynes, "DC/DC converters are getting smaller with higher performance and lower cost", PCIM, Mar 1990.

23   J. Bassett, "Improving DC/DC converters", Electronics Engineering Times, Mar 26, 1990.

24   T. Harbert, "DC/DC converters work in varied environments", EDN Product News, Oct 1986.

25   J.H. Mayer, "Compact DC/DC converters serve distributed power applications", Computer Design, Nov 1, 1988.

26   E. Marchese, "High-density converters", Powertechniques, Mar 1990.

HIGH FREQUENCY MAGNETICS

27   A.F. Goldberg, J.G. Kassakian and M.F. Schlecht, "Issues related to 1-10 MHz transformer design", IEEE Trans. Power Electronics, 4 (1), Jan 1989.

28   V. Gregory, "Building magnetics with flexible circuits", Powertechniques Magazine, Feb 1989.

29   K.K. Sum and E. Herbert, "Novel low profile matrix transformers for high density power conversion", PCIM, Sept 1988.

30   H.A. Savisky, "New core materials will improve high frequency switchers in the 90s", PCIM, Mar 1990.

INTEGRATED INTELLIGENT POWER SILICON

31   E. Balbonni, G. Pietroban, D. Rossi and C. Vertemara, "Bipolar-CMOS-DMOS smartpower IC drives brushless DC motor", PCIM, Jan 1989.

32   M. Mehler, "Power Ics: No longer the new kids on the block", Electronic Engineering Times, Feb 20, 1989.

33   F. Longo, M. Pesce and I. Wilson, "Increasing system reliability with power Ics", Powertechniques Magazine, Dec 1988.

34   D. Pryce, "Smart-power Ics", EDN, Mar 31, 1988.

35   B. Dunn and R. Frank, "Future cars to employ more smart power Ics", PCIM, Feb 1989.

36   T. Hopkins, "Smart power - technologies for all applications", PCIM, Feb 1990.

THERMAL MANAGEMENT

37   M.E. Jacobs, "Practical limits of forced-air cooling", Powertechniques Magazine, Feb 1990.

38   M.E. Jacobs, "Limiting fan and blower noise", Powertechniques Magazine, Mar 1990.

39   R.W. Johnson, R. Weeks, D.C. Hopkins, J. Muir and J.R. Williams, "Plated copper on ceramic substrates for power hybrid circuits", IEEE Trans. Components, Hybrids & Manuf. Technol., 12 (4), Dec 1989.

40   H. Fick, "Thermal management of surface mount power devices", PCIM, Aug 1987.

41   A. Pshaenich and D. Hollander, "Managing heat dissipation in DPAK surface-mount power packages", Powertechniques Magazine, Dec 1988.

42   H. Fick, A. Pshaenich and D. Hollander, "Metal-backed boards improve performance of power semis", PCIM, Sep 1989.

POWERLINE QUALITY

43   J.M. Salzer, "Worldwide review of power disturbances", IEEE APEC Conference, 1987.

RECENT DEVELOPMENTS

IN THERMAL TECHNOLOGY

FOR ELECTRONICS PACKAGING

RICHARD C. CHU

ROBERT E. SIMONS

5.1. Introduction

For more than 50 years, the development of thermal technology for electronics packaging has been an important arena for the application of advanced heat transfer techniques. This has been especially true in the evolution of electronics packaging for digital computers. Since the development of the first electronic digital computers in the 1940s, the removal of heat has played a major role in ensuring their reliable operation. Early digital computers such as ENIAC (Electronic Numerical Integrator and Computer) used vacuum tubes as the basic logic element building blocks [1,2]. These physically massive machines were cooled by forced air and by today's standards were unreliable. The invention of the transistor by Bardeen, Brattain, and Shockley at Bell Laboratories in 1947 foreshadowed the development of generations of computers yet to come. As a replacement for vacuum tubes, the miniature transistor generated less heat, was much more reliable, and promised lower manufacturing costs. At the time it was even thought that the use of transistors would greatly reduce, if not totally eliminate, cooling concerns. This thought was short-lived as engineers sought to improve computer speed and storage capacity by packaging more and more transistors, first on printed circuit boards (PCBs), and then on ceramic substrates.

The trend towards higher packaging densities gained momentum during the 1960s with the introduction of monolithic circuit technologies fully integrated in the body of silicon chips [3]. As shown in Figure 5.1, the trend continued through the 1970s with the development of LSI (Large Scale Integration) technologies offering hundreds to thousands of devices per chip, and then through the 1980s with the development of VLSI (Very Large Scale Integration) technologies offering thousands to tens of thousands of devices per chip. This trend has generally been accompanied by increased power dissipation at the chip level. Demands to further increase packaging density and reduce signal delay between communicating circuits has also led to the development of multi-chip modules. As shown in Figure 5.2, the net effect of these developments has been a rapid growth in module level heat flux within mainframe computers, especially over the past 10 years [4]. This trend may be expected to continue. Projected heat fluxes for the early 1990s are in excess of 100 W/cm^2 at the chip level, 25 W/cm^2 at the module level, and 10 W/cm^2 at the PCB [5].

If these trends were not matched by the development and application of more efficient cooling techniques the result would be increased circuit operating temperatures. These higher operating temperatures would in turn lead to an increase in failure rate [6] and a

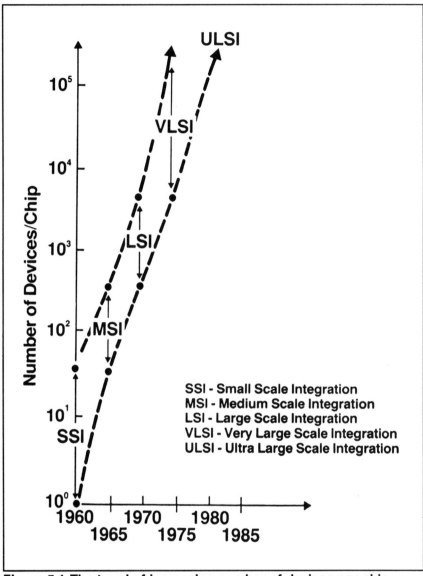

Figure 5.1 The trend of increasing number of devices per chip.

reduction in reliability. The power densities in some of today's mainframe computers are so high that without adequate cooling physical damage to the powered components would soon occur [7]. The remainder of this chapter will discuss the role of thermal management in electronics packaging and some of the recent thermal technology developed to provide adequate temperature control for electronic modules used in computers.

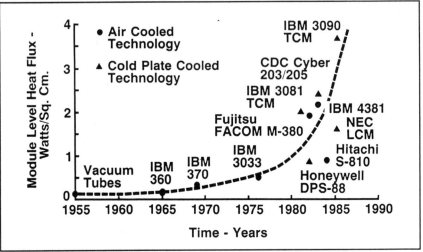

Figure 5.2 The trend in heat flux at the module level.

5.2. Thermal Management

Thermal management of electronic packages encompasses all the heat transfer processes and technologies required to remove and transport heat from the individual components to the system thermal sink in a controlled manner. In this context, thermal management has two primary objectives. The first is to ensure that the temperatures of all components are maintained within both their functional and maximum allowable temperature limits. Functional temperature limits define the temperature range within which the electrical circuits of a component are expected to meet their specified performance requirements. Operation outside this range can result in degraded machine performance or logic errors. The maximum allowable temperature limit is the highest temperature which a component or part of a component may be safely exposed to. Operation above the maximum allowable temperature limit may result in physical damage to the component or irreversible changes in its operating characteristics. The second objective of thermal management is to ensure that the distribution of component operating temperatures will be such that the aggregate of the resulting component failure rates will satisfy the overall system or product reliability objectives. It has already been noted that higher operating temperatures can lead to increased failure rates. The strong effect temperature can have on failure rate is illustrated in Figure 5.3, which provides normalized failure rates as a function of device operating temperature [8]. Based upon a reference temperature of 85°C and activation energies of 0.8 and 1.0 eV, a 15°C increase in operating temperature above the reference temperature results in about a 3 to 4 times increase in failure rate. Consequently, although a failure to meet product reliability temperature requirements may not be immediately obvious, over the long run it can become both apparent and costly.

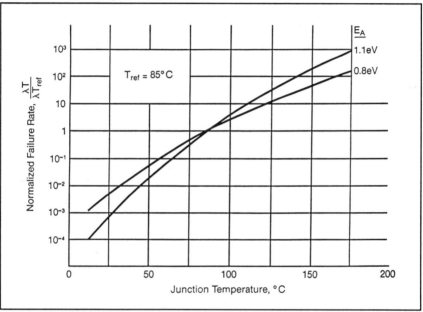

Figure 5.3 The effect of component operating temperature on component failure rate.

In addition to the two primary objectives of thermal management, there are a number of important secondary objectives. These objectives include:

1) Safety - the temperature of exposed surfaces must be kept within safe operating limits;

2) Cooling Reliability - the cooling subsystem design must be consistent with the overall system reliability and availability objectives;

3) Cooling Compatibility - the cooling system design must be consistent with the customer's environment and heat rejection capability; and

4) Cooling Cost - the cost of cooling cannot be excessive in relation to

the overall system cost.

Complex analytical and numerical models are often used to predict the degree to which the primary thermal management objectives may be achieved. However, relatively simple thermal models based upon the concept of thermal resistance can provide considerable insight into the nature and magnitude of the electronics cooling challenge.

5.3. Thermal Resistance

Many heat transfer situations may be modeled using an electro-thermal analog circuit approach with temperature difference providing a driving potential to induce heat flow through a thermal resistance. The heat transfer characteristics of a microelectronic package can then be characterized by a thermal resistance defined as:

$$R = \frac{\Delta T}{Q}$$

[5.1]

where
$$\Delta T = \text{temperature drop (°C)}$$
$$Q = \text{heat flow rate (W)}$$
$$R = \text{thermal resistance (°C/W)}$$

The relatively low-powered packages found in consumer electronics, communications equipment, and personal computers, often exhibit a relatively high total thermal resistance (e.g. 100°C/W) from the component heat source to the external cooling medium. In contrast, some of the electronic modules with high power dissipations and high power densities which are used in mainframe computers often exhibit a total thermal resistance less than 10°C/W [9]. The total thermal resistance of a package is sometimes broken down into an internal and an external thermal resistance. The internal thermal resistance (R_{int}) indicates the temperature rise which will occur within the package per unit of power dissipation. The external thermal resistance (R_{ext}) indicates the temperature rise which will occur between the external package surface and the cooling fluid per unit of power dissipation. These parameters may be used in the fundamental junction temperature equation:

$$T_J = \Delta T_{J\text{-}C} + P_C R_{int} + P_M R_{ext} + \Delta T_F + T_{F_i}$$

[5.2]

where

T_J = chip junction temperature (°C)
$\Delta T_{J\text{-}C}$ = junction to chip temperature rise (°C)
P_C = chip power (W)
R_{int} = package internal thermal resistance (°C/W)
R_{ext} = package external thermal resistance (°C/W)
P_M = module power (W)
ΔT_F = cooling fluid temperature rise (°C)
T_{Fi} = cooling fluid inlet temperature (°C)

This equation is often used to predict junction temperature of devices on an integrated circuit chip for known values of thermal resistance and power. Conversely, the equation may also be used to determine acceptable values of internal and external thermal resistance for specified values of power and junction temperature. Application of this equation to determine the combination of maximum acceptable internal and external thermal resistances to cool increasing chip powers for specified temperature constraints is shown in Figure 5.4. As might be expected, the acceptable total thermal resistance ($R_{int} + R_{ext}$) diminishes rapidly as chip power increases. For the example shown in Figure 5.4, in order to limit the junction temperature to 85°C for a 10 W chip, the total thermal resistance from chip to cooling fluid cannot be allowed to exceed 5.5°C/W. Accordingly, the thermal management task becomes one of selecting or developing, in concert with other packaging objectives and constraints, the proper combination of cooling mechanisms and packaging alternatives to minimize total thermal resistance and maximize thermal dissipation capability.

5.4. Cooling Options

Perhaps the two most fundamental decisions facing the thermal engineer are the choice of coolant and the mode of heat transfer. These choices will determine the magnitude of heat transfer coefficient which may be obtained. The heat transfer coefficient is especially important since it characterizes the effectiveness with which heat can be removed from the external surface of an electronics package. Consequently it has a first order effect on the magnitude of external thermal resistance.

For air-cooled packages the thermal engineer has the option of choosing natural or forced convection flow. Natural convection flow

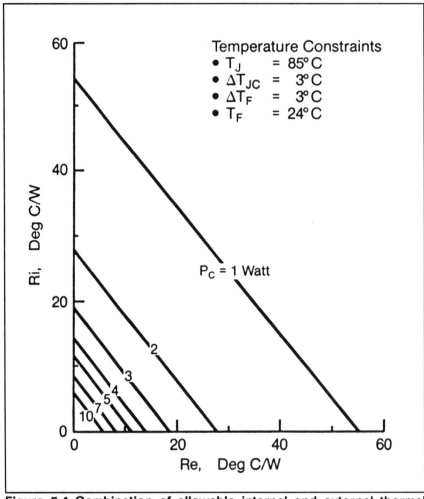

Figure 5.4 Combination of allowable internal and external thermal resistances at several chip power levels.

offers the advantage of not requiring a fan or blower to move the air. Movement of the air is induced by density differences resulting from the heat given off by the electronic components. However, this advantage is usually achieved at the cost of lower heat transfer coefficients. By incorporating a fan or blower, the thermal engineer may take advantage of forced convection heat transfer and usually obtain substantially higher heat transfer coefficients.

For components requiring the highest heat transfer coefficients available, direct liquid immersion cooling using forced convection [10] or boiling [11] may be employed.   The magnitude of heat transfer coefficients which may be obtained is shown in Figure 5.5.   Although water offers the best heat transfer characteristics, the first consideration must be the chemical and electrical compatibility of the coolant with the circuit chips and other packaging materials exposed to the cooling liquid. For this reason interest in direct immersion liquid cooling within the electronics industry has focused upon dielectric coolants such as the fluorocarbon liquids.

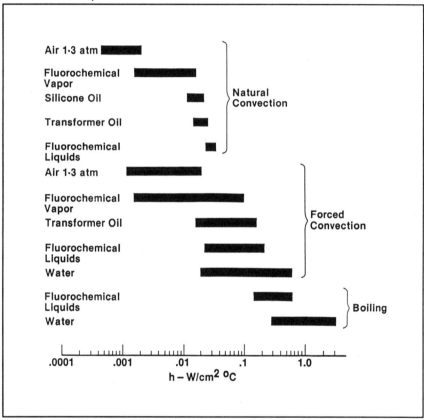

Figure 5.5 Magnitudes of heat transfer coefficients for various coolants and modes of convection.

5.5. Air Cooling

Air cooling has been, and continues to be, the most widely used method of cooling electronics packages.   To achieve the cooling capability required by some recent electronic module packages and still retain air cooling has required the development of enhancement techniques to minimize both internal and external thermal resistances. It has also been necessary to give consideration to minimizing thermal stresses within a module, and to reducing the resistance to air flow across a module package. Air-cooled multi-chip modules developed by Hitachi, IBM, and Mitsubishi serve as examples of electronics packages designed with these considerations in mind.

The Hitachi SiC RAM module design [12] shown in Figure 5.6 illustrates a number of desirable thermal design features. Silicon chips on a silicon carrier results in a good match of thermal expansion coefficients. The silicon carbide ceramic, with a thermal conductivity of about 14 times that of alumina acts as a heat-spreader and provides a good thermal expansion match to the silicon. To lower the external thermal resistance of the module, a low flow impedance/high airflow heat sink is attached to the module with a layer of thermal silicone rubber.

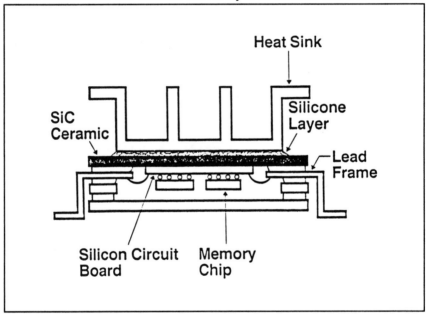

Figure 5.6 Cross-sectional view of Hitachi air-cooled SiC multi-chip module.

The Mitsubishi module design [13], shown in Figure 5.7, provides an example of a multi-chip module with improved thermal paths inside the module to reduce internal thermal resistance, and a heat sink outside the module to reduce external thermal resistance. The improved internal thermal paths are provided by bonding a heat-spreader plate to the back of each chip with soft solder. This is done in such a way that only a small gap remains between the spreader plate and the cap. The thermal path within the module is further improved by filling the module with hydrogen gas.

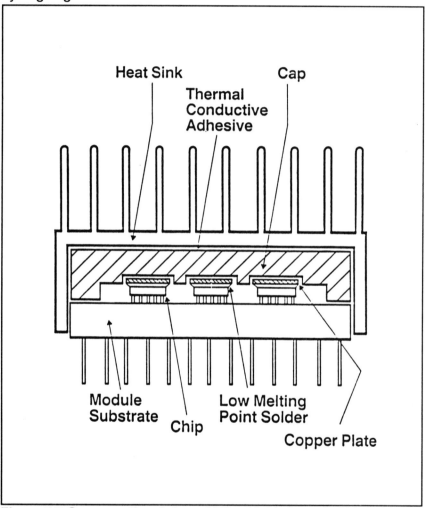

Figure 5.7 Cross-sectional view of Mitsubishi air-cooled high thermal conduction module.

A larger air-cooled multi-chip module was developed by IBM for use in the IBM 4381 processor [14]. This module package supports up to 36 chips mounted on a 64 mm x 64 mm multi-layer ceramic substrate.

A direct thermal path from the back of the chip to a ceramic cap is provided by using a unique thermal paste developed at IBM. This paste has a thermal conductivity of 1.25 W/m-K, is electrically nonconductive, and is easily removable for chip removal and replacement.   To adequately remove heat from the module, a novel heat sink designed to be cooled by an impinging air flow is used. As shown in Figure 5.8, the heat sink is made of multiple segmented fin structures attached to the ceramic cap. These structures allow air to readily flow through the heat sink, and provide sufficient compliance to take up the thermal expansion mismatch between the heat sink and the ceramic cap.

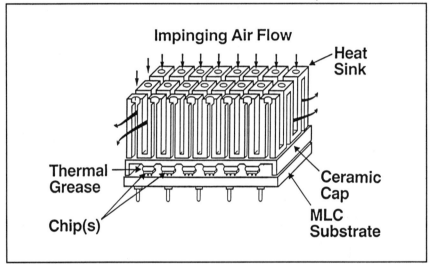

Figure 5.8 Impingement air-cooled multi-chip module used in IBM 4381 processor.

An even larger air-cooled multi-chip module is used in the IBM 9370 Model 90. All the processor logic, cache, memory, and control storage is contained in a single air-cooled Thermal Conduction Module (TCM) [15]. This package is an offshoot of the original water-cooled TCM (to be discussed in the next section). For this particular application a large straight fin heat sink occupying approximately 10 cm x 10 cm x 10 cm was designed to cool the TCM with air.

In addition to improvements in module cooling technology, improvements have been made in air cooling systems technology to support higher chip and module powers. Parallel air flow cooling systems have been used to reduce the cooling air temperature rise across a machine. With previous serial flow cooling systems the same airstream passed over a number of PCBs arranged in series. As a result, each PCB received cooling air which had been heated by the PCB preceding it. Depending on total PCB power and the air flow rate, serial air flow sometimes resulted in a substantial rise in the temperature of the cooling air from inlet to outlet. This cooling air temperature rise was directly reflected in increased circuit operating temperatures. As shown in Figure 5.9, parallel air flow systems of the type used in some Amdahl [16] and Fujitsu [17] computers, supply multiple parallel streams of air to cool the PCBs. As a result, each of the PCBs receive the same "fresh" cooling air. As shown in Figure 5.10, the IBM 4381 processor discussed earlier, also utilizes parallel air flow cooling to supply inlet air to each module.

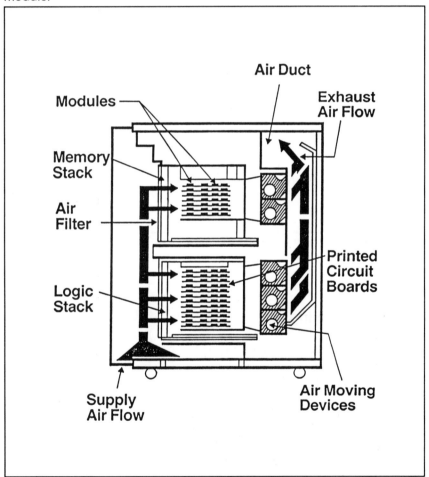

Figure 5.9 Parallel air flow system cooling configuration.

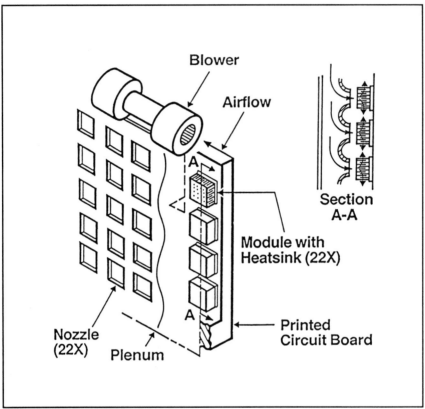

Figure 5.10 Parallel impingement air flow cooling scheme used in the IBM 4381 processor.

5.6. Water Cooling

As substantial as the improvements in air cooling have been, air cooling can not support the level of packaging density found in some of today's large mainframe computers. In these computers, it has been necessary to utilize water cooling via cold plates attached to the electronic modules. One of the earliest implementations of such a cooling technology was for the TCM shown in Figure 5.11 [18]. The first TCMs used in the IBM 3081 processor, contained a 90 mm x 90 mm multi-layer ceramic substrate designed to hold more than 100 chips, and as many as 45,000 circuits per module. As a result of these high packaging densities, peak heat fluxes of 20 W/cm^2 and 4 W/cm^2 respectively, were encountered at the chip and module levels. By contrast, earlier air-cooled IBM processors such as the System/370 Model 168 and the IBM 3033 had typical peak heat fluxes of 1.5-2.5 W/cm^2 at the chip level and 0.3-0.6 W/cm^2 at the module level.

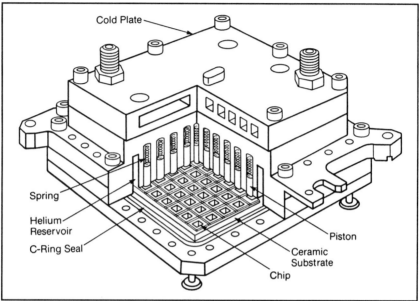

Figure 5.11 IBM thermal conduction module (TCM) with water-cooled cold plate.

The TCM was designed to bring the water-cooled cold plate as close as possible to the chip heat sources and to allow for variations in chip heights and locations resulting from the manufacturing process. In addition, allowances had to be made for nonuniform thermal expansion or contraction across whatever thermal path was provided. This led to the concept of a spring-loaded aluminum piston touching the back of the chip to provide a thermal path from the chip through the piston and into an aluminum housing. The thermal path created from the back of the chip to the water-cooled cold plate is shown in Figure 5.12. Thermal conduction across the interface gaps between the chip and the piston, and between the piston and the housing, is enhanced by filling the module with helium gas. Helium is used because it has a thermal conductivity which is about six times greater than that of air and is inert.

The Liquid Cooled Module (LCM) used in the NEC SX Supercomputer [19] is another example of a large multi-chip module which uses enhanced internal thermal paths and a water-cooled cold plate. As shown in Figure 5.13, the NEC LCM houses 36 chips mounted

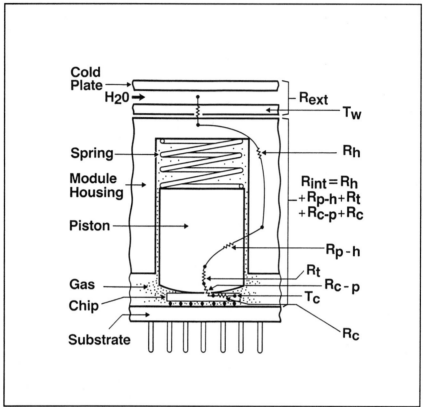

Figure 5.12 Heat flow in the TCM.

within individual Flipped Tab Carriers (FTC) on a 100 mm x 100 mm multi-layer ceramic substrate. The maximum power dissipation per chip is 7 watts and the maximum module power dissipation is 250 watts.

The thermal path within the LCM from an individual FTC to cooling water is shown in Figure 5.14. Heat is conducted from the chip within the FTC to the top surface of the FTC. Heat is then conducted from the FTC across a thermal compound to the face of an aluminum stud. The heat flow path continues along the stud and radially outward across the cylindrical surface of the stud into the surrounding aluminum Heat Transfer Block (HTB). Heat is removed from the HTB by the water-cooled cold plate bolted to the surface of the HTB.

Although it is also water-cooled, the multi-chip packaging technology used in the Fujitsu FACOM M-780 computer differs markedly from the IBM and NEC packages previously discussed. The Fujitsu packaging approach utilizes single chip modules mounted on a large

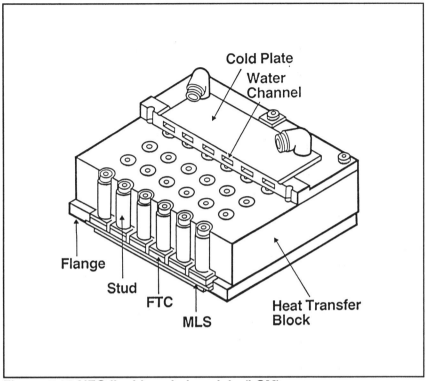

Figure 5.13 NEC liquid cooled module (LCM).

PCB [20]. A total of 336 single chip modules are mounted on both sides of a 540 mm x 488 mm PCB. The maximum chip power is 9.5 watts and the total board power dissipation is 3000 watts.

A section of the PCB, with single chip modules sandwiched in between two water-cooled cold plate assemblies, is shown in Figure 5.15. These assemblies are comprised of water-cooled bellows which contact each module. A compliant material consisting of a binder (e.g. silicone rubber) with a metal oxide filler (e.g. alumina) is used at the tip of each bellows, apparently to ensure adequate thermal contact with the surface of each module. High convective heat transfer rates are achieved within each bellows by using an impinging water jet. The cold plates are attached to the PCB in the factory, and are not meant to be separated in the field for replacement of the electronic package. Both the IBM TCM and NEC LCM cold plates are removable in the field.

Perhaps the ultimate water-cooling approach for application to electronic packages is that proposed by Tuckerman and Pease [10]. They proposed to eliminate almost all of the package thermal conduction

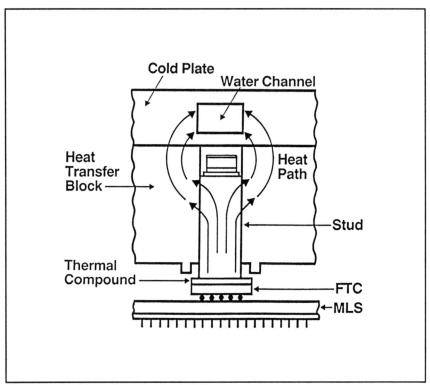

Figure 5.14 Heat flow in the LCM.

resistance by making the cold plate part of the integrated circuit chip as shown in Figure 5.16. To demonstrate the cooling performance that could be achieved, 50 micron wide channels were chemically etched 300 microns deep in a one centimeter square silicon test chip, and water was pumped directly through the channels. For a heat flux of 790 W/cm^2 the chip temperature rise above the cooling water temperature was 71°C. The real challenge in adopting this cooling approach is in how to provide the required chip and module cooling structure and flow interconnections in a manner which would be both manufacturable and reliable.

5.7. Immersion Cooling

Immersion cooling has been of interest as a possible method to cool high heat flux components for many years. Unlike the water-cooled cold plate approaches which utilize physical walls to separate the coolant from the chips, immersion cooling brings the coolant in direct physical contact

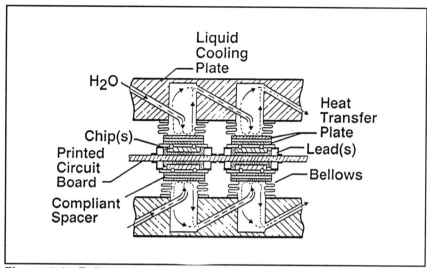

Figure 5.15 Fujitsu water-cooled bellows cold plate cooling scheme.

with the chips. As a result, except for the conduction resistance from the device junction to the surface of the chip in contact with the liquid, most of the contributors to internal thermal resistance may be eliminated. The dominant thermal resistance will then be the external (i.e. chip to liquid) resistance, which is inversely proportional to the heat transfer coefficient. As noted earlier very high heat transfer coefficients may be obtained with liquids in either the forced convection or boiling mode of heat transfer. It is essential that the liquid used has dielectric characteristics which will not adversely affect circuit delay, and is chemically compatible with the electronics packaging materials it contacts. The fluorocarbon liquids (e.g. FC-72, FC-77) meet these requirements and are considered to be the most suitable liquids for immersion cooling of electronics.

The Cray-2 supercomputer provides a good example of forced convection immersion cooling of electronics [21]. As shown in Figure 5.17, the multi-chip module used in the Cray-2 consists of a stack of 8 PCBs on which are mounted arrays of single chip modules. Although specific details of chip power and thermal resistance have not been reported in the literature, the module power dissipation has been reported to be 600 to 700 watts [22].

If 700 watts were distributed uniformly over the surfaces (i.e. top and bottom) of the PCBs within the multi-chip module, the average surface heat flux would be 0.21 W/cm^2 This heat flux is within the range of air cooling. However, considering the number of multi-chip modules (i.e. 14 columns x 24 modules/column = 336) in the Cray-2, the total

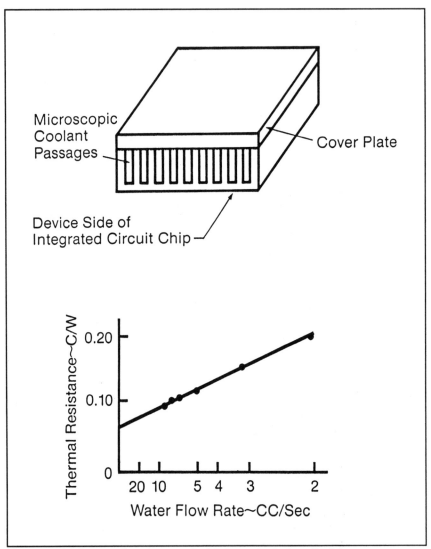

Figure 5.16 Integral water-cooled integrated circuit chip.

volumetric flow rate of air which would be required to remove the heat would not be practical. The problem is nicely solved in the Cray-2 by immersion cooling with FC-77 liquid coolant. Fluorocarbon coolants offer a higher thermal conductivity and higher density than air. Both of these properties are important in determining a coolant's ability to remove heat from a surface and transport it away without an excessive temperature rise at the surface or in the coolant. The FC-77 in the Cray-2 is

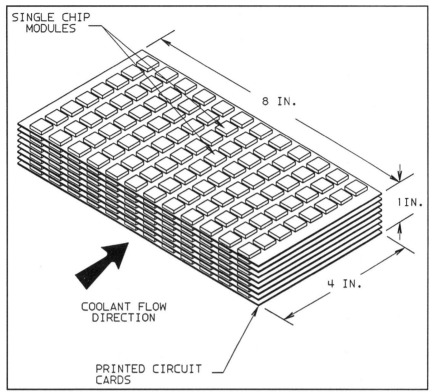

Figure 5.17 Liquid immersion-cooled CRAY-2 multi-chip module package.

distributed vertically between the stacks of multi-chip circuit modules, and flows horizontally through the modules and over the chip packages at a velocity of 1 in./s. A total volumetric FC-77 flow rate of about 200 gpm is required to cool the 194 Kw dissipated in the Cray-2.

The IBM Liquid Encapsulated Module (LEM), which preceded the TCM [23], provides an example of an immersion cooled electronics package utilizing pool boiling. As shown in Figure 5.18, the substrate with chips was attached to a module-cooling assembly to form a sealed unit containing FC-72 coolant. Boiling at the chip surfaces resulted in high heat transfer coefficients (0.17-0.57 W/cm^2-°C) with which to meet the chip cooling requirements. The heat was transported from the FC-72 coolant to internal fins and was then transferred to water flowing through the external cold plate. Although replaced in production machines by the TCM cooling approach, the LEM was proven capable of cooling chip powers as high as 4 watts in a 300 watt module.

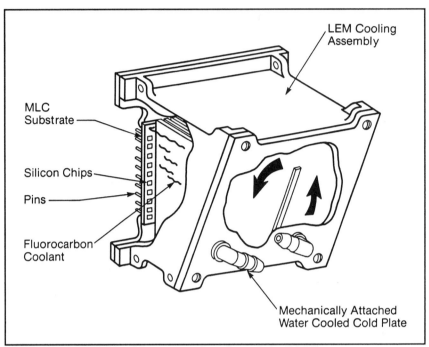

Figure 5.18 IBM Liquid encapsulated module (LEM).

Some aspects of the immersion cooling technology developed for the LEM continue to be used in IBM today [24]. Prior to final module assembly, each TCM substrate with its chips undergoes an electrical test. To obtain the required electrical measurements the surface of the substrate area around each chip site must be accessible for electrical probing. To fulfill this requirement and provide satisfactory cooling the substrate assembly is placed in a test chamber filled with fluorocarbon coolant as shown in Figure 5.19. As shown in Figure 5.20, a flexible bellows is attached to one wall of the test chamber so that the test probe can be moved horizontally and vertically to test each chip on the substrate. A remote cooling and circulating unit supplies subcooled fluorocarbon coolant through spray nozzles in a chamber above the test chamber. The spray provides both direct contact condensation of vapor from the test chamber and a continuous supply of liquid to the test chamber.

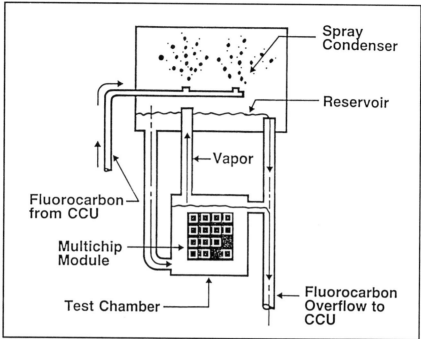

Figure 5.19 Direct immersion liquid cooled electrical test system for TCM substrates with chips.

5.8. Summary

Significant advances have been made in electronics packaging technology since the days of vacuum tubes and discrete transistors. Many of these advances would not have been possible without corresponding advances in thermal technology. The driving forces have been and continue to be demands for:

1) higher circuit density,
2) higher circuit power, and
3) improved reliability.

As described in this chapter, these demands have been met to date by improvements in internal package thermal paths, enhanced air cooling designs and technology, the development of water-cooled cold plate technology, and limited applications of immersion cooling technology. The need for further increases in cooling capability is generally recognized throughout the electronics industry, and continued research and development is needed to meet this challenge.

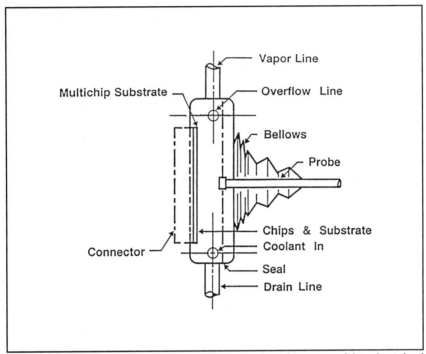

Figure 5.20 Side view of test chamber showing movable electrical probe assembly.

5.9. Acknowledgement

The authors wish to thank their colleague, T.M. Anderson, for his review of this chapter, and his assistance with the final preparation of the manuscript.

5.10. References

1   D. Hanson, The New Alchemists, (Avon Books, New York, 1982), pp. 57-61.

2   C.J. Bashe, et al, "The Architecture of IBM's Early Computers," IBM Jour. Res. and Dev. 25, (5), 363-365 (1981).

3   E.J. Rymaszewski, J.L. Walsh, and G.W. Leehan, "Semiconductor Logic Technology in IBM," IBM Jour. Res. and Dev. 25, (5), 603-607 (1981).

4   R.C. Chu, "Heat Transfer in Electronic Equipment," Proc. of 8th Intl. Heat Trans. Conf., San Francisco, CA, 1986, pp. 293-305.

5   F.P. Incopera (Editor), Research Needs in Electronic Cooling, Proc. of NSF/Purdue Univ. Sponsored Workshop, Andover, MA, 1986, p. 76.

6   A.D. Kraus and A. Bar-Cohen, Thermal Analysis and Control of Electronic Equipment, (McGraw Hill Co., New York, 1983), pp. 34-35.

7   R.E. Simons, "Thermal Sensing and Control for a Large-Scale Digital Computer," Proc. of SEMI-THERM1 Conf., Phoenix, AZ, 1984.

8   R.E. Simons, "Thermal Management of Electronic Packages," Solid State Technology 28, (10), 131 (1983).

9   V.W. Antonetti, S. Oktay, and R.E. Simons, "Heat Transfer in Electronic Packages,"in Microelectronics Packaging Handbook, edited by R.R. Tummala and E.J. Rymaszewski, (Van Nostrand Reinhold, New York, 1989), pp. 169-170.

10   D.B. Tuckerman and R.F. Pease, "High Performance Heat Sinking for VLSI," IEEE Elec. Device Lett. EDL-2, (5), (1981).

11   R.E. Simons, "Direct Liquid Immersion Cooling: Past, Present, and Future," Proc. of the 1987 Intl. Symp. on Microelectronics, Minneapolis, MN, 1987, pp. 186-197.

12   K. Okutani, K. Otsuka, K. Sahara, and K. Satoh, "Packaging Design of a SiC Ceramic Multi-Chip RAM Module," Proc. of 4th Ann. Intl. Elec. Pack. Soc. (IEPS) Conf., Baltimore, MD, 1984, pp. 299-304.

13   M. Kohara, S. Nakao, K. Tsutsumi, K., H. Shibata, and H. Nakata, "High Thermal Conduction Package Technology for Flip Chip Devices," IEEE Trans., CHMT-6, (3), 267-271 (1983).

14   S. Oktay, B. Dessauer, and J.L. Horvath, "New Internal and External Cooling Enhancements for the Air-Cooled IBM 4381 Computer," IEEE Intl. Conf. on Computer Design: VLSI in Computers, Port Chester, NY, 1983.

15   P.J. Dvorak, "Packaging Computers to Serve the Real World," Machine Design, 60, (12), 70-76 (1988).

16   P.B. Wesling, "Packaging for the 580-Series Amdahl Computer," 31st IEEE Electronic Components Conference, Atlanta, GA, 1981.

17   T. Murase, H. Hirata, and S. Ueno, "High Density Three Dimensional Stack Packaging for High Speed Computer," IEEE CH1781-4/82, 1982, pp. 448-455.

18   R.C. Chu, U.P. Hwang, and R.E. Simons, "Conduction Cooling for an LSI Package: A One-Dimensional Approach," IBM Jour. Res. and Dev., 26, (1), 45-54 (1982).

19   T. Mizuno, M. Okano, Y. Matsuo, and T. Watari, "Cooling Technology for the NEC SX Supercomputer," Proc. Intl. Symp. on Cooling Technology for Electronic Equipment, Honolulu, HI, 1987, pp. 110-125.

20   H. Yamamoto, Y. Udagawa, and M. Suzuki, "Cooling System for FACOM M-780

Large Scale Computer." Proc. Intl. Symp. on Cooling Technology for Electronic Equipment, Honolulu, HI, 1987, pp. 110-125.

21   R.D. Danielson, N. Krajewski, and J. Brost, "Cooling a Superfast Computer," Electronic Packaging and Production, 26, (7), 44-45 (1986).

22   S.R. Cray, "Immersion Cooled High Density Electronic Assembly," U.S. Patent No. 4,590,538 (May 1986).

23   R.C. Chu and R.E. Simons, "Evolution of Cooling Technology in Medium and Large Scale Computers - An IBM Perspective," Proc. Intl. Symp. on Heat Transfer in Electronic and Microelectronic Equipment, Dubrovnik, Yugoslavia, 1988.

24   U.P. Hwang and K.P. Moran, "Boiling Heat Transfer of Silicon Integrated Circuits Chip Mounted on a Substrate," ASME HTD - Vol. 20, pp. 53-59 (1981).

HEAT SINKS IN FORCED

CONVECTION COOLING

GARY L. LEHMANN

6.1 Background

The need for reliable, simple and cost-effective designs to provide adequate cooling of electronic packages is well documented [1]. The relative simplicity of direct air cooling makes it an attractive design provided sufficient cooling can be achieved.

Cooling capacity is described in terms of the total heat removal, q, or alternately in terms of the heat transfer per unit area, q/A, termed the heat flux. The heat transfer, for a given design, can be characterized by its thermal resistance, R, to heat flow. The actual heat flow between two surfaces of temperatures T_1 and T_2, is proportional to the temperature difference, $\Delta T = T_1 - T_2$, available to drive the heat flow:

$$q = \frac{\Delta T}{R}$$

[6.1]

For direct air cooling ΔT is constrained by the maximum chip temperature and the ambient temperature in which the system operates. Therefore improved cooling performance is achieved by decreasing the total thermal resistance.

The limits for which adequate cooling can be achieved through direct air cooling were estimated by Kraus and Bar-Cohen [2] and are quoted here as follows:

"... for a typical allowable temperature difference of 60°C between the component surface and the ambient, "natural" cooling in air -relying on both free convection and radiation- is effective only for heat fluxes below approximately 0.05 W/cm^2. Although forced convection cooling in air offers approximately an order-of-magnitude improvement in the heat transfer coefficient, this thermal configuration is unlikely to provide a heat removal capability in excess of 1 W/cm^2 even an allowable temperature difference of 100°C. To facilitate the transfer of moderate and high heat fluxes from component surfaces, the thermal designer must choose between the use of finned air-cooled heat sinks and direct or indirect liquid cooling."

A heat sink is essentially a device for spreading a given heat load over a larger surface area and thus hopefully allowing one to reduce the heat flux to a range in which adequate air cooling can be achieved. This chapter identifies recent research pertinent to the use of finned air-cooled heat sinks.

Heat sink use can be divided into two common applications. First consider the traditional arrangement of a rack of circuit cards, each card populated with an array of components. Furthermore allow for the presence of a limited number of high power components in a system of predominantly low power components. In this case heat sinks could be attached to achieve local cooling enhancement of these sparsely located "hot" modules. At the other end of the spectrum is the use of heat sinks to significantly increase the operating window of air cooling for high performance modules. In 1990 Hanneman [3] reported an up-to-date discussion on estimating the upper limits of air cooling. His analysis is inclusive of the use of extended surfaces, as packaging volume permits, to increase the effective heat transfer area. This is the area that Hanneman [3] is in particular addressing.

To illustrate how finned heat sinks improve cooling performance consider, as a simple model for convective air cooling, one dimensional heat transfer through a plane wall. In Figure 6.1, a heat load is applied to the "inner" surface of a solid wall.

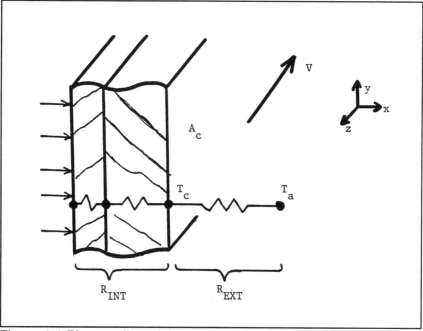

Figure 6.1 Plane wall model for convective air cooling.

The heat is drawn through the wall and convectively removed by an air flow adjacent to the "external" surface. The external surface temperature is T_c and the fluid temperature far from the surface is T_a. For a fixed temperature difference T_c-T_a, the heat removal depends on the external thermal resistance

$$R_{EXT} = \frac{(T_c - T_a)}{q} = \frac{1}{hA_c}$$

[6.2]

where A_c is the wetted surface area and h is the heat transfer coefficient. Improved cooling through the reduction of R_{EXT} is achieved though an increase in h or A_c. In general the heat transfer coefficient will be a function of the air speed, the air properties and the thermal boundary conditions at the surface.

The addition of an external surface is considered in Figure 6.2.

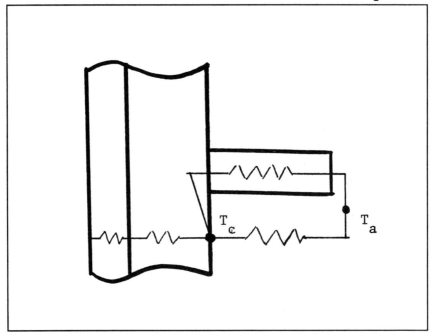

Figure 6.2 Heat flow path of a convectively cooled extended surface (fin).

The external resistance is now calculated as

$$R_{EXT} = \frac{(T_c - T_a)}{q} = \frac{1}{UA_s}$$

[6.3]

where A_s is the total wetted surface and U is an overall heat transfer coefficient. Roughly speaking the value of U will be nearly equal to the h felt in the absence of the fin, while the fin significantly increases the wetted surface area (i.e. $A_s > A_c$). By comparing Equations 6.2 and 6.3 it is clear that the increase in surface area provides a reduced R_{EXT}.

Extended surface analysis is a combined-mode heat transfer problem. The coupled conduction-convection problem is often termed the conjugate problem. The coupling occurs at the solid-fluid interface where the heat conduction to a differential area dA must equal the convective heat removal from dA. This boundary condition is stated as

$$\frac{dq}{dA} = -k_{fin}\nabla t \cdot \underline{n} = -k_{air}\nabla T \cdot \underline{n}$$

[6.4]

where t is the fin material temperature, T is the air temperature and $\underline{n}$ is the local unit vector normal to surface dA, or equivalently

$$\frac{dq}{dA} = -k_{fin}\nabla t \cdot \underline{n} = h(T_s - T_{ref})$$

[6.5]

where T_s is the surface temperature. In Equation 6.5 the convective heat transfer is represented through the use of "Newton's Cooling Law"[1]. By assuming a value of h the conduction problem can be uncoupled from the convection problem. Assuming a uniform value of h_{AVG} the net heat transfer from a fin is calculated as

---

[1]The definition of the heat transfer coefficient, $h(x) = q/A/[T_s - T_{ref}]$, is dependent on the choice of T_{ref}. In heat transfer literature, particularly as applies to air cooling, a number of choices for T_{ref} have been employed. The reader is cautioned that consistency in the meaning of T_{ref} is essential in the application of h correlations. A review of this point and its consequences has been provided by Moffat et al. [4].

$$q = h_{AVG}A_{EFF}(T_b - T_a)$$

[6.6]

where T_b is the temperature at the base of the fin and A_{EFF} is the effective fin area behaving as if it were at temperature T_b. This is the approach commonly used in fin analysis. A thorough review of extended surface analysis is presented by Kraus [5].

Accurate knowledge of R_{EXT} is required to predict operating temperatures and thus evaluate cooling designs. The primary source of uncertainty in the analysis is due to the difficulties in predicting the value of h_{AVG} through either analysis or application of empirical correlations. Related to this is the uncertainty in T_{ref} if it is taken to be the local effective air temperature.

The presence of one or more component mounted heat sinks, in an array of card-mounted components will affect the heat transfer rate, as described by Equation 6.6, by changing the values of h_{AVG} and T_a felt at both the heat sink site and the neighboring component sites. These changes occur because the presence of the heat sink modifies the air flow pattern. The heat sinks can introduce increased flow resistance and the flow pattern will adjust to follow the path of least resistance, possibly by-passing the heat sink. Also introduction of a heat sink can modify the turbulence level. Lehmann et al [6-7] postulate that turbulence level modification is responsible for observed increases and decreases in heat transfer at components downstream of a single finned heat sink. Lau and Mahajan [8-9] summarize heat sink use and design as follows:

"Fins provide increased contact area between the heat
source and the cooled air, which results in increased
heat transfer. On the other hand, obstruction due to the
fins slows down the air flow, and thus decreases the
heat transfer rate. By balancing the tradeoff between the
surface area and the flow velocity, an optimum fin design
for a given cooling system can be achieved."

While it may be possible, in principle, to speak of an optimum design, there remain many applications in which the required analysis is simply to difficult to allow a systematic approach to optimization.

The remainder of this chapter is devoted to highlighting the approaches and results of investigations of the application of finned heat sinks in air cooled systems. It is useful to divide the material into two configurations. In the first, the heat sink is treated as an isolated system. This allows the prescription of the flow inlet conditions. The second configuration is to mount one or multiple component mounted heat sinks to a component in an array of card-mounted components. Here the local flow behavior, at the heat sink, and the overall system behavior, determine the mass flow and inlet temperatures felt by a particular heat sink. In the system approach both the behavior at the heat sink site and its impact on neighboring sites have been of interest.

6.2 Isolated Heat Sinks

In this section we restrict the discussion to the behavior of a single heat sink. The approach flow and temperature conditions are prescribed. The heat sink consists of an array of equally spaced fins, aligned parallel to the flow direction. Two configurations have been studied and are identified as:

1. an infinite array of fins
2. a finite width (number of fins) heat sink.

6.2.1 Infinite array heat sinks.

Figure 6.3 shows the cross-section of the geometry under consideration. By symmetry the mass flow rate in each passage is identical. Therefore not only is the total approach flow prescribed but so is the mass flow in each passage.

Far from the passage inlet, conditions of fully developed flow are achieved if the overall passage length in the flow direction, L, is sufficiently large. Computational studies have been reported [10-12] for assumed conditions of hydrodynamically and thermally fully developed flows. These studies allowed for natural variation of the heat transfer coefficient around the passage perimeter by directly obtaining a solution to the coupled temperature field (the conjugate problem). These studies have provided detailed investigation of the effect of fin tip clearance from the opposite wall. Sparrow, Baliga and Patankar [10] first considered this problem for laminar flow conditions. An approximation of one-dimensional temperature within the fins was made, limiting the study to relatively thin fins. Later Kadle and Sparrow [11] extended the analysis to turbulent flow and reported both numerical and experimental work. Their calculations were well supported by the experimental data. In 1989 Lau, Ong and Han [12] expanded on the laminar flow computations of Sparrow et al [10] by allowing for two-dimensional heat flow:

1) in the fins, and
2) through the wall to which the fins are attached.

This allowed Lau et al to consider a range of fin spacing and aspect ratios. The tip clearance effect is significantly different for short stubby fins than for tall thin fins. The performance was based on the reported

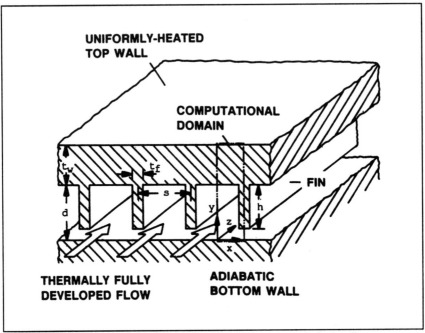

Figure 6.3 Schematic cross-sectional view of an infinite fin array heat sink. Reproduced from [12]

values of heat transfer per unit pumping power.

Karki and Patankar [13] have reported numerical computations for developing laminar, mixed convection in a horizontal flow. The effect of fin tip clearance is discussed. Their study shows that buoyancy forces can significantly alter the heat transfer and flow patterns in the laminar flow regime.

Lau and Mahajan [9] have reported the results of a set of experiments to determine the effects of fin density and tip clearance on heat transfer rates and resulting fluid flow pressure drop. Figure 6.4 provides a sketch of their laboratory configuration.

Figures 6.5 through 6.8 show their measurements of thermal resistance, Θ^2, and pressure loss P. They have defined the total thermal resistance as the ratio of the maximum wall temperature rise over the inlet temperature to the heat transfer:

---

[2]The use of θ as a thermal resistance is to be consistent with the figures reproduced from reference [9]. It is not a departure from the concepts introduced for R.

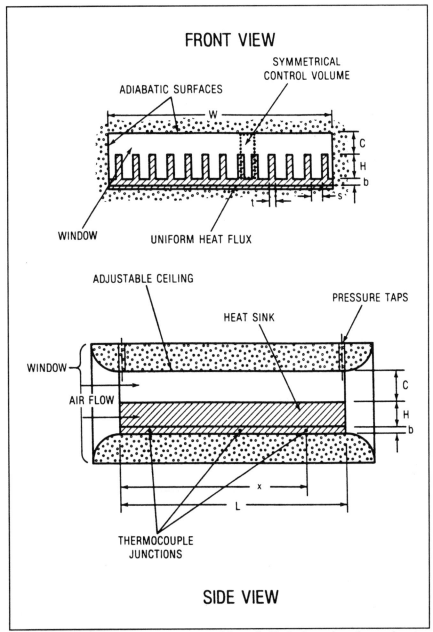

Figure 6.4 Laboratory apparatus of Lau & Mahajan showing a heat sink with rectangular fin array. Reproduced from [9]

$$R_{EXT} = \Theta = \frac{(T_{W,max} - T_{in})}{q}$$

[6.7]

Fin densities of 1.3, 4.6 and 5.6 fins per cm of width correspond to the 13, 46 and 56 fin configurations. Figure 6.5 shows that thermal resistance decreases with increases in mass flow rate and fin density. However the data of Figure 6.6 show that a penalty is paid for this increase in cooling ability by higher pressure losses as both mass flow rate and fin density are increased.

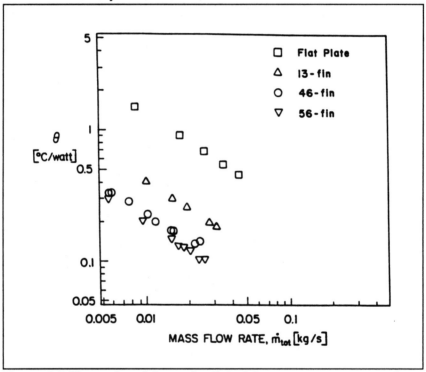

Figure 6.5 Thermal resistance for flat plate (1 cm channel height) **and finned** (1 cm fin height, zero tip clearance) **heat sinks.** Reproduced from [9]

Figures 6.7 and 6.8 show the effect of varying the tip clearance on thermal resistance, Θ, and pressure drop, P, respectively. At a fixed fin density, thermal resistance decreases as the tip clearance is reduced, with the maximum value at zero clearance. The reduction in heat transfer with increasing tip clearance is a result of the reduction in mass flow rate

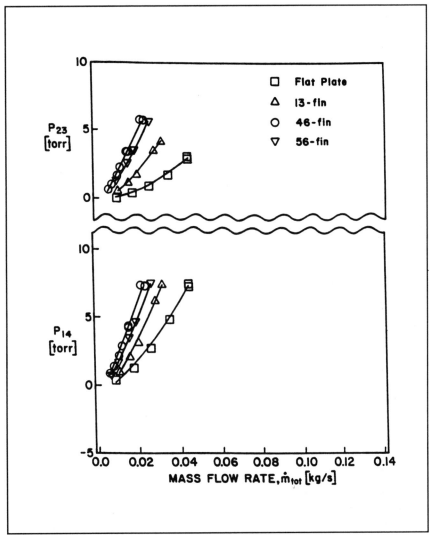

Figure 6.6 Pressure drop data for flat plate and finned heat sinks.
Reproduced from [9]

remaining in the fin-to-fin passages.   As would be expected, the
configuration of largest heat transfer (zero clearance) also produces the
largest pressure loss, at a fixed mass flow rate.

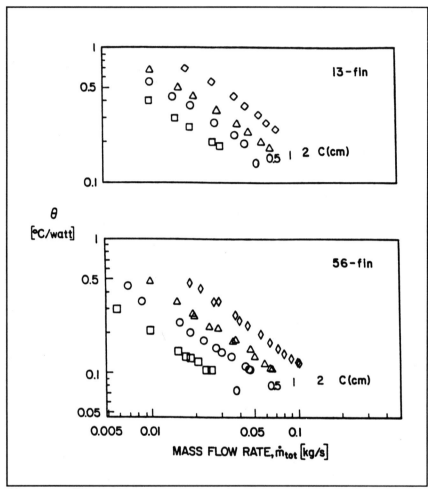

Figure 6.7 Thermal resistance data for 13- and 56-fin heat sinks for various values of tip clearance C. Reproduced from [9]

6.2.2 Finite width heat sinks.

Consider a rack of circuit cards placed parallel to one another, forming low aspect ratio channels for the coolant air to pass through. Furthermore let a heat sink be mounted to a single component. The heat sink footprint area will be nearly equal to that of the component. The cross-section of a typical heat sink is similar to that of Figure 6.3. The length of each fin-to-fin passage will be relatively short. This will tend to emphasize any entrance region effects in the fin-to-fin passages. The

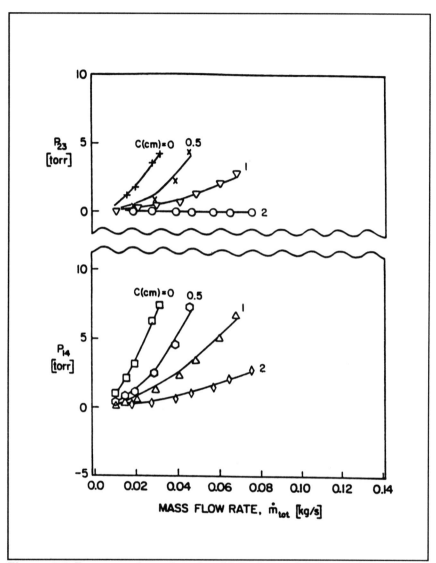

Figure 6.8 Pressure drop data for 13-fin heat sink for various values of tip clearance. Reproduced from [9]

finite width of the heat sink and any tip clearance from the heat sink to the adjacent card will allow by-pass of fluid flow around the heat sink. As a result of the by-pass one can only prescribe the approach flow rate. Analysis or experimental measurements are required to determine the actual mass flow rate passing through the heat sink.

Ashiwake, Nakayama and Daikoku [14] have reported a study of the resistance to heat flow from finned LSI packages. An analysis to predict the external resistance is presented. Their modeling technique is based on correlations of thermal and hydraulic parameters (e.g. heat transfer and friction factor coefficients). The analysis estimates the fraction of the prescribed approach flow which actually will pass through the fin-to-fin passages. Based on this, the effective heat transfer coefficient is determined which in turn allows calculation of the fin efficiency. The heat transfer correlations account for the relatively short passage lengths (axial length/hydraulic diameter ~7), but are based on the mixed mean temperature. The desired external resistance is defined in terms of the approach temperature. Therefore a method to relate the mixed mean temperature rise, in the fin-to-fin passage, to the approach temperature is used to obtain the required external resistance prediction. Figure 6.9 shows a comparison of the model predictions and their own experimental data. The agreement is good.

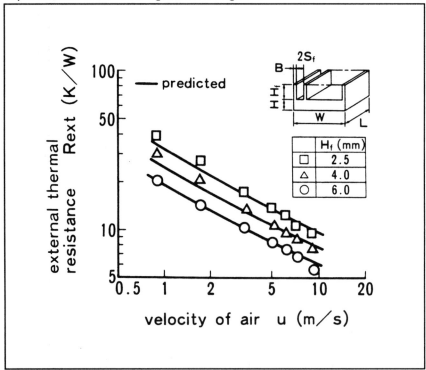

Figure 6.9 External thermal resistance between the finned package and the incoming air. (W=L=20mm H=3mm B=1mm S_f=1.4mm Al fin)
Reproduced from [9]

The figure contains a sketch of the heat sink used. The external resistance decreases with increases in the approach air speed and increasing fin height at a fixed fin density. This model was subsequently used to consider the effect of fin density as discussed by Nakyama [15]. The results are summarized in Figure 6.10. For a fixed approach velocity, identified as V_o on Figure 6.10, there is a fin spacing which provides the maximum heat transfer. This optimum spacing shifts as the fin height is varied.

Additional experience on the effects of by-pass around a finite width heat sink are reported in references [16-17]. Yokono, Sasaki and Ishizuka [16] report experiments in which a finite width array of fins is attached to a large smooth wind tunnel wall. Their fin-to-fin passages are relatively short. They observed that the total heat removal is nearly constant, for a constant surface area, independent of the fin height. Their data was well correlated in the form:

$$\frac{h_{AVG}S}{k} = c_1\left[\left(\frac{V_oS}{v}\right)\frac{S}{L}\right]^{c_2}$$

[6.8]

where S is the fin-to-fin spacing, L is the fin length in the flow direction, V_o is the approach air speed, and k and v are the air thermal conductivity and viscosity respectively. The correlation constants are quoted as $c_1 = 0.33$ and $c_2 = 0.63$. The heat transfer coefficient is based on the unheated approach temperature (i.e. $T_{ref} = T_o$).

6.3 Heat Sink(s) in a System

References [18-19] present experimental measurements of the heat transfer performance when an array of heat sinks is attached to one surface of a channel flow. This models the situation in which an entire array of card-mounted components each has a heat sink attached.

Lehmann and co-workers [6-7] have considered the effect that a single heat sink has on the heat transfer rates at locations downstream. The presence of a heat sink modifies the flow pattern, seen at positions downstream, as compared with that when the heat sink is absent. In references [6-7] heat transfer data are presented with the heat sink as a thermally passive (unheated) flow disturbance. Measurements were made at positions downstream of the heat sink, one component at time, with and without a heat sink present in row i as sketched in Figure 6.11.

Figures 6.12 and 6.13 provide a comparison of the Nusselt

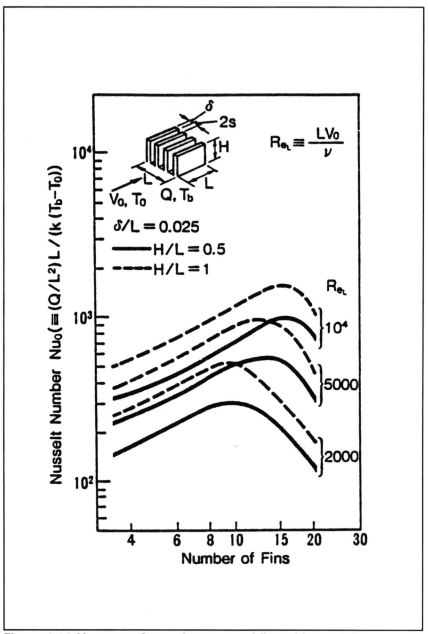

Figure 6.10 Heat transfer performance of finned heat sinks convectively cooled in open environment. Reproduced from [15]

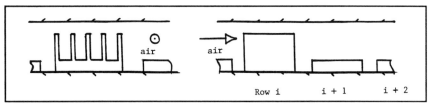

Figure 6.11 Air flow in a channel over a simulated array of components. Reproduced from [8]

number at row i, Nu_i, with a heat sink at $i = 6$, to that with the heat sink absent Nu_{oAVG} for laminar channel flow and turbulent channel flow respectively. A value of Nu_i/Nu_{oAVG} less than one or greater than one represents a relative increase or decrease in heat transfer respectively, due to the flow pattern modification caused by the presence of the heat sink in row 6. In the case of laminar flow a large reduction in heat transfer occurs one row downstream of the heat sink, as shown on Figure 6.12. Further downstream the heat transfer rate rapidly regains the value it would have in the absence of the heat sink. Figure 6.13 shows a quite different effect when the channel flow is turbulent. In this case the heat transfer rate one row downstream of the heat sink is enhanced. However at positions three and four rows downstream the heat transfer rate is degraded by the flow pattern modification of the heat sink located at row 6. A postulated explanation is provided in references [6-7]. Briefly it is felt that the heat transfer enhancement is due to a local flow acceleration which results from the flow area restriction of the heat sink. The degradation further downstream is thought to be a result of a reduction in the turbulence level experienced when the flow passed through the relatively narrow fin-to-fin passages.

6.4 References

1.	F. P. Incropera, "Convection Heat Transfer in Electronic Cooling," Trans. ASME J. Heat Transfer, 110, 1097-1111, Nov. 1988.

2.	A. D. Kraus and A. Bar-Cohen, Thermal Analysis and Control of Electronic Equipment, Hemisphere Publishing Company, 1983.

3.	R. Hannemann, "Thermal Control for Mini- and Microcomputers: The Limits of Air Cooling," in Heat Transfer in Electronic and Microelectronic Equipment, edited by A. E. Bergles (Hemisphere Publishing Corp., New York, 1990), pp. 61-82.

4.	R. J. Moffat and A. Anderson, "Applying Heat Transfer Coefficient Data to Electronics Cooling," ASME Publication HTD-Vol. 101, 33-43, 1988.

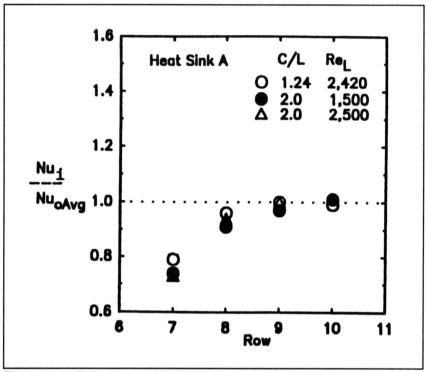

Figure 6.12 Relative heat transfer downstream of a finned heat sink in row 6; laminar channel flow. Reproduced from [8]

5.   A. D. Krauss, "Analysis of Extended Surface," Trans. J. Heat Transfer, 110, 1071-1081, Nov. 1988.

6.   G. L. Lehmann and S. J. Kosetva, "A Study of Forced Convection Direct Air Cooling in the Downstream Vicinity of Heat Sinks", Trans. ASME J. of Electronics Packaging, September, 1990.

7.   G. L. Lehmann and J. Pembroke, "Forced Convection Air Cooling of Simulated Low Profile Electronic Components: Part 2 Heat Sink Effects," accepted for publication in Trans. ASME J. Electronic Packaging.

8.   K. S. Lau and R. L. Mahajan, "Convective Heat Transfer from Longitudinal Fin Arrays in the Entry Region of Turbulent Flow," Trans. ASME J. Electronic Packaging, 111, 213-219,Sept. 1989.

9.   K. Lav and R. L. Mahajan, "Effects of Tip Clearance and Fin Density on the Performance of Heat Sinks for VLSI Packages," IEEE Trans. on Components, Hybrids, and Manufacturing Technology, 12 (4), 757-765, December 1989.

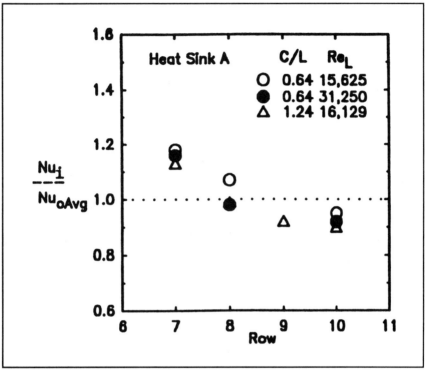

Figure 6.13 Relative heat transfer downstream of a finned heat sink in row 6; turbulent channel flow. Reproduced from [8]

10.   E. M. Sparrow, B. R. Baliga, and S. V. Patankar, "Forced Convection Heat Transfer from a Shrouded Fin Array With and Without Tip Clearance," Trans. ASME J. Heat Transfer, 100, 572-579, Nov. 1978.

11.   D. S. Kadle and E. M. Sparrow, "Numerical and Experimental Study of Trubulent Heat Transfer and Fluid Flow in Longitudinal Fin Arrays," Trans. ASME J. Heat Transfer, 108, 16-23, Feb. 1986.

12.   S C. Lau, L. E. Ong, J. C. Han, "Conjugate Heat Transfer in Channels With Internal Longitudinal Fins," AIAA J. Thermophysics, 3 (3), 303-308, July 1989.

13.   K. C. Karki and S. V. Patankar, "Heat Transfer Augmentation Due to Buoynacy Effects in the Entrance Region of a Shrouded Fin Array", Numerical Heat Transfer, vol. 14, pp. 415-428, 1988.

14.   N. Ashiwake, W. Nakayama, and T. Daikoku, "Forced Convective Heat Transfer from LSI Packages in an Air-Cooled Wiring Card Array," Heat Transfer Engineering, 9, (3), 76-84, 1988.

15.   W. Nakayama, "Thermal Management of Electronic Equipment: A Review of Technology and Research Topics," Appl. Mech. Rev. 39, (12), 1847-1860, Dec. 1986.

16.   Y. Yokono, T. Sasaki and M. Ishizuka, "Small Cooling Fin Performances for LSI Packages," Cooling Technology for Electronic Equipment, edited by W. Aung (Hemisphere Publishing Corp., New York, 1988) pp. 211-218.

17.   Y. Yokono and M. Ishizuka, "Thermal Studies on Finned LSI Packages Using Forced Convection," IEEE Trans. on Components, Hybrids, and Manufacturing Technology, 12, (4), 753-756, December 1989.

18.   W. Nakayama, H. Matsushima and P. Goel, "Forced Convection Heat Transfer from Arrays of Finned Packages," in Cooling Technology for Electronic Equipment, edited by W. Aung (Hemisphere Publishing Corp., New York, 1988) pp. 195-210.

19.   Y. P. Gan, S. Wang, D. H. Lei and C. F. Ma, "Enchancement of Forced Convection Air Cooling of Block-like Electronic Components in In-line Arrays," in Heat Transfer in Electronic and Microelectronic Equipment, edited by A. E. Bergles (Hemisphere Publishing Corp., New York, 1990) pp. 223-233.

DIAMOND THIN FILMS:

APPLICATIONS IN

ELECTRONICS PACKAGING

SUI-YUAN LYNN,

D. GARG & D.S. HOOVER

7.1 Introduction

Diamonds may be a girl's best friend, but scientists have long known that diamond is much more than a sparkling gemstone. Besides being the hardest material known to humanity, diamond scores high on almost all the categories of engineering properties such as thermal conductivity, electrical resistivity, and dielectric constant as summarized in Table 7.1. The combination of these outstanding properties makes diamond very attractive as the engineering material of tomorrow.

Table 7.1 Properties of Type IIa natural diamond.

Hardness	9,000 Kg/mm^{-2}*
Thermal Conductivity	20 W/cm.K*
Electrical Resistivity	10^{16} Ω.cm
Thermal Expansion Coefficient (@ 293 K)	1.7 x10^{-6} /K**
Bulk Modulus	5 x10^{11} N/m^2**
Refractive Index (@ 590 nm)	2.41
Dielectric Constant (@ 1 MHz)	5.5
Optical Transmissivity	225nm - Far IR
Chemical Reactivity	extremely low
* Highest among all materials ** Lowest among all materials	

The electronics packaging field offers a major opportunity for application of diamond. Most electronic devices consist of an integrated chip mounted on a heat-spreading material, which is placed at the top of a metallic flange. The heat-spreading material transfers heat generated by the chip quickly and efficiently to the metallic flange, thereby maintaining the chip temperature at or below the design temperature. The metallic flange facilitates the transfer of heat to the surrounding atmosphere.

A pervasive problem in electronics packaging is quick and efficient dissipation of heat generated by the transistors on the integrated chips. The heat dissipation efficiency of the package generally determines the packing density of the transistors on the integrated chips as well as the spacing of the chips on printed circuit boards used in computers and other electronic equipment. The heat dissipation problem is particularly acute in devices operating at radio and microwave frequencies such as laser diodes. Without adequate heat dissipation, these high-frequency devices do not operate efficiently and reliably and their useful life generally decreases substantially.

Because of heat dissipation limitations, microelectronic design engineers spend a lot of time designing and configuring VLSI devices. In addition, recent trends toward micro-miniaturization and higher-and-higher packing densities have compounded the heat dissipation problems. These changes in the technology are forcing microelectronic design engineers and materials scientists to look for new materials with higher heat dissipation capabilities.

To overcome the heat dissipation problem, the heat-spreading material used in electronics packaging must have attributes such as high thermal conductivity and electrical resistivity as listed in Table 7.2. This table also compares properties of diamond, beryllium oxide (the currently used material), and aluminum nitride (one of the future candidates). Among the attributes listed in Table 7.2, the most desirable properties of heat-spreading materials are high thermal conductivity and high electrical resistivity. Interestingly, diamond is the only material that possesses the rare combination of high thermal conductivity and high electrical resistivity. Because diamond conducts heat about five times better than copper at room temperature (see Figure 7.1) and has high electrical resistivity ranging from 10^{10} to 10^{16} Ω-cm, its use as a heat-spreading material in electronics packaging is becoming increasingly important.

7.2 General Overview

7.2.1 Diamond Growth Under High-Pressure and High-Temperature Conditions

Ever since man discovered diamond, its synthesis has presented scientists with a big challenge. Diamond and graphite are both thermodynamically stable allotropes of carbon, but only diamond is stable at high pressure and high temperature (HPHT), as shown in Figure 7.2. Therefore, scientists made countless attempts to synthesize diamond in the thermodynamically stable region (HPHT) at the beginning of the

Table 7.2 Attributes of heat-spreading material.

Attributes	Req't	Type IIa Natural Diamond	BeO	AlN
Resistivity (Ω.cm)	High	10^{16}	10^{14}	10^{14}
Thermal Conductivity (W/cm.K)	High	18 - 20	3.7	3.2
Dielec Constant	Low	5.7	6.9	8.8
Dielectric Loss	Low	0.001	0.001	0.001
Dielec Strength (V/cm)	High	10^6	10^5	10^5
CTE match with chip material Si: $3.5 \times 10^{-6}/^\circ$C	Good Match	2.8×10^{-6}	7.2×10^{-6}	4.1×10^{-6}

century. Researchers at General Electric were the first in the world to successfully produce synthetic diamond reproducibly by using a HPHT process in the mid-1950s [1]. Subsequently, molten-metal catalysts were developed and used to increase growth rates, as shown in Figure 7.3. Because of advances in diamond growth technology, synthetic diamonds have become readily available from powder to 1 mm size particles and are very useful in the tool and electronics industries. For example, synthetic diamonds have been extensively used as abrasives (polishing and grinding media) as well as to produce a wide variety of cutting tools. Additionally, the electronics industry uses them as heat-spreading material for laser diodes and high-temperature semiconductor applications where performance rather than cost is the important factor.

7.2.2 Diamond Growth Under Metastable Conditions

Instead of synthesizing diamond in a thermodynamically stable regime, many scientists followed a different, unconventional route for producing diamonds. As early as 1911 they began exploring the possibility of synthesizing diamonds from hydrocarbons in an operating regime where graphite is thermodynamically stable and diamond is, in fact, metastable [2]. This operating regime is characterized by low temperature (less than 1,100°C) and low pressure (atmospheric to sub-atmospheric pressure). In this operating regime the free energy of diamond formation is only

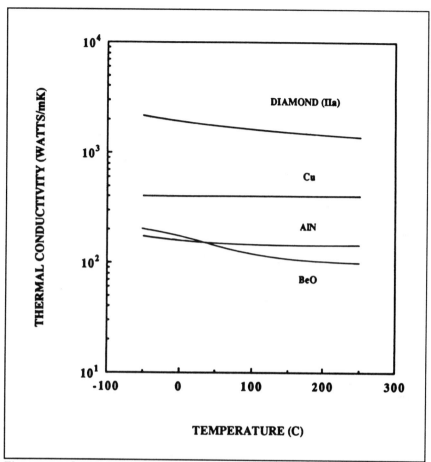

Figure 7.1 Comparison of thermal conductivities of BeO, AlN, Cu, and diamond.

slightly higher than that of more stable graphite, but the free energies of carbon in the form of methane, acetylene, carbon monoxide, and carbon vapor are considerably higher than those of diamond and graphite. Therefore, scientists thought that the use of the metastable operating regime with these hydrocarbons might result in the formation of diamond. Unfortunately, however, only limited results were reported by scientists between early 1900 and the mid-1950s, known as the Von Bolton-Ruff-Tammann Era [3].

It was not until the late 1950s and early 1960s that scientists started systematically studying ways to synthesize diamond in the metastable region.   During   that   time   frame,   known   as   the

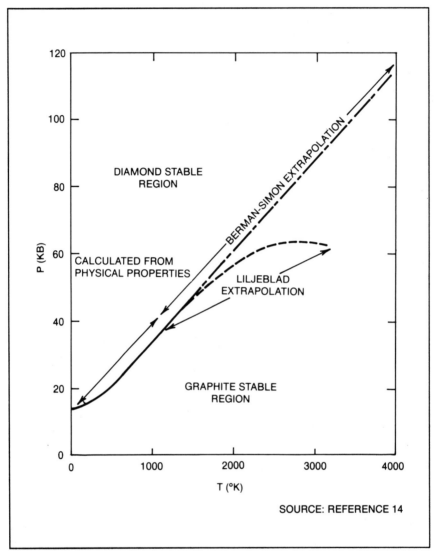

Figure 7.2 Carbon phase equilibrium diagram [14].

Deryagin-Eversole-Angus Era [3], scientists recognized the presence of two competing reactions, one for the formation of diamond and the other for graphite, in the metastable operating regime. They also realized that atomic hydrogen was needed to etch or dissolve graphite from the diamond/graphite mixture. Synthesis of diamond in the metastable region got its biggest boost from the work conducted by Matsumoto and his group at the National Institute for Research in Inorganic Materials

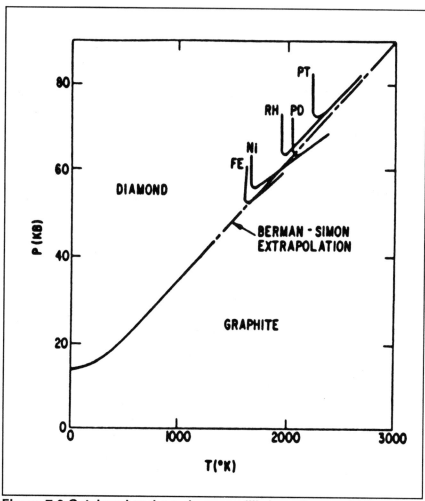

Figure 7.3 Catalyzed carbon phase equilibrium diagram [14].

(NIRIM) in the late 1970s [4,5].

7.2.3 Diamond Growth by CVD Processes

Since the pioneering work of Matsumoto, et al., a number of chemical vapor deposition (CVD) methods have been developed to grow diamond in the metastable operating regime. In all of these methods a gaseous feed containing a mixture of hydrocarbon (such as methane, ethane, or acetylene) and hydrogen is energetically activated to produce hydrocarbon free radicals which are responsible for diamond growth. All of these methods are basically the same but differ mainly in the way the feed gas is activated.

One of the popular methods is a hot-filament assisted CVD (HFCVD) in which feed gas is thermally activated by a filament made of a refractory metal selected from tungsten, tantalum, molybdenum, or rhenium and heated to 1800-2300∘C. A typical HFCVD reactor design is shown in Figure 7.4. The other methods utilize the energy of plasma generated by either radio frequency, direct current, or microwaves to activate feed gas and are commonly referred as plasma-assisted CVD (PACVD) methods. These methods differ only in the way plasma is generated. A typical schematic of a microwave plasma CVD reactor is shown in Figure 7.5.

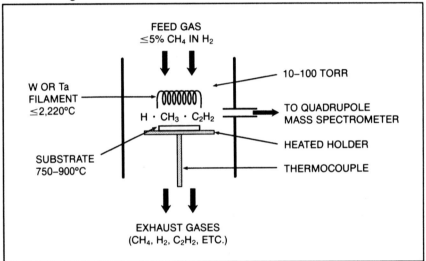

Figure 7.4 Typical schematic of HFCVD reactor.

7.2.4 Characterization of Diamond

Laser Raman spectroscopy and X-ray diffraction techniques have generally been used to characterize natural as well as synthetic diamonds. Laser Raman spectroscopy provides a powerful tool to differentiate diamond from amorphous carbon and graphite. It provides a sharp and distinct peak at ~1332 cm^{-1} for crystalline diamond and a broad and diffused peak at ~1580 cm^{-1} for amorphous carbon. Since chemically vapor deposited diamond films contain a mixture of crystalline diamond and amorphous carbon, the relative proportions of diamond and amorphous carbon as well as the quality of these films have been

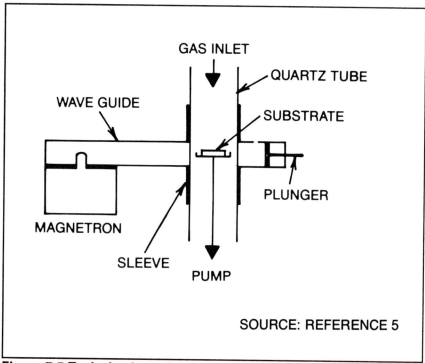

Figure 7.5 Typical schematic of microwave PACVD reactor [5].

effectively determined by Laser Raman Spectroscopy.

Laser Raman spectrographs of diamond films deposited by CVD containing different amounts of amorphous carbon are shown in Figure 7.6. They show that the peak height or area at ~1580 cm^{-1} increases with an increasing amount of amorphous carbon in the film. Unlike Laser Raman Spectroscopy which is used to determine the presence of diamond and amorphous carbon and film quality, the X-ray diffraction technique has been used to identify size, structure, and orientation of crystals in natural and synthetic diamonds as well as chemically vapor deposited diamond films (Figure 7.7).

7.3 Diamonds in Electronics Packaging

Attempts to use diamond as a heat-spreading material began over 20 years ago. Because of its excellent thermal and electrical properties, Swan used it for the first time as a heat-spreading material in a microwave avalanche diode [6]. It was, as expected, instrumental in

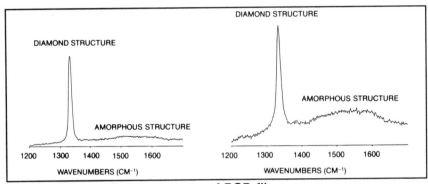

Figure 7.6 Laser Raman spectra of PCD films.

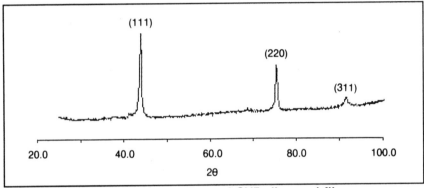

Figure 7.7 X-ray diffraction pattern of CVD diamond film.

increasing power output by a factor of two relative to a conventional heat sink. Similar improvements have been reported with Ge IMPATT diodes and GaAs junction lasers [7,8]. Simplified schematics of the use of diamond in these devices are shown in Figure 7.8.

The heat-spreading material used in microwave avalanche diodes, Ge IMPATT diodes, and GaAs junction lasers was Type IIa natural or HPHT synthetic diamonds. The major drawbacks of using single-crystal diamond heat-spreading material in these devices are cost and size limitations. Because of the cost and size limitations of single-crystal diamond, beryllium oxide (BeO), which is electrically insulative but has thermal conductivity less than half that of copper, has been commonly used as a heat-spreading material by the industry. BeO is mounted on a high thermal conductivity material, such as copper metal, copper/tungsten, or copper/molybdenum composite materials, to dissipate heat to the surrounding atmosphere. However, BeO is presently being phased out as a heat-spreading material due to its

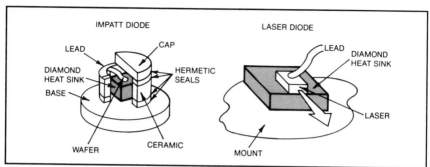

Figure 7.8 Applications of diamond heat-spreading material in electronic devices.

toxicity. Therefore, the electronics industry has been actively pursuing development of alternative heat spreading materials such as aluminum nitride and CVD diamond.

The advent of CVD diamond coatings has provided scientists and engineers with a significant opportunity to use diamond as a heat-spreading material without size and cost limitations. A proposed scheme for using CVD diamond instead of BeO as heat-spreading material is shown in Figure 7.9. In this scheme, a thin diamond film is deposited on high thermal conductive metal or composite material by CVD. The thin diamond film must, however, meet a number of critical requirements discussed in detail below before it can be exploited commercially.

7.3.1 Adhesion, Surface Finish, and Uniformity

Since diamond film is deposited directly on metals or composite materials by CVD, good bonding and adhesion between the substrate and diamond film is extremely critical. The adhesive force between the substrate and diamond film must be high enough to withstand stresses generated by the differences in coefficients of thermal expansion and keep film intact during rapid thermal cycling, which is generally used to simulate changes in atmospheric conditions. The root mean square surface finish of the film must be less than 1.0 μm in order to effectively metallize the surface, obtain good bonding between the metallized layer and chip, and prevent cracking and breaking of the chip. Finally, the film must be fairly uniform and free of defects such as pin holes to provide consistent properties over the entire surface.

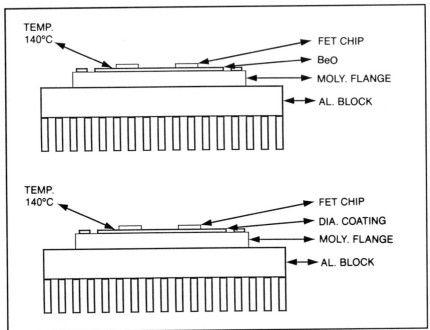

Figure 7.9 Current (BeO) and proposed (diamond) schemes for heat-spreading material.

7.3.2 Dielectric Constant

Single-crystal natural and synthetic diamonds have the lowest dielectric constant among all the heat-spreading materials currently used or being considered for future use. The CVD diamond film, therefore, must have a dielectric constant close to natural diamond. Also, it must have high dielectric strength. Both of these properties are required to reduce signal delays and to sustain a high electric field in high-frequency and high-power devices.

7.3.3 Electrical Resistivity

Single-crystal Type IIa natural and synthetic diamonds have electrical resistivity ranging from 10^{10} to 10^{16} Ω-cm [9]. CVD diamond film, on the other hand, is believed to require electrical resistivity $>10^8$ Ω-cm for the heat-spreading application. This high electrical resistivity is required to prevent breakdown of the film in high-frequency and high-power devices.

7.3.4 Thermal Conductivity

High thermal conductivity is the most important property for a heat-spreading material. Thermal conductivity of Type IIa natural diamond is known to be between 18 and 20 W/cm-°K at room temperature, almost five times that of copper and seven times that of BeO and AlN. The thermal conductivity of CVD diamond film must therefore be far superior to current and future heat-spreading materials such as BeO and AlN. In fact, a thermal conductivity value between those of copper metal and Type IIa natural diamond would be preferable for CVD diamond films.

7.4 Recent Developments in CVD Diamond Heat-Spreading Material Technology

To demonstrate the feasibility of using thin diamond films as heat-spreading material in electronics packaging, the authors have deposited them directly on metallic substrates such as molybdenum by a hot-filament assisted chemical vapor deposition (HFCVD) technique. The film thickness and quality determined by sectioning the deposited specimens and analyzing them by Laser Raman Spectroscopy are consistent and uniform (see Figures 7.10 and 7.11). These films have high electrical resistivity (Figure 7.12) and breakdown voltage greater than 1,000 volts. Their surface finish is good enough to facilitate metallization and to provide good die (chip) bonding. They also exhibit a dielectric constant value close to 6 at low (1 MHz) and high (35 GHz) frequencies. This value is slightly higher than that of the best quality natural diamond. The thermal conductivity value of freestanding CVD diamond film prepared by other researchers and measured between 100 and 130°C has been reported to be approximately 10 W/cm-°K [10,11]. Although this value is approximately 80% that of natural diamond in this temperature range, it is still two and a half times that of copper metal. All of these properties are well within the desired range, demonstrating the feasibility of using thin diamond film for spreading heat in electronics packages. In fact, metallic substrates deposited with CVD diamond films are currently being field-tested.

7.5 Future Challenges

Although considerable research has been done to understand deposition of diamond, the mechanisms involved during nucleation and growth of diamond are not yet well understood. It is believed that the nature and

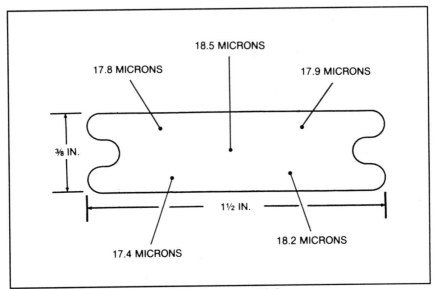

Figure 7.10 Variations in CVD diamond thickness.

type of substrate, composition of feed gas, formation of intermediate reactive species (free radicals of methyl, ethyl, and acetylene [12,13]), atomic hydrogen, surface preparation technique, substrate temperature, and operating pressure play key roles during nucleation and growth of diamond. A focused and well-defined research effort is therefore needed to understand and describe the mechanism involved during nucleation and growth of diamond.

Considerable progress has also been made in depositing thin diamond films by CVD on metallic substrates with good surface finish and electrical properties such as dielectric constant, electrical resistivity, and breakdown voltage. Additional work is needed to successfully utilize them as heat-spreading material in electronic devices, especially those operating at high frequencies. The use of CVD diamond in these devices is constrained by low capacitance requirements; the thickness of CVD diamond film must be in the range of 100-150 μm to meet these requirements. Thus, researchers face the challenge of producing 100-150 μm thick diamond films on metallic substrates with good adhesion and surface finish in a cost-effective way.

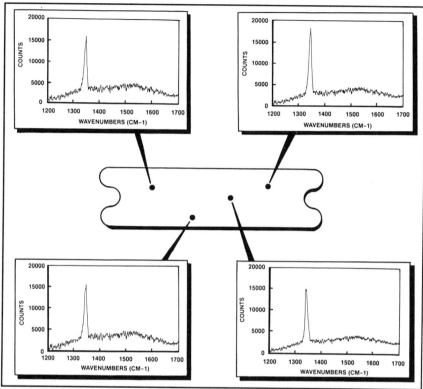

Figure 7.11 Variations in CVD diamond quality.

7.6 Summary

The field of electronics packaging represents the largest single area of application for thin diamond films. In an attempt to penetrate this market in the last five years, researchers have made significant progress in using CVD to deposit diamond films, with excellent physical and material properties, uniformly and reproducibly over large areas. Preliminary test results indicate that thin diamond films will provide solutions to the heat dissipation problems experienced by microelectronics design engineers. Although a few technical hurdles have yet to be overcome before using thin diamond films as heat-spreading material commercially, their future looks very bright. In summary, thin diamond films deposited by CVD are expected to play an important and significant role in microelectronics devices of the future.

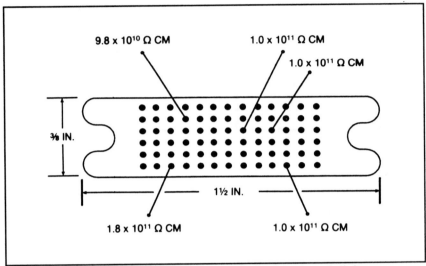

Figure 7.12 Variations in CVD diamond electrical resistivity.

7.7 Acknowledgements

The authors wish to acknowledge Air Products and Chemicals, Inc. for funding this research. They also wish to thank Vyril A. Monk, Robert L. Iampietro, Carl F. Mueller, Ernest L. Wrecsics, Gary L. Johnson, and Richard F. Hamilton for technical assistance.

7.8 References

1   F.P. Bundy, H.T. Hall, H.M. Strong and R.H. Wentorf, "Man-made diamonds," Nature 176, 51 (1955).

2   W. Von Bolton, "Uber die ausscheidung von Kohlenstoff in form von diamant," Z. Elektrochem. 17, 971 (1911).

3   R.C. DeVries, "Synthesis of diamond under metastable conditions," General Electric Corporate R & D Technical Report 86CRD247 (1987).

4   S. Matsumoto, Y. Sato, M. Tsutsumi and N. Setaka, "Growth of diamond particles from methane-hydrogen gas," J. Mater. Sci. 17, 3106 (1982).

5   M. Kamo, Y. Sato, S. Matsumoto and N. Setaka "Diamond synthesis from gas phase in microwave plasma," J. Cryst. Growth 62, 642 (1983).

6   C.B. Swan, "The importance of providing a good heat sink for Avalanching transit

time oscillator diodes," Proc. IEEE (Letters) $\underline{55}$, 451 (1967).

7   D.R. Decker and A.J. Schorr, "High-power IMPATT diodes on diamond heat sinks," IEEE Trans. Electron Devices, $\underline{ED-17}$, 739 (1970).

8   J.C. Dyment and L.A. D'Asaro, "Continuous operation of GaAs junction lasers on diamond heat sinks at 200°K," Appl. Phys. Lett. $\underline{11}$, 292 (1967).

9   J.E. Field, The Properties of Diamond, (Academic Press, London, 1979).

10   A. Ono, T. Baba, H. Funamoto and A. Nishikawa, "Thermal conductivity of diamond films synthesized by microwave plasma CVD," Jap. J. Appl. Phys. $\underline{25}$, L808 (1986).

11   D.T. Morelli, C.P. Beetz and T.A. Perry, "Thermal conductivity of synthetic diamond films," J. Appl. Phys. $\underline{64}$, 3063 (1988).

12   M. Tsuda, M. Nakajima and S. Oikawa, "Epitaxial growth mechanism of diamond crystal in CH4-H2 plasma," J. Amer. Chem. Soc. $\underline{108}$, 5780 (1986).

13   M. Franklach and K.E. Spear, "Growth mechanism of vapor-deposited diamond," J. Mater. Res. $\underline{3}$, 133 (1988).

14   R. Berman and F. Simon, "Graphite-diamond equalibrium," Z. Electrochem. $\underline{59}$, 333 (1955).

LOW DIELECTRIC CONSTANT

MATERIALS FOR PACKAGING

HIGH SPEED ELECTRONICS

MELVIN P. ZUSSMAN

8.1 Introduction: The impact of dielectric constant on electrical performance

Dielectric materials used for electronics packaging serve mechanical, thermal and electrical functions. In functioning as electrical insulation, a dielectric can be characterized by its dielectric constant, with the related properties of volume resistivity and dissipation factor, and by its dielectric strength. The dielectric constant (as well as dissipation factor and volume resistivity) describe the interaction of a dielectric material with an electric field; the dielectric strength defines a limiting electric field gradient above which the interaction of the dielectric with the field causes an irreversible change in the dielectric. Values of these properties for commercially important dielectric materials are shown in Table 8.1.

Table 8.1 Dielectric properties of insulating materials [1-5].

Material	Dielectric Constant	Dissipation Factor	Volume Resistivity (Ω-cm)	Dielectric Strength (V/mil)
Ceramics				
Silica	3.78	0.0005	10^{12}	380-640
Alumina	8.81	0.00035	10^{14}	250-400
Organics				
PTFE (Teflon®)	2.1	0.0002	10^{18}	430
Polyimide (Kapton®)	3.4	0.01	10^{18}	7,000
Epoxy	3.4-3.6	0.024-0.032	10^{14}-10^{15}	
Silicone	2.9-3.7	0.003-0.005	10^{15}-10^{19}	300-700
Polyethylene	2.2	0.0003	10^{16}	500
Composites				
FR-4 Epoxy/ glass	5.8	0.045	$>10^{11}$	400
Polyimide/glass	4.8	0.01	10^{14}	700
Teflon®/glass	2.5	0.001	10^{13}	800

Since dielectric properties determine the interaction of a material with an electric field, it follows that these properties will impact circuit performance. The lower the dielectric constant, the smaller the interaction of the field with the material: the limiting case is for a field propagating in a vacuum, for which the dielectric constant is 1. One example of the impact of dielectric constant k' on circuit performance is the effect on propagation velocity V_p:

$$V_p = \frac{c}{\sqrt{k'}}$$

[8.1]

Dielectric properties impact circuit impedance, cross-talk, power loss, and signal distortion, as well as propagation delay. A knowledge of the dielectric is vital in the design of the circuit; the use of low k' dielectrics permits closer spacing of signals to ground planes at constant impedance, smaller signal to signal spacings and thinner dielectric between signal layers. The result is higher density circuits. This chapter will focus on the properties of low dielectric constant materials, on the relationship of structure to mechanical and dielectric properties and on the impact of mechanical, thermal and processing attributes on the selection of a dielectric material for a particular packaging job.

8.2 Fundamentals of dielectrics

Imagine two conductive parallel plates separated at distance d by vacuum[*]. Applying a voltage V to this capacitor creates an electric field E oriented perpendicular to the plates:

$$V = Ed$$

[8.2]

The plates are now oppositely charged with a charge density σ such that

$$E_{vac} = 4 \pi \sigma = D$$

[8.3]

---

[*]A full development of the physics of dielectrics is given in references [6-7].

where D is defined as the electric displacement. If this imaginary capacitor is now filled with a dielectric material, the voltage and electric field strength between the plates is found to drop, and

$$E = \frac{E_{vac}}{k_s} , \ (k_s \! > \! 1)$$

[8.4]

The lowering of the electric field is caused by the polarization P of the dielectric; charge is displaced in the dielectric in response to E_{vac} until equilibrium is reached at the field strength E.   The electric displacement is then the sum of the electric field and the polarization:

$$D = E + 4\pi P$$

[8.5]

k_s in equation 8.4 is the relative, static dielectric constant of the material and is the ratio of the permittivity of the dielectric to ϵ_o, the permittivity of free space (= 8.85 x 10^{-12}F/m). Some authors use ϵ_r' for the relative permittivity or dielectric constant.

Useful electronic devices contain circuits with fluctuating fields (i.e. signals) so the static dielectric constant is of limited interest. If a sinusoidal voltage is applied to the imaginary, vacuum-filled capacitor described above, the voltage will be 90° out of phase with the induced current: at maximum voltage the current is zero and at maximum current the voltage is zero. This is described mathematically as

$$I_c = i\omega C_{vac} V$$

[8.6]

where I_c is the charging current, i is $\sqrt{-1}$, C_{vac} is the capacitance, and ω is the angular frequency at which the voltage V is changed. Filling the capacitor with a dielectric increases its capacitance by a factor k′ and introduces a loss current I_l in phase with the voltage. The total current I is then

$$I = I_c + I_l = (i\omega C + G)V = (i\omega k' + \omega k'') C_{vac} V$$

[8.7]

where G is the conductance of the dielectric, and k' and k" are the real and "imaginary" parts of the dielectric constant. k" is also called the relative loss factor and the loss tangent is given by

$$\tan \delta = \frac{k''}{k'}$$

[8.8]

To understand the origin of dielectric loss, it is helpful to examine the sources of electric dipoles in dielectric materials. All dielectrics are composed of atomic nuclei and electrons: the properties of the dielectric depend on the composition of the nuclei and the spatial organization of the components. Electron locations relative to other electrons as well as to positively charged nuclei are described by quantum mechanic orbitals; many of these orbitals are oriented relative to the nuclei with which the electron is associated. Electron transitions from one orbital to another, with the accompanying change in polarization, take about 10^{-15} seconds to complete (frequency equals 1,000,000 GHz). Vibrational transitions of nuclei occur over times on the order of 10^{-12} seconds (1,000 GHz). Since electronic and vibrational polarization transitions occur at such high frequencies, their contributions to the dielectric constant are nearly constant over the range of frequencies important for electronic devices (i.e. <10 GHz) and is denoted k'_∞. Multiatomic segments of the dielectric take longer to move; polarization changes which require the rearrangement of groups within molecules or the reorientation of entire molecules can take place over time periods of seconds. Each polarization mechanism can be characterized by a relaxation time τ which is temperature dependent. The static dielectric constant k'_s is reached only if the electric field is held constant for times much greater than the longest relaxation time. The Clausius-Mossotti formula, (Equation 8.9) relates k'_s of the bulk dielectric to the sum of the microscopic polarizations [4]

$$\frac{k'_s-1}{k'_s+1} = \frac{1}{3\epsilon_o}\frac{N_o\rho_d}{M}\alpha = \frac{1}{3\epsilon_o}\Sigma N_i\alpha_i$$

[8.9]

where N_o is Avogadro's number, ρ_d is the density of the dielectric, M is the molecular weight of the dielectric and α is the polarizability. Since the value of k'_∞ ignores long τ polarizations, k'_s is always greater than

k'_∞.

To illustrate the sources of polarization in molecules, examine two cases: PTFE (polytetrafluoroethylene) and Kapton® polyimide (poly(pyro-melitimido-4,4'-oxydiphenylene)). As shown in Figure 8.1, PTFE consists of repeating $-CF_2-$ groups, each with a dipole moment of 1.66 Debye units (see Figure 8.1B). PTFE is a semi-crystalline polymer; this means that long segments of the molecular chains take on a regular, repetitive structure and that the three dimensional ordering of chains is high in the crystalline regions. X-ray studies show that in the crystalline regions the PTFE chain assumes a helical configuration[8]. The dipole moment for this regular helix (see Figure 8.1D) is zero because the dipole moments of the individual $-CF_2-$ groups cancel out; in this case k'_s should be very nearly equal to k'_∞. As a result, the most significant contributors to dielectric loss in PTFE are impurities and kinks in the chain. A crystal-crystal transition occurs in PTFE near room temperature; perhaps because of this transition, the loss tangent for crystalline PTFE is thought to be higher than that of amorphous PTFE [9]: 3.0×10^{-4} vs. 0.2×10^{-4} at 1 MHz. As the crystalline fraction in PTFE goes down, the density also drops, since the packing efficiency is lowered. Consequently, amorphous PTFE has a slightly lower k' than crystalline PTFE.

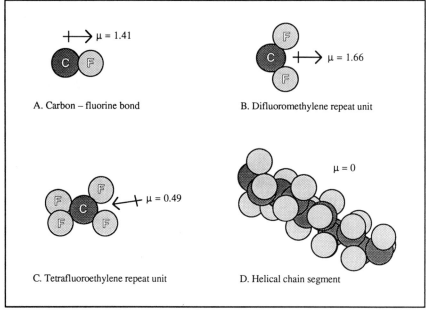

A. Carbon – fluorine bond $\mu = 1.41$

B. Difluoromethylene repeat unit $\mu = 1.66$

C. Tetrafluoroethylene repeat unit $\mu = 0.49$

D. Helical chain segment $\mu = 0$

Figure 8.1 Molecular conformation and dipole moment in Polytetra-fluoroethylene.

Figure 8.2 illustrates the symmetric nature of the dipoles in the Kapton® polyimide backbone. Each repeat unit has an apparent zero dipole moment The high frequency dielectric constant of Kapton® is 3.2; structure-property studies have shown that intermolecular dipolar interactions contribute significantly to polyimide dielectric constants [10,11]. Water absorption also leads to an increase in k' in polyimides; Beuhler et. al. [12] used the Clausius-Mossotti equation to determine that absorbed water was only weakly bound in polyimides.

Dielectric analysis can reveal subtleties in material structure and composition, as well as the suitability of the material for use as a dielectric. The dielectric constant of insulators exposed to changing electric fields has a value between k'_s and k'_∞. Debye showed that this value depends on the relaxation time of the dominant, low frequency relaxation mode τ_{dom}:

$$k'(\omega) = k'_\infty + \frac{k'_s - k'_\infty}{1 + \omega^2 \tau^2_{dom}}$$

[8.10]

$$k''(\omega) = \frac{(k'_s - k'_\infty)\omega\tau}{1 + \omega^2\tau^2_{dom}}$$

[8.11]

$$\tan\delta(\omega) = \frac{k''}{k'} = \frac{(k'_s - k'_\infty)\omega\tau}{k'_s + k'_\infty\omega^2\tau^2_{dom}}$$

[8.12]

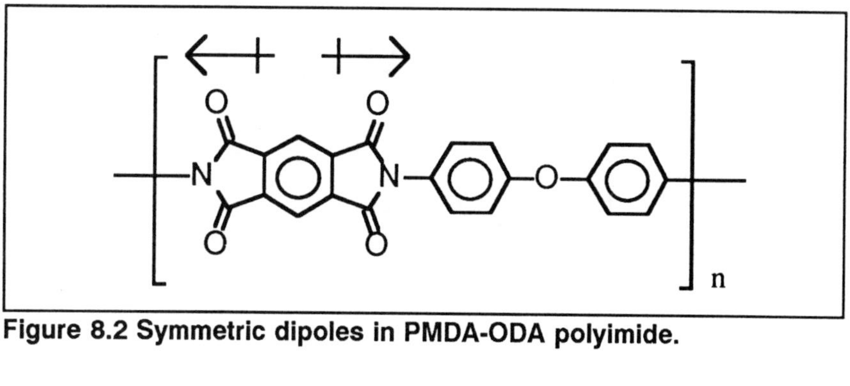

Figure 8.2 Symmetric dipoles in PMDA-ODA polyimide.

Analysis of these formulae shows that as ω increases, k' goes down. At $\omega\tau = 1$, the rate of change of k' is at a maximum and k'' also peaks; tan δ reaches a maximum value at a somewhat higher frequency. Materials which have several important low frequency polarization mechanisms show a series of k' steps accompanied by peaks in k'' and tan δ.

Figure 8.3 shows that the effect of frequency on tan δ is clearly greater than its effect on k'. Analysis of the tan δ data illuminates the effects of processing on dielectric properties. Note the difference in tan δ at low frequency for oven dried versus room temperature dried PTFE/quartz laminates. These laminates are typically made by coating an aqueous PTFE emulsion on quartz fibers and then removing the water. The dielectric data suggest the presence of residual water or possibly surfactant in the material dried at room temperature; at elevated temperature either of these impurities is removed. Raising the cure temperature of DC996 silicone from 150 to 250°C lowers the dissipation factor by decreasing the number of small, mobile molecular segments. The Micarta® cresol-formaldehyde/paper composite shows a series of transitions that reflect the presence of several phases in the material.

Plots of dielectric properties versus frequency bear a striking resemblance to dynamic mechanical spectra. In fact, these techniques are complementary methods for obtaining information about molecular motions. The connection between these two methods is that the low frequency polarizations are temperature-dependent kinetic events. In addition to the variation of τ with temperature, it should be recognized that k_s' and k_∞' are also temperature dependent .

This discussion shows that the dielectric properties of real materials are not constant, but depend on the temperature and field frequency as well as the composition. Low k' is found when dielectrics contain few polar groups (avoid hydroxyls, amines, sulfones, salts), when the polar groups present are oriented antiparallel by the molecular structure (e.g. polyimides), and when intermolecular interactions are minimized [13]. Other factors, such as a processing history which may bring about non-equilibrium molecular orientation or cause molecular decomposition, or the presence of unintended impurities (e.g. air, moisture) must also be considered when evaluating dielectrics.

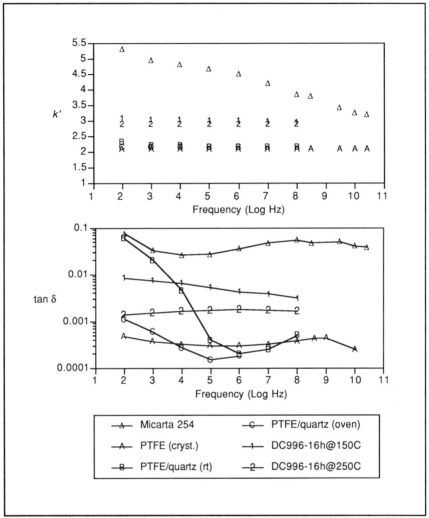

Figure 8.3 Dielectric spectra of packaging materials.

8.3 Why low dielectric constant materials are important for electronics packaging

The two major reasons for interest in low k' materials are increasing circuit density and faster circuit operation. These two motivations are highly interrelated; however, it is possible to distinguish between the desire for less expensive, higher functionality per volume systems and systems designed for higher reliability and increased functionality per unit time. The trend to increasing chip functionality supports both of these goals; the job of the packaging components, IC dielectrics, chip packages, printed wiring boards, connectors, etc. is to enable the IC devices to operate at their full capacity.

Control of impedance is critical in the design and operation of circuits. For microstrip and stripline circuitry, the impedance Z_o is determined by the geometry of the circuit and by the dielectric constant [14,15]:

$$Z_o \sim \ln\left(\frac{h}{w+t}\right)\frac{1}{\sqrt{k'}}$$

[8.13]

where h is the dielectric thickness from signal to ground plane, w is the line width and t is the line thickness. For a given impedance, the use of a lower k' dielectric requires either a thinner dielectric or increased line width or thickness: higher density is achieved by reducing dielectric thickness. One result of this reduced signal to ground spacing is a reduction in crosstalk between adjacent signal lines. Circuit modelling using the ICONSIM[SM] computer-aided interconnection simulation system reveals the increased density achievable using low k' materials[16]. Figure 8.4 shows microstrip circuits at 50Ω impedance and constant cross-talk. Changing from k' of 3.7 to 2.7 reduces the circuit cross section by sixty percent. One consequence of the use of dielectrics with reduced thickness is a need for tighter control over dielectric thickness. For example, a 0.0005" thickness variation is acceptable in a 0.020" dielectric and catastrophic in a 0.002" dielectric.

The speed at which a system can operate is determined by the sum of the logic device switching time, the propagation delay (time for a signal to travel from the near-end device to the far-end device), and the rise time of the far-end device. The propagation delay T_d of an electric field along a conductor of length $l_{circuit}$ surrounded by a dielectric is:

$$T_d = l_{circuit}\sqrt{\frac{k'_{eff}}{c}}$$

[8.14]

The effective dielectric constant k'_{eff} varies somewhat from the material dielectric constant depending on the circuit construction; for stripline circuits the values are approximately equal. The use of a low k'

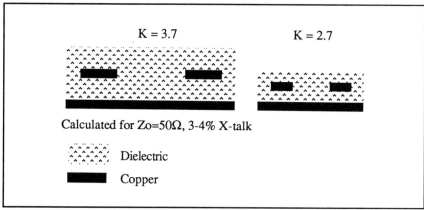

Figure 8.4 Increased circuit density through reduced dielectric constant.

dielectric contributes to reduced T_d both by enabling higher circuit density (smaller $l_{circuit}$) and by speeding the propagation.

The dielectric also has an impact on the rise time of the far-end device. The power loss or signal attenuation a_d (in dB/in) for a stripline circuit is

$$a_d = 27.3 \frac{f}{c} \sqrt{k'} \tan\delta \qquad [8.15]$$

where f is the electric field frequency [17]. This equation shows that power loss increases linearly with frequency and tan δ and to the square root of the dielectric constant; consequently for microwave circuitry where f is necessarily high, low tan δ is more critical than low k'. The relationship between this attenuation formula and signal distortion/rise time degradation in digital circuits derives from the way digital waveforms are generated. A square wave pulse commonly used in digital processing is equivalent to a series of cosine waves at frequencies which are odd integer multiples of (pulse width)$^{-1}$. Figure 8.5 shows an example of a 60 ns wide pulse. The lowest frequency component of this pulse is 8.3 MHz and the calculated rise time (10% to 90% of signal intensity) is 27 ns. By adding higher frequency pulses according to Equation 8.16,

$$V(t) = \frac{A}{\pi} \sum_{n=1,3,5,7,...} i^{(n-1)} \cos \frac{(n\omega_o t)}{n} \qquad [8.16]$$

the rise time is progressively shortened, until at 224.1 MHz (n=27) the agreement of the calculated pulse with the desired square wave is quite good and the rise time is only 2 ns.

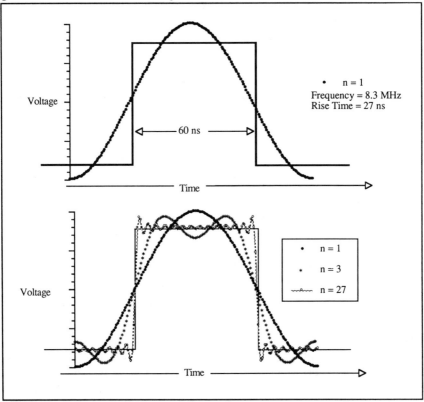

Figure 8.5 Sinusoidal components of a square wave pulse.

The highest frequency of importance in defining the signal rise time is called the band width BW and is generally given [18] by the equation

$$BW = \frac{0.35}{T_r} \qquad [8.17]$$

Each frequency up to BW will undergo attenuation according to Equation 8.15; a significant rise in tan δ at frequencies less than the band width will lead to significant attenuation of the high frequency signal with a consequent increase in rise time delay. This analysis demonstrates the importance of measuring material dielectric properties at frequencies well into the GHz region for systems which incorporate high speed logic devices. Remember, however, that if the interconnect distance is sufficiently short, the propagation delay and rise time degradation due to the dielectric will be less significant than delays due to the IC components.

8.4 Measurement techniques

To determine k' and k" over the frequency range of interest for electronic devices, generally from >10 KHz to >10GHz, requires a variety of test methods. At frequencies below 200 MHz, the electrode-dielectric combination is treated as a lumped circuit [19]. At higher frequencies, transmission line and resonance methods are generally used [20,21]

Table8.2 Measurement techniques for dielectric properties.

Method	Frequency Range	Reference
Stripline Resonator ASTM D3380	9 - 11 GHz	[22]
Waveguide Cavity Perturbation	1 - 10 GHz	[23,24]
Time domain reflectometry	1 - 10 GHz	[25]
Q-Meter	0.02 - 70 MHz	[26,27]
LCR meter	0.01 - 10 MHz	[28]

8.5 Low k' packaging materials

This section will review the menu of low k' materials from which a designer can choose to construct a packaging and interconnection system for high speed electronics. This menu is not all-inclusive, in the sense that not all materials that have been explored for use in high speed packaging are described, nor is the menu restrictive in the sense of including only currently available, commercial products. The discussion below focuses on the chemistry and some key properties of these materials; detailed property and processing information will be found in the references. Materials are divided into three categories:

- Thermoplastics - uncrosslinked polymeric materials
- Thermosets - polymeric materials which are cross-linked in the use condition
- Composites - combinations of materials with enhanced properties

Low k' ceramic materials will not be described in this Chapter; interested readers should consult references [29-31] for more information.

8.5.1 Thermoplastics

While polyethylene has very good electrical properties, its use in electronics packaging is generally limited by its poor mechanical properties and restricted thermal capabilities. To extend the serviceability of polyethylene, the resin is generally filled with glass and lightly crosslinked by irradiation [32]. Such materials have been used in microwave packaging. The principal use of polyethylene in electronics is as cable insulation.

Fluoropolymers have been widely used in packaging. As a class of materials they exhibit unusually low dielectric constant and loss, can be used over a wide temperature range, are generally not affected by solvents, acids or bases and are flame retardant. The chemical structures of several important fluoropolymers are shown in Figure 8.6 [33]. Key properties of several fluoropolymers are shown in Table 8.3 [34]. The semicrystalline fluoropolymers (including PTFE, PFA, FEP, and ETFE) are generally employed either as adhesive layers or as composites with fibrous or particulate reinforcements. The semicrystalline fluoropolymers exhibit cold flow and high coefficients of thermal expansion which reduce the dimensional stability of unreinforced laminates. Composite materials based on PTFE will be described later in this chapter.

High Tg, amorphous fluoropolymers have been recently introduced by Du Pont under the tradename Teflon® AF[35,36]. These copolymers of tetrafluoroethylene and perfluorodimethyldioxole have the lowest dielectric constants of any known polymers. Because of their amorphous nature, Teflon® AF polymers exhibit some solubility (3 - 15%) in perfluorinated ethers. As a result, thin films of Teflon® AF can be spin-, spray-, brush- or dip-coated onto substrates. Teflon® AF is currently used as a cladding in fiber optic cables because of its low refractive index and optical clarity. Potential electronics packaging applications include passivation layers and inter-layer dielectrics.

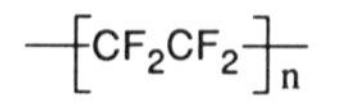

Polytetrafluoroethylene
(Teflon® PTFE,
Hostaflon® TF, Fluon®)

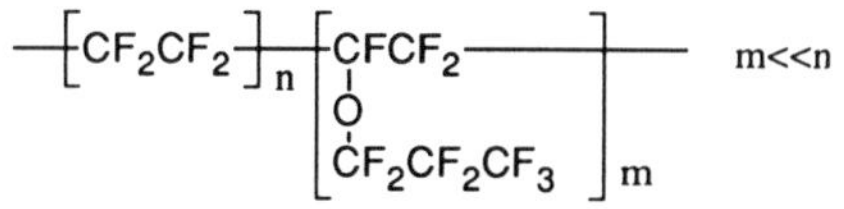

Poly(tetrafluoroethylene-co-perfluoropropylene vinyl ether)
(Teflon® PFA)

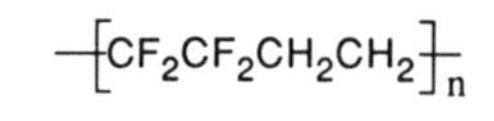

Poly(ethylene-co-tetrafluoroethylene)
(Tefzel®, Hostflon® ET)

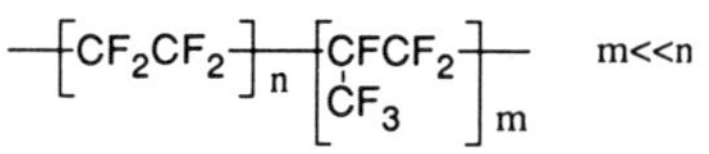

Poly(tetrafluoroethylene-co-hexafluoropropylene)
(Teflon® FEP)

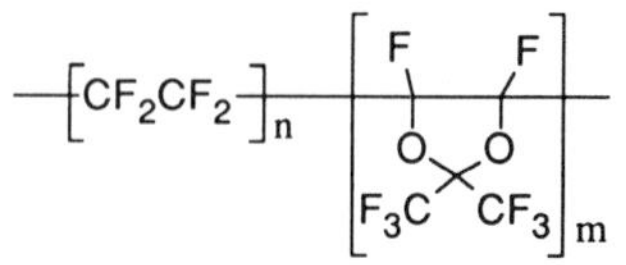

Poly(tetrafluoroethylene-co-perfluorodimethyldioxole)
(Teflon® AF)

Figure 8.6 Structures of thermoplastic fluoropolymers.

Parylene is a unique packaging material in that the polymer is formed by simultaneous condensation and polymerization of the unstable para-xylelene monomer onto the substrate surface[37]. As shown in Figure 8.7, the monomer is prepared by in-situ pyrolysis of a paracyclophane. The linear polymer that forms is highly crystalline, with a melting point above 300°C and a glass transition temperature of 80°C; there are no known solvents for parylene. The most common use of parylenes is as conformal coatings for circuit boards[38]. Parylene has also been employed as an insulation layer over semiconductor devices [5]. The elevated temperature (>250°C) oxidative stability of parylene is poor because of the high number of benzylic hydrogens. Taking advantage of this oxygen sensitivity, efforts to utilize parylene as an IC interlayer dielectric have focused on reactive ion etching with oxygen plasma.

Linear polyimides are used at all levels of electronics packaging. Thin films may be spin cast or sprayed onto silicon or ceramic substrates for use as IC passivation layers or inter-layer dielectrics. Thicker films are used as substrates for TAB (tape automated bonding) or for flexible laminates. Polyimides exhibit excellent mechanical and electrical properties as well as exceptional high temperature properties (up to

Table 8.3 Selected properties of fluoropolymers.

		PTFE	PFA	FEP	ETFE	AF 2400
Property	units					
Max Use Temp.	°C	260	260	205	>150	285
Melting Point	°C	327	305	290	270	NA
Tensile Modulus	Kpsi	76	62	75	120	950-2150
Coeff. Therm. Exp.	ppm/°C	151	184	182	113	80-100
Density	gm/cc	2.2	2.15	2.15	1.7	
Flammability	(UL94)	V0	V0	V0	V0	V0
Dielectric Constant	1 MHz	2.05	2.05	2.1	2.6	1.90
Tan δ	1 MHz	0.00009	0.0002	0.0004	.005	0.00008

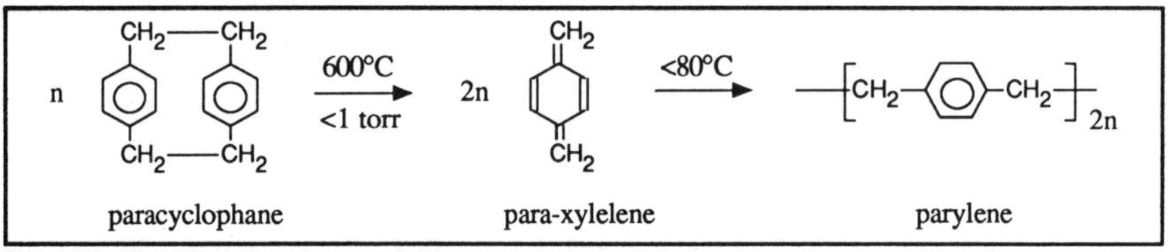

Figure 8.7 Formation of paralene polymers.

500°C for short times).   Most commonly, polyimide films are cast as polyamic acid solutions which are thermally dehydrated to form the final polymer: proper "curing" is necessary to obtain the best mechanical and electrical properties.   The cured polymers are generally insoluble, intractable materials; however, they are not truly crosslinked. This distinction is important because polyimides with modified chemical structure can be soluble.

Structure-property relationships in polyimides are much studied. St. Clair et. al. evaluated chain constituents which decrease intermolecular charge transfer, such as fluoroalkyl groups, meta-catenated aromatic diamines, and -O-, -S-, and -SO$_2$- linked aromatic groups[10,39,40]. These constituents are effective at reducing the UV absorption of polyimides and, with the exception of the polar -SO$_2$- groups, lower the dielectric constant. Studies at Du Pont [41,42] and elsewhere [43] show that the use of fluorinated anhydrides reduces k′ and moisture absorption; however, when the diamine portion of the polyimide is also fairly flexible, the resulting polyimide has lower T_g and higher solvent sensitivity than is desirable . By using the inflexible para-phenylene diamine with biphenyl dianhydride, high T_g (450°C) is restored at a small penalty in k'. An added benefit of this composition is that it yields films with low in-plane coefficients of thermal expansion (~3 ppm/°C) which reduces interfacial stress for films coated on low expansion inorganic substrates. The use of siloxane-containing diamines or dianhydrides results in lower k' and reduced moisture absorption at the expense of lower T_g [44,45]. The presence of dimethylsiloxane linkages also leads to reduced thermal stability; the use of an aromatic siloxane dianhydride restores much of the thermal stability of the polyimide. Low k' polyimides are being used, or evaluated for use, principally in applications which justify the higher materials costs: inter-layer dielectrics and passivation layers in IC devices and inter-layer thin film dielectrics for multichip modules (Figure 8.8).

8.5.2 Thermosets

Epoxy resins dominate the market for thermosets in electronics packaging; they are used in printed wiring boards (both rigid and flex), as encapsulants, as potting resins and for coatings [46]. Most epoxy resin systems have dielectric constants around 3.5. Figure 8.9 shows the formation of polar -OH and -NH groups during "cure" by the reaction of amine curing agents with the epoxy ring and the subsequent addition of secondary hydroxyls to the ring. The flexible aliphatic portions of the epoxy chain result in fairly low T_g's, typically between 120 and 150°C. Higher Tg's, up to 180°C have been obtained by using multifunctional epoxy systems[47].

The use of anhydride curing agents results in the formation of less polar ester linkages, such that epoxy resins with dielectric constants as low as 2.8 may be formed from cycloaliphatic epoxies and anhydrides. The resins,and prepregs made from them, have limited shelf life, so their use is mostly in potting applications rather than in the larger volume printed wiring board market.

Thermosetting polyimide resins have been developed with thermal and electrical properties superior to the epoxies. These resins consist of preformed imide segments end-capped with functional groups which cross-link through addition chemistry. Most common is the use of bismaleimides which are chain extended with diamines in the B-stage and then crosslinked by further reaction with amines accompanied by chain polymerization of the double bonds at higher temperatures [48]. The dielectric constant of these resins is generally near 3.2.

A route to thermoset polyimides with lower dielectric constants

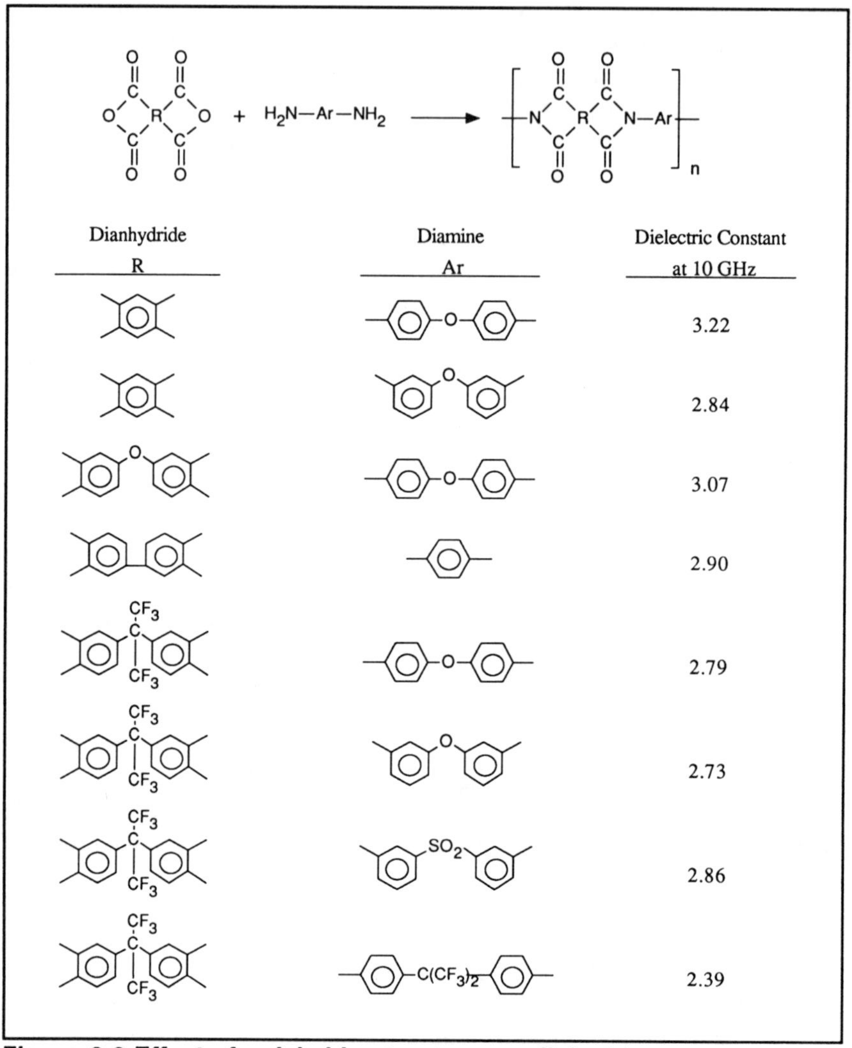

Figure 8.8 Effect of polyimide structure on dielectric constant.

is to prepare oligoimides end-capped with acetylenic groups. Capo and Schoenberg described the preparation and properties of oligoimides based on a fluorinated dianhydride and an ether linked aromatic diamine with propargyl end-caps[49,50]. This resin can be processed from solution and cured at temperatures above 180°C to give thermally stable, crosslinked resins with dielectric constant ranging from 3.35 at 1 KHz to 2.95 at 8.5 GHz [51]. Suggested applications for these resins are as passivation coatings, interlayer dielectrics and encapsulants.

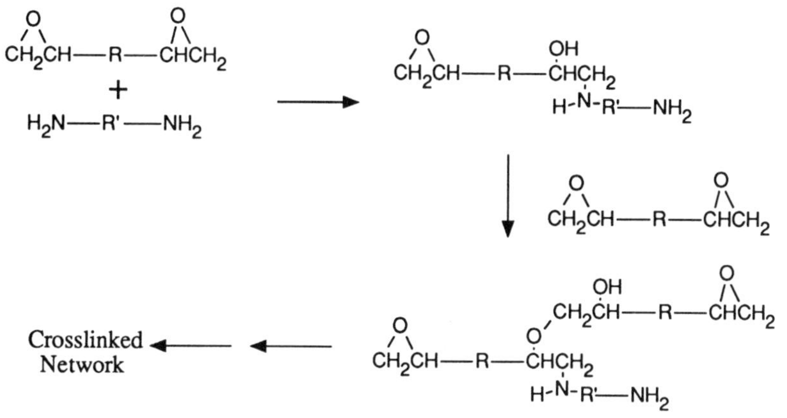

O
/\
CH$_2$CH——R——CHCH$_2$
\/
O

+

H$_2$N——R'——NH$_2$

⟶

O
/\
CH$_2$CH——R——CHCH$_2$
                    |
                    OH
H-N-R'——NH$_2$

↓

O   O
/\   /\
CH$_2$CH——R——CHCH$_2$

OH   O
| /\
CH$_2$CH——R——CHCH$_2$
O
|
O
/\
CH$_2$CH——R——CHCH$_2$
          |
H-N-R'——NH$_2$

Crosslinked ⟵ ⟵
Network

Figure 8.9 Chemistry of amine-cured epoxies.

During the past few years considerable interest has developed in the use of cyanate ester resins for printed wiring board substrates. The resin preparation is based on the reaction of bisphenols (or polyphenols) with cyanogen halides. Mitsubishi Gas Chemical introduced blends of cyanate ester and bismaleimide (BT resins) as an improvement on the existing bismaleimide systems [48]. Hi-Tek Polymers and Dow Chemical have introduced several cyanate resin systems for use in electrical as well as structural composites [52-54]. In general, these resins show a good balance of low dielectric constant, high T_g, moderate toughness, and facile processing (Figure 8.10). Many of the cured resin properties have been attributed to the regularity and symmetry of the crosslinked network [55]. The dielectric constant of the cyanate ester resins is fairly flat over the frequency range from 1 kHz to 10 GHz and shows a slight increase with temperature up to 200°C. Cyclotrimerization of cyanate esters proceeds in the presence of active hydrogen species (alcohols, acids, amines, etc.) and is catalyzed by transition metals. While small amounts of active hydrogen compounds are necessary, too many protonic impurities, i.e. too much water, leads to degradation of the resin through formation of carbamate groups and subsequent loss of carbon dioxide to form amines. The results are lower T_g and higher dielectric constant. Cyanate esters can be co-reacted with epoxies to permit lower temperature cure [56], or blended with thermoplastics to improve toughness [57].

Silicone resins are primarily used as IC encapsulants to protect devices from moisture, ionic contaminants and radiation[61,62]. Silicone gels are lightly cross-linked elastomers; as a result, they are very soft and put little stress on surfaces which they contact[63]. This reduces

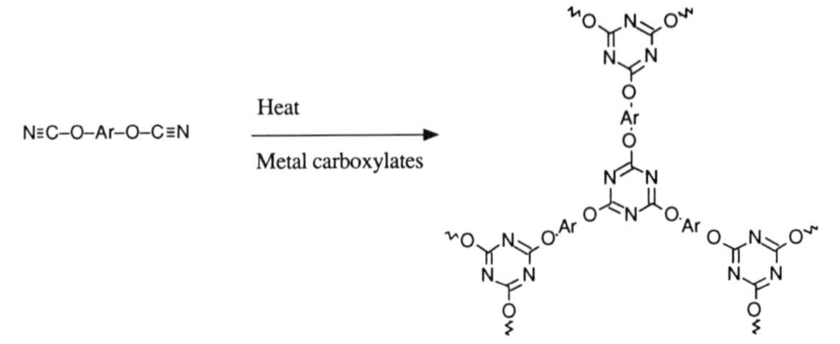

Ar	k' (1 MHz)	T_g (°C, TMA)	Water absorption (%)
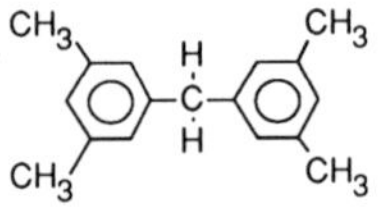	2.75	244	1.2
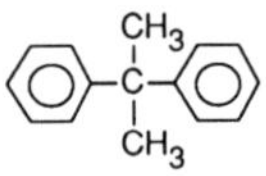	2.91	257	1.6
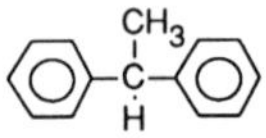		259	1.3
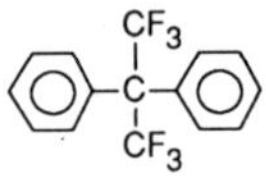	2.66	265	1.2
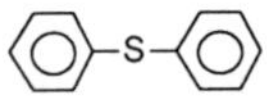	3.11	270	1.4
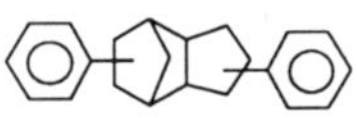	2.8	223-250	1.4

Figure 8.10 Structure and cured resin properties of cyanate ester. [47,58-60]

problems of mismatched thermal coefficient of expansion on the chip surface as well as stresses on bond wires encapsulated by the resin. Two crosslinking mechanisms are widely used: room temperature condensation and elevated temperature vinyl addition (see Figure 8.11). The dielectric constant for thermally cured addition type silicones is in the 2.7 to 2.9 range with tan δ below 0.002.

Dow Chemical has introduced a family of resins which crosslink through benzocyclobutene groups [64,65]. The structures of two compounds being promoted for use in electronics applications are shown in Figure 8.12 along with a suggested two stage curing mechanism.

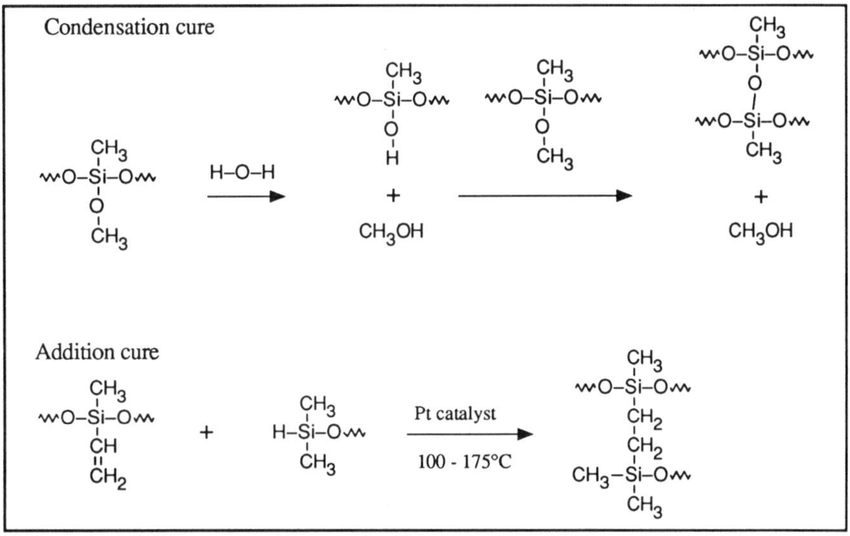

Figure 8.11 Cure reactions of silicones.

Crosslinking of BCBs (benzocyclobutenes) is believed to occur both through cyclo-addition reactions and through linear addition across the conjugated o-xylylene tautomer [66]. BCBs are under current evaluation as chip passivation coatings and inter-layer dielectrics [67]. For both of these applications, the monomeric BCB is oligomerized and the B-staged material is coated from a high solids (typically 35 - 55%) solution. The dielectric constant of cured BCBs (T_g>350°C) is 2.7±0.1 over a wide range of frequencies and temperatures, with tan δ values of 0.0005 to 0.001. The hydrocarbon structure of the BCBs results not only in low k' and tan δ, but also low moisture absorption. The cross-linked structure of BCBs is reminiscent of the parylene structure; the benzylic protons in the structure increase oxygen sensitivity so that elevated temperature (>250°C) processing of the BCB must be carried out in low oxygen environments [68].

Crosslinking versions of high vinyl polybutadiene [69,70] and polyphenylene oxide [71] have been proposed for use in printed wiring board laminates. Processing difficulties limit the utility of these materials in this application; the polybutadiene systems seem to have greater potential in potting/encapsulation and in radome composites [72].

8.5.3 Composites

Composite materials are generally used for circuit board substrates, connectors and plastic chip carriers. The dielectric constant of a

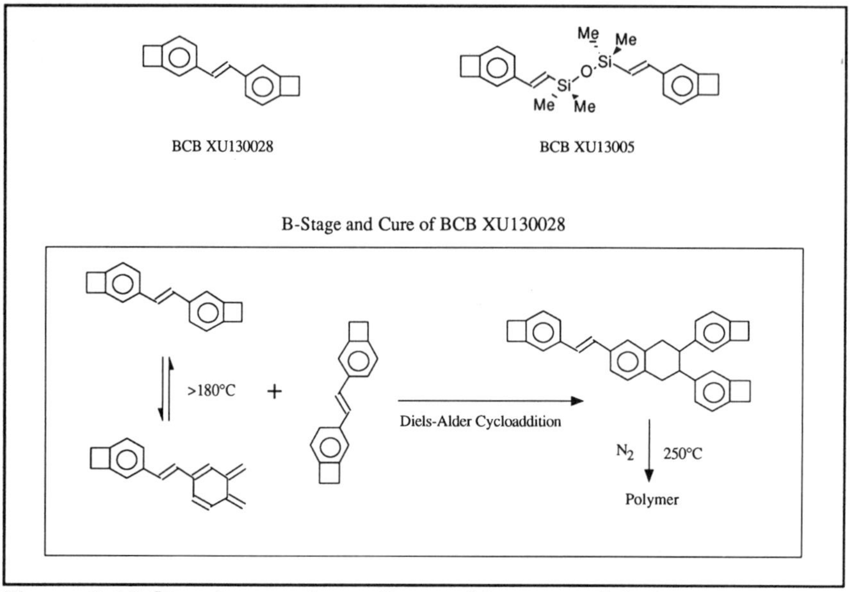

Figure 8.12 Structure and reactions of benzocyclobutenes.

composite k_T' is determined both by the dielectric constants of the component materials (k_1', k_2',...) as well as by the volume fractions ($\underline{v}_1$, $\underline{v}_2$, ...) of the components. For a fiber reinforced laminate, the dielectric constant is generally taken as a linear function:

$$k_T' = k_1'v_1 + k_2'v_2 + ...$$

[8.18]

Mumby and Yuan showed that epoxy/fiberglass laminates obeyed this linear relation for resin volume fractions from 0.55 to 0.75 [73]. Various other theoretical equations have been developed to relate the dielectric constant of a composite to k' of the components [74]. Over the practical range of compositions for most composites, it is difficult to determine whether the curve is better described by the linear formula Equation 8.18 or by an alternate expression that has k' as a non-linear function of the resin content. Moreover, the dielectric constants of the reinforcements and resins found in the literature are frequently derived from composite data. As a result, Equation 8.18 is a good guide for predicting composite dielectric constants, but is no replacement for making actual measurements.

The most common reinforcement for electrical composites is a lime-alumina-borosilicate glass commonly called E-glass; a substantial infra-structure exists for the production of yarn and a variety of fabrics from this glass for use in both electrical and structural applications [75]. Important electrical and physical properties of E-glass and other reinforcing fibers are given in Table 8.4. Of the materials shown, E-glass has the highest k' and tan δ. Notice that the inorganic reinforcements have isotropic thermomechanical properties while the organic fibers all display negative CTEs along the fiber axis and positive transverse CTEs.

Table 8.4 Properties of reinforcing fibers. [18,48,76,77]

Material	k' (1 MHz)	tanδ (%) (1 MHz)	CTE (ppm/°C)	Modulus (Mpsi)	UpperUse Temp. °C
E-Glass	5.8 {6.3}	0.1 {0.37}	5.0	10.5	841
S-Glass	4.53 {6.0}	0.2 {0.2}	2.9	12.4	860
D-Glass	3.56 {4.6}	0.05 {0.15}	3.1	7.5	771
Fused Quartz	3.8 {3.7}	0.01 {0.02}	0.5	10	1670
Kevlar® 49	{3.8}	{0.2}	-5.0/55	18.5	300
e-PTFE	2.0	0.02	<-5/>100	1.3	280
Spectra®	2.0 - 2.3	0.02 - 0.04	<-5/>100	17	120

Dielectric values in brackets are calculated from composite data. CTE values for organic fibers are shown parallel/perpendicular to the fiber axis.

Woven glass reinforced PTFE laminates are commonly used in microwave circuit boards. Materials are available with k'/tan δ ranging from 2.2/0.0008 to 2.6/.002 depending on the volume fraction glass. These laminates typically have in-plane CTE (coefficient of thermal expansion) near 10 ppm/°C with out-of-plane CTEs greater than 120 ppm/°C. While the electrical properties of these materials are excellent, their use in printed wiring boards for digital applications has been limited by several processing and performance problems [17,78]:

- Recall that PTFE undergoes a crystal to crystal phase change around room temperature; as a result, the PTFE based laminates tend to show dimensional instability over this critical temperature range. This instability makes the registration of multiple board layers quite difficult, unless the feature sizes are large.
- The high out-of-plane CTE is thought to reduce the reliability

of vias due to thermally induced stresses on the plated copper vias.

• Adhesion of plated copper to fluoropolymers is problematic. Fluoropolymer surfaces generally need to be etched with organo-sodium solutions to ensure good bonding.

• Multilayer lamination using fluoropolymer adhesive layers must be carried out at temperatures higher than are typical for board manufacturers. Care is required to ensure that void free laminates are produced without distortion of inner layers.

Rogers Corp. produces PTFE laminates reinforced with glass fibers oriented randomly in-plane. These laminates are reported to have lower dielectric loss and smoother surfaces than laminates with woven glass reinforcements, but inferior dimensional stability [79,80]. Thin laminates (thickness 0.001 - 0.01") with a dielectric constant of 2.5 are being promoted for use as low k' substrates for non-dynamic flexible circuitry [81,82]. These laminates are clad with copper or are used as adhesive layers between Kapton® film and copper [83,84].

Flexible fluoropolymer laminates based on Kapton® film as well as Kevlar® fabric reinforcements have been prepared by Du Pont [85]. The Kevlar® reinforced material (T-Lam™) has lower dielectric constant, better dimensional stability, and higher mechanical strength than the Kapton® reinforced laminates (T-Flex™); however T-Flex™ has better dynamic flexural properties. The properties of these fluoropolymer flexible laminates are compared to a standard acrylic/polyimide material in Table 8.5.

Our experience is that the fluoropolymer composites used for flexible circuits do not have sufficient dimensional stability for use in higher layer count (>8 layers) multilayer circuit boards. Rogers has recently introduced a fluoropolymer composite (RO2800) reinforced with amorphous silica powder and small amounts of glass fiber [86,17,48,87]. At roughly 50% silica by weight, the dielectric constant of RO2800 is 2.8 to 2.9. The incorporation of randomly dispersed, unoriented silica gives these laminates fairly isotropic expansion properties, with in-plane CTEs of 16 ppm/°C and out-of-plane CTE of 24 ppm/°C. These expansion properties, along with reported dimensional stability of 0.05%, support fabrication of high layer count multilayer printed wiring boards. These excellent end-use properties come at the expense of non-standard processing [18,88] and high materials cost; finished board costs are estimated as two to three times the cost of the same board prepared from epoxy/glass.

W. L. Gore has introduced a fluoropolymer composite designed to permit standard processing conditions: the laminates use expanded PTFE (e-PTFE) as the reinforcement and standard thermoset resins [89].

Table 8.5 Properties of fluoropolymer flex laminates.

Properties	Acrylic/ Kapton®	GRF**	GRF/ Kapton®	PTFE/ Kevlar®	PFA/ Kapton®
Tensile strength, kpsi	24	5	8	7	23
Elongation,%	40	4	25	29	80
Flex endurance, cycles	6000	1400		400-600	8600
Dimensional stability,%	0.1	0.15	0.18	0.1	<0.6
Peel strength, lb/in	9 - 12	8	8 - 12	13	13
Chemical resistance, %	>80	100	85-100	100	90-100
Water Absorption, %	6.0	0.4	1.0	0.2	0.6-0.9
Propagation velocity, 10^8 m/s	1.7	1.92		2.02	1.90
** GRF = glass microfiber reinforced fluoropolymer					

To enable wetting and adhesion of the resins, the e-PTFE is treated with alkali naphthelene prior to impregnation [90]. The dimensional stability of copper clad laminates based on e-PTFE/thermosets is inadequate for the preparation of high layer count multilayers. Gore has suggested two solutions to this limitation: co-weaving of quartz fibers with the e-PTFE and use of e-PTFE only as bond ply. Quartz containing laminates with epoxy resin are reported to have k′ of 2.8, dimensional stability less than 0.05%, good electrical and mechanical properties and fairly standard processing. The second approach, which Gore has named the "mixed dielectric approach" utilizes cyanate ester/e-PTFE prepreg (k′ = 2.7) as the bonding ply and conventional glass-reinforced inner layers in a dual stripline construction as shown in Figure 8.13 [91-94]. Since the electric field is concentrated between the signal line and the neighboring ground plane, the effective dielectric constant is nearly equal to that of the bond ply.

Quartz reinforced composites have received attention as PWB substrates both because of the low k′ of the quartz, and because of the

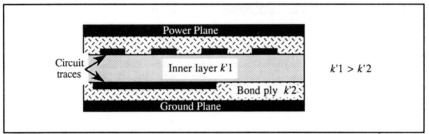

Figure 8.13 Dual stripline construction with mixed dielectrics.

low CTE achievable [95,96]. While standard bismaleimide resins have been most studied, acetylene-terminated polyimides [97] and cyanate esters have also been examined [53,98]. In our study of quartz fabric reinforced cyanate ester, we obtained laminates with k' of 3.1 to 3.2 and other properties as shown in Table 8.6. Two impediments to widespread use of woven quartz as a reinforcement are that it is more difficult to drill than E-glass and that it is much more expensive than E-glass. Composites reinforced with Kevlar® aramid fabrics are also known to have lower dielectric constant than the corresponding glass reinforced materials. The principal application of woven Kevlar® reinforced laminates is for control of in-plane CTE in printed wiring boards with surface mounted ceramic chip carriers [99]. Non-woven reinforcements have the potential to expand the market for aramids in electronics packaging. Non-woven aramid reinforcements offer a route to low k' laminates with:

- Lower cost than woven quartz or Kevlar®
- Improved surface smoothness which assists in the definition of fine line circuitry and reduced drill wander,
- More uniform distribution of the reinforcement which leads to reduced drill wear and improved thermal cycling performance vs. woven aramids, and
- Tailorable thickness and CTE properties.

Table 8.6 compares the properties of cyanate ester laminates reinforced with woven and non-woven materials.

8.6 Conclusion

Low dielectric constant materials help to meet the packaging needs for controlled impedance, high circuit density, high clock-speed electronic systems. Recent developments in composite materials, in new cross-linking chemistries and in designed molecular architecture demonstrate

Table 8.6 Properties of cyanate ester laminates.

Property	Reinforcements			
	E-glass, style 108**	Quartz, style 525	Kevlar® style 108	Modified non-woven aramid
Dielectric Constant (1 MHz)	3.7 - 3.8	3.1 - 3.2	3.4 - 3.5	3.1 - 3.3
Dissipation Factor (1 MHz)	0.003	0.006	0.008	0.007
Tg(°C, TMA)	240 - 250	240	240	225
Resin Content (wt %)	62 - 68	46.5	66	58
CTE (ppm/°C):				
X,Y	11	8 - 10	6 - 9	15 - 17
Z	52	35	110 - 120	50 - 60
Copper Peel Strength (pli)	10 - 12	6 - 7	6	6 - 7
Dielectric Thickness (in)		0.0062	0.0055	0.0072
Dimensional Stability (%)		0.025	0.02	0.04

** Data for E-glass reinforced material taken from refs 53 and 96, other data from experimental laminates prepared at Du Pont.

the response of materials scientists and chemists to the needs of the electronics industry. It seems likely that the packaging materials of the future will be application- and process-specific and that dialogue between materials suppliers and system designers will be the key to successful new products.

8.7 Acknowledgement

I am grateful to my colleagues at Du Pont Electronics for providing background information, useful insights, and patient discussion of the issues discussed in this Chapter. It is a pleasure to acknowledge the help of A.T. Murphy, J.P. Currilla, Y.Belopolsky, D.L. Goff, J.D. Craig, W.J. Lautenberger and H.S. Hartmann.

8.8 References

1   I. A. Blech, "Properties of Materials" in <u>Electronics Engineers Handbook 3'rd Edition</u>, edited by D. G. Fink and D. Christiansen, (McGraw-Hill, New York, 1989) Sec. 6

2   "Di-Clad® 522 and 527 woven PTFE composite laminates", Keene Corporation, Technical bulletin code #1065

3   "G-30 Polyimide laminates and prepreg for multilayer", Norplex/Oak product Bulletin No. 2002 5/1/87

4   R. C. Buchanan, In <u>Ceramic Materials for Electronics</u>, edited by R. C. Buchanan (Marcel Dekker: New York, 1986), Chapter 1.

5   J. J. Licari, <u>Plastic Coatings for Electronics</u>, (Robert E. Krieger: Malabar, FL 1981) Chapter 2

6   H. Fröhlich, <u>Theory of Dielectrics - Dielectric Constant and Dielectric Loss</u>, (Oxford University Press, Oxford, 1958)

7   A. R. Von Hippel, in <u>Dielectric Materials and Applications</u>, edited by A. R. Von Hippel (MIT Press: Cambridge, MA, 1954), Chapter 1.

8   T. W. Bates, in <u>Fluoropolymers</u>, edited by L. A. Wall (John Wiley & Sons:New York, 1972) Chapter 14.

9   R. E. Schramm, A. F. Clark, and R. P. Reed, in <u>A Compilation and Evaluation of Mechanical, Thermal and Electrical Properties of Selected Polymers</u>, (U.S. National Bureau of Standards, 1972) p. 17.

10   A. K. St. Clair, T. L. St. Clair and W. P. Winfree, "Process for preparing low dielectric polyimides", European Patent Appl. 0 299 865 A2 (January 1989).

11   E. D. Wachsman, P. S. Martin and C. W. Frank, "Cure studies of PMDA-ODA- and BTDA-ODA-based polyimides by fluorescence spectroscopy", in <u>Polymeric Materials for Electronics Packaging and Interconnection</u>, edited by J. H. Lupinski and R. S. Moore, (Amer. Chem. Soc., Washington, DC, 1989), Chapter 2.

12   A. J. Beuhler, N. R. Nowicki and J. M. Gaudette, "Dielectric characterization of water in polyimide and poly(amide-imide) thin films", in <u>Polymeric Materials for Electronics Packaging and Interconnection</u>, edited by J. H. Lupinski and R. S. Moore, (Amer. Chem. Soc., Washington, DC, 1989), Chapter 5.

13   J. M. Butler, R. P. Chartoff, and B. J. Kinzig, "Some approaches to low dielectric constant matrix resins for printed circuit boards", presented at 15th National SAMPE Technical Confer.,Cincinatti OH, October 1983.

14   J. R. Paulus, "High performance laminate systems for high speed electronics applications", Circuit World <u>15</u>(4), 19-24 (1989)

15   A. L. M. Angstenberger, "The impact of microwave theory on MLBs for high speed digital applications - design, production, qualification", Printed Circuit World Convention IV, June, 1987; WCIV-31.

16   E. Abramson, "Improved high speed interconnects with Du Pont flexible fluoropolymer laminates", Electronics Packaging & Production, in press.

17   D. J. Arthur, "Electrical and Mechanical Characteristics of Low Dielectric Constant Printed Wiring Boards", presented at IPC 29th annual meeting, April 1986, Boston, MA, IPC-TP-585

18   S. J. Mumby,"An overview of laminate materials with enhanced dielectric properties", J. Electronic Materials 18(2),241-250 (1989).

19   R. F. Field, "Lumped circuits", in Dielectric Materials and Applications, edited by A. R. Von Hippel (MIT Press: Cambridge, MA, 1954), Chapter II.A.1.

20   W. B. Westphal, "Distributed circuits" in Dielectric Materials and Applications, edited by A. R. Von Hippel (MIT Press: Cambridge, MA, 1954), Chapter II.A.2.

21   R. K. Hoffmann, Handbook of Microwave Integrated Circuits, translated by G. A. Ediss and N. J. Keen, edited by H. H. Howe, Jr. (Artech House, Norwood, MA, 1987) Chapter 7.

22   ASTM D3380-89 "Standard test method for elative permittivity (dielectric constant) and dissipation factor of polymer based microwave circuit substrates", in 1989 Annual Book of ASTM Standards vol. 10.02, (Amer. Soc. Test. Matls., Philadelphia, 1989) pp 323 - 332.

23   ASTM D2520 "Standard test methods for complex permittivity (dielectric constant) of solid electrical insulating materials at microwave frequencies and temperatures to 1650iC", in 1989 Annual Book of ASTM Standards vol. 10.02, (Amer. Soc. Test. Matls., Philadelphia, 1989) pp 210 - 225.

24   D. C. Dube, M. T. Lanagan, J. H. Kim and S. J. Jang, "Dielectric measurements on substrate materials at microwave frequencies using a cavity perturbation technique", J. Appl. Phys. 63(7) pp. 2466-2468 (1988).

25   H. Fellner-Feldegg, "The measurement of dielectrics in the time domain", J. Phys. Chem. 73(3) pp 616-623 (1969).

26   "Q Meter 4342A Operating and Service Manual", Yokogawa-Hewlett Packard, Tokyo, Japan 1970.

27   ASTM D 150-87 "Standard test methods for A-C loss characteristics and permittivity (dielectric constant) of solid electrical insulating materials", in 1989 Annual Book of ASTM Standards vol. 10.02, (Amer. Soc. Test. Matls., Philadelphia, 1989) pp 17-35.

28   "4262A Digital LCR Meter, Operating and Service Manual", Yokogawa-Hewlett Packard, Tokyo, Japan 1977.

29   H. S. Hartmann and C. L. Booth, "Development of a crystallizable low-k low-loss
        low temperature multilayer interconnect system", presented at ISM meeting, May
        1990, Tokyo.

30   B. Schwartz, "Review of multilayer ceramics for microelectronic packaging", J.
        Phys. & Chem. Solids 45(10), pp. 1051-1068 (1984).

31   N. Kamehara, K. Niwa and K. Murakawa, "Packaging material for high speed
        computers", IMC Proceedings, Tokyo, May 1982, pp. 382-393

32   T. S. Laverghetta, Microwave Materials and Fabrication Techniques (Artech House:
        Dedham, MA 1984) p. 26.

33   D. C. England, R. E. Uschold, H. Starkweather and R. Pariser, Proceedings of The
        Robert A. Welch Conferences on Chemical Research XXVI. Synthetic Polymers,
        Houston,TX, 1982 pp. 203-215.

34   S. V. Gangal, in Kirk-Othmer Encyclopedia of Chemical Technology, Third Ed. 11
        (John Wiley & Sons, 1980) pp. 1 - 35.

35   P. R. Resnick, "The preparation and properties of a new family of amorphous
        fluoropolymers: Teflon® AF", Polymer Preprints 31(1) p 312 (1990).

36   "Teflon® AF Amorphous Fluoropolymer", Du Pont product brochure H-16577-1,
        12/89.

37   M. Szwarc, "Poly-para-xylelene: its chemistry and application in coating
        technology", Polymer Eng. Sci. 16(7), pp 473-479 (1976).

38   W. F. Beach and T. M. Austin, "Update: parylene as a dielectric for the next
        generation of high density circuits", SAMPE Journal 24(6) pp 9-12 (1988).

39   A. K. St.Clair, T. L. St.Clair and W. P. Winfree, "Low dielectric polyimides for
        electronic applications", Polym. Mater. Sci. Eng. 59, p. 28-32 (1988).

40   D. M. Stoakley and A. K. St. Clair, "Effect of diamic acid additives on dielectric
        constants of polyimides", in Polymeric Materials for Electronics Packaging and
        Interconnection, edited by J. H. Lupinski and R. S. Moore, (Amer. Chem. Soc.,
        Washington, DC, 1989), Chapter 7.

41   D. L. Goff, E. L. Yuan, H. Long and H. J. Neuhaus, "Organic dielectric materials
        with reduced moisture absorption and improved electrical properties", in Polymeric
        Materials for Electronics Packaging and Interconnection, edited by J. H. Lupinski
        and R. S. Moore, (Amer. Chem. Soc., Washington, DC, 1989), Chapter 8.

42   C. C. Schuckert, G. B. Fox and B. T. Merriman, "The evolution of packaging
        dielectrics" presented at the Second Du Pont Symposium on High Density
        Interconnect Technology; Wilmington, DE, 1988.

43   A. J. Beuhler, M. J. Burgess, D. E. Fjare, J. M. Gaudette and R. T. Roginski,
        "Moisture and purity in polyimide coatings", Mat. Res. Soc. Symp. Proc. 154 pp.

73-90 (1989).

44   P. P. Policastro, J. H. Lupinski and P. K. Hernandez, "Siloxane polyimides for interlayer dielectric applications", in Polymeric Materials for Electronics Packaging and Interconnection, edited by J. H. Lupinski and R. S. Moore, (Amer. Chem. Soc., Washington, DC, 1989), Chapter 12.

45   C. A. Arnold, Y. P. Chen, D. H. Chen, M. E. Rogers and J. E. McGrath, "Low dielectric, hydrophobic polyimide homopolymers and poly(siloxane imide) segmented copolymers for electronics applications", Mat. Res. Soc. Symp. Proc. 154 pp. 149-160 (1989).

46   C. A. May, "Epoxy resins", in Engineered Materials Handbook™ Volume I (ASM International, 1987) pp 66-77.

47   D. R. McGowan, P. C. Fabrication, August 1984.

48   D. W. Wang, "Advanced materials for printed circuit boards", Mat. Res. Soc. Symp. Proc 108 p 125 (1988).

49   D. J. Capo and J. E. Schoenberg, "Acetylene-terminated AT fluorinated polyimide", SAMPE Journal, March/April 1987.

50   D. J. Capo and J. E. Schoenberg, "An acetylenic-terminated fluorinated polyimide, properties and applications", presented at 18th International SAMPE Technical Conf., October, 1986.

51   "Thermid® FA-7001 Product Data", National Starch and Chemical, Data Sheet 16286.

52   D. A. Shimp, J. R. Christenson and S. J. Ising, "Cyanate esters - an emerging family of versatile composite resins", presented at 34'th International SAMPE Symp., May 8, 1989, Reno, NV.

53   G. W. Bogan, M. E. Lyssy, G. A. Monnerat and E. P. Woo, "Unique polyaromatic cyanate ester for low dielectric printed circuit boards", SAMPE J. 24(6) pp. 19-25 (1988).

54   D. A. Jarvie, "Characterization of dicyclopentadiene based cyanate ester resin mechanical and electrical properties of unreinforced resin castings and composites", 33'rd International SAMPE Symposium, Volume 33, March 1988, p. 1405.

55   V. A. Pankratov, S. V. Vinogradova, and V. V. Korshak, "The synthesis of polycyanates by the polycyclotrimerisation of aromatic and organoelement cyanate esters", Russian Chem. Rev. 46(3),pp 278-295 (1977), translated from Uspekhi Khimii 46, pp 530 - 564 (1977).

56   D. A. Shimp, F. A. Hudock and S. J. Ising, "Co-reaction of epoxide and cyanate resins", 33'rd International SAMPE Symposium, Volume 33, March 1988, p. 754.

57   D. A. Shimp, "Polycyanate esters of polyhydric phenols blended with thermoplastic polymers", EP 0 311 341 A2 (12/4/89).

58   S. J. Ising and D. A. Shimp, "Flammability resistance of non-brominated cyanate ester resins", presented at 34'th International SAMPE Symp., May 8, 1989, Reno, NV.

59   S. J. Ising, D. A. Shimp and J. R. Christenson, "Cyanate cure behavior and the effect on physical and performance properties", 3'rd Int'l SAMPE Electronic Conf., June 1989, Los Angeles, CA.

60   "Dow XU-71787 Product Brochure", Dow Chemical Form No. 296-00851-589-A&L, , April 1989.

61   C. P. Wong, "High performance silicone gel as integrated-circuit-device chip protection", in Polymeric Materials for Electronics Packaging and Interconnection, edited by J. H. Lupinski and R. S. Moore, (Amer. Chem. Soc., Washington, DC, 1989), Chapter 19.

62   C. P. Wong, "Integrated circuit device encapsulants", in Polymers for Electronic Applications, edited by J. H. Lai, (CRC Press, Boca Raton, FL, 1989). Chapter 3.

63   G. J. Kookootsedes, "Silicone gels for semiconductor applications", in Polymeric Materials for Electronics Packaging and Interconnection, edited by J. H. Lupinski and R. S. Moore, (Amer. Chem. Soc., Washington, DC, 1989), Chapter 20.

64   R. A. Kirchoff, "Polymers derived from poly(arylcyclobutenes)", U. S. Patent 4,540,763, September 10, 1985.

65   S. F. Hahn, P. H. Townsend, D. C. Burdeaux and J. A.Gilpin, "The fabrication and properties of thermoset films derived from bis-benzocyclobutene for multilayer applications", Polym. Mater. Sci. Eng. 59 pp 190 -194 (1988).

66   L-S. Tan and F. Arnold, "New high-temperature thermoset systems based on bis-benzocyclobutene", ACS Polymer Preprints 26(2) pp 176-177 (1985).

67   P. Garrou, "Dow Chemical polymer and ceramic material developments for VLSI and beyond", presented to IEEE Computer Packaging Workshop, January 26, 1988 Oiso, Japan.

68   T. G. Tessier, G. M. Adema and i. Turlik, " Polymer dielectric options for thin film packaging applications", Technical Report TR89-46, Microelectronics Center of North Carolina, October 23, 1989.

69   M. P. Zussman, "Polybutadiene-epoxy-anhydride laminating resins", U. S. Patent 4,601,944 July 22, 1986.

70   N. Sawatari, I. Watanabe, H. Okuyama and K. Murakawa; IEEE Trans. Elec. Insul., EI-18 (2), 131 (1983).

71   M. Itoh, S. Maeda, T. Heiuchi, T. Ozeki and T. Sakamoto, "Thermosetting PPO

laminates for high frequency circuits", presented at IPC Fall Meeting, Anaheim, CA, October 1988, IPC-TP-750.

72   "Ricotuff and Ricotuff L.V., Technical Bulletin", Colorado Chemical Specialties, 1987.

73   S. J. Mumby and J. Yuan, "Dielectric properties of FR-4 laminates as a function of thickness and the electrical frequency of the measurement", J. Electronic Materials 18(2) pp 287-291 (1989).

74   L. K. H. van Beek, "Dielectric behavior of heterogeneous systems", Progress in Dielectrics 7 pp 69 -113 (1967).

75   D. J. Vaughan and C. G. Herschdorfer, "The manufacture of woven glass fabric" in The Multilayer Printed Circuit Board Handbook, edited by J. A. Scarlett,(Electrochemical Publications Ltd, Ayr, Scotland, 1985), Chapter 3.

76   I. Englen, R. Hengl, G. Hinricksen, "Thermal expansion and Young's modulus of uniaxially drawn polytetrafluoroethylene in the temperature range from 100 to 400K", Colloid and Polymer Sci. 262 pp 780 -787 (1984).

77   G. C. Weedon and T. Y. Tam, Modern Plastics 63(3) pp 64,66-68 (1986).

78   R. J. Bonfield, "A high reliability soft substrate for high speed digital and microwave applications", presented at IPC Fall Meeting, October 1987, Chicago, IPC-TP-658.

79   "RT/duroid®" product bulletin, Rogers Corporation, 1982, brochure 9341-026-MG-5.0.

80   J. R. Carroll, L. W. McGinnis, T. L. Miller and M. B. Norris, "Glass fiber reinforced fluoropolymeric circuit laminate", US 4,886,699 December 12, 1989.

81   S. Gazit and S. C. Lockard, " Low dielectric constant circuit materials for high speed digital systems", presented at Printed Circuit World Convention IV, June, 1987, Tokyo, Japan, paper WCIV-33.

82   "RO2500™ Preliminary Data Sheet", Rogers Corporation, 1986, brochure 0330-067-3.0A.

83   S. Gazit and T. S. Kneeland, "Laminated circuit material", UK Patent GB 2 162 124, June 2, 1988.

84   S. Gazit and C. A. Fleischer, "Flexible circuit laminate and method of making the same", US 4,634,631 January 6, 1987.

85   "T-Flex™/T-Lam™ Flexible Laminates" data sheet, Du Pont Electronics, 1989, brochure H-05339 2/89 10M.

86   D. J. Arthur, J. C. Mosko, C. S. Jackson and G. R. Traut, "Electrical substrate material", US 4,849,284, July 18, 1989.

87   "RO2800™ Preliminary Data Sheet", Rogers Corporation, 1989, brochure 2296-089-5.0A.

88   D. B. Noddin, "Processing high performance multi-layer boards of fluoropolymer composite materials" IEPS: Proc. Tech. Confer. 6'th Ann. Int. Elec. Pack. Confer., Wheaton, IL, Nov. 1986 pp. 318 - 330.

89   "Laminates and pre-pregs of expanded PTFE for high speed digital multilayers", W. L. Gore and Assoc., GCGP1813-3/87.

90   D. D. Johnson, "Dielectric materials having low dielectric constants and methods for their manufacture", UK Patent GB 2 171 356 B, August 23, 1989.

91   J. Mosko, "The mixed dielectric approach: improving speed and density with Gore-Ply® precision dielectric prepreg", presented at 1989 IEPS meeting, San Diego, CA.

92   W. W. Snyder, "A new low-dielectric material for PCB's", PC Fab, March 1990, pp. 41-42,43,48,50.

93   "Gore-Ply® Precision Dielectric Pre-pregs", W. L. Gore & Assoc., Technical Note 3126, November 1989.

94   J. LÜsch, "Neue Perspektiven in der Multilayer-Technologie", Elektronik 37(24) pp. 58-62 (1988).

95   A. M. Ibrahim, "Surface mount technology (SMT) substrate material requirements - a brief review", Mat. Res. Soc. Symp. Proc. 108pp 159-168 (1988).

96   L. E. Gates, Jr. and W. G. Reiman, "An alternate printed wiring board material for LCC's D polyimide-quartz fabric", Proc. 2'nd Ann. Int. Elec. Pack. Cong. pp. 605-612 (1982).

97   A. M. Ibrahim, "Acetylene-terminated polyimide composites for advanced electronic packaging applications", Mat. Res. Soc. Symp. Proc. 154 pp 3-10 (1989)

98   "N-8000 Laminate and Prepreg for High Speed Circuitry", NELCO product bulletin, 1989.

99   "Kevlar® Controlled Expansion Substrates for Leadless Chip Carriers", E. I. Du Pont bulletin E-69540, 1984.

INTEGRATED OPTICAL DEVICES BASED ON SILICA WAVEGUIDE TECHNOLOGY

TADASHI MIYASHITA

9.1 Introduction

9.2 Waveguide formation on silicon

9.3 Waveguide circuit elements

9.4 Optical chip integration

9.5 Integrated optical devices
    9.5.1 Branching device
    9.5.2 WDM devices
    9.5.3 Frequency control devices
    9.5.4 Switching and interconnection devices

9.6 Packaging

9.7 Conclusion

9.8 References

9.1 Introduction

Light was more familiar to us than electricity before the modern industrialized age. However, the great progress in electrical and electronic technologies has opened up a new era in which electricity and electronics have become more useful in our lives. In the early 1960's, the laser was developed, which provided a stable source of coherent light, and low-loss optical fiber has been developed to make it possible to transmit light signals over long distances. Today, the transmission and processing of signals carried by light has become a topic of great interest.

To manipulate light for signal processing and transmission, many optical components must be developed, including lenses, mirrors, prisms, light sources and detectors, modulators and so on. Figure 9.1 shows the developmental progress in optical components, in comparison with electronic devices. As is well known, bulk optical components such as mirrors and prisms have been widely used for a long time. Then, micro optics technology based on miniaturized bulk materials was developed. In the late 1960's, the concept of the optical integrated circuit (OIC) emerged, in which light signals are transmitted and processed by waveguides.

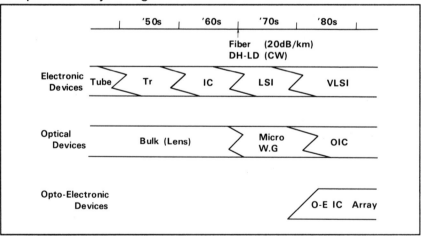

Figure 9.1 Progress in electrical and optical devices.

The current stage of optical components technology is rather behind that of the advanced electronic device technology. But the integrated optics approach to signal transmission and processing may offer advantages in both performance and cost compared to electrical and electronic methods.

Currently, optical integrated circuit technologies are being researched by using various kinds of materials and methods, including dielectric crystals, polymers, glasses and semiconductors [1]. With the shift of optical fiber applications from long distance telecommunications to short distance applications in a wide variety of industries, optical device technology has been aimed at practical uses with the primary interest of commercialization. However, big challenges in performance and cost-effectiveness remain for its introduction into the real market place.

An optical fiber compatible silica glass planar waveguide technology based on a combination of flame hydrolysis deposition with photolithographic and reactive ion etching processes has been developed which provides high performance and cost effective waveguide circuits [2]. A novel technique for hybridizing optical fibers and chips onto the waveguide has been researched based on guiding groove structures, which are formed simultaneously with the waveguide by using a single photomask [3]. This technique provides a wide range of functions in integrated optical devices.

In the following sections, silica glass waveguide technology is described first. Second, construction and performance of various kinds of integrated optical devices, including passive and functional devices based on the silica waveguide technology, are summarized [4].

9.2 Waveguide Formation on Silicon.

Material configurations, planar glass layer deposition and waveguide patterning methods are described. The waveguide material is silica based glass, which is widely used for conventional optical fibers, and the substrate material is silicon. A combination of silica and silicon is one of the most stable materials.

Figure 9.2 shows the silica glass waveguide fabrication process. The silica glass layers are formed on the silicon substrate by flame hydrolysis deposition. Porous-glass, produced by flame hydrolysis of halide materials such as $SiCl_4$, $TiCl_4$, $GeCl_4$, etc., is deposited directly onto silicon wafers. After deposition, the wafers are heated to high temperature for consolidation. The refractive index can be controlled precisely by changing the flow rate of the starting halides. The thickness of the glass layer is also controlled in the range of 10-200 μm within 2%.

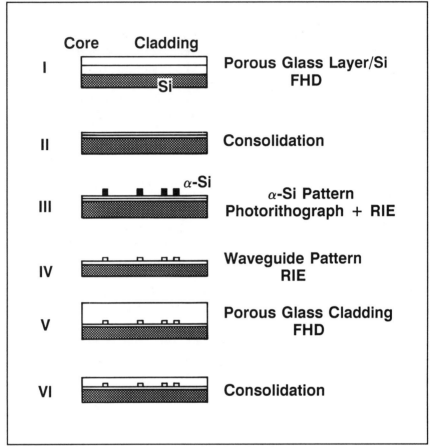

Figure 9.2 Silica planar waveguide fabrication process.

A channel waveguide is formed by a combination of photolithographic pattern recognition and reactive ion etching. Amorphous silicon is used as a mask material for etching silica glass. Ridge waveguides with vertical and smooth side-walls are obtained by selecting suitable etching conditions. Thanks to the isotropic property of the glass layer and the photolithographic technique, any desired circuit configuration, including circular patterns, can be easily fabricated, and many identical circuits can be simultaneously processed on a single substrate and repeatedly duplicated. Figure 9.3 shows a scanning electron microscope photograph of a patterned ridge waveguide. The patterned ridge waveguide is ordinarily embedded by the second flame hydrolysis deposition process.

Figure 9.3 SEM observation of a patterned silica waveguide.

The structural parameters and optical characteristics of the waveguide thus fabricated are listed in Table 9.1. The structural parameters are selected to fit to conventional optical fibers. Propagation losses are as low as 0.1 dB/cm. Butt-coupling losses of silica waveguides with conventional optical fibers are also listed here. The relatively high coupling loss in the multimode system is due to the geometrical mismatch between the square waveguide core and circular fiber core. Fortunately, coupling losses even lower than 0.1 dB/point can be attained in the single mode system.

It should be noted that the present waveguide technology has the advantage of mass producibility and reproducibility in addition to high performance with low loss and excellent structural controllability.

9.3 Waveguide Circuit Elements.

Prior to device fabrication, the fundamental circuit elements shown in Figure 9.4 are formed and evaluated.

The cross is very useful for optical circuitry. Unlike a conventional electrical circuit, there is no crosstalk between two waveguides if an

Table 9.1 Performance of silica waveguides.

Wave-guide	Core size (μm)	Index diff. Δ (%)	Loss (dB/cm)	Fiber-coupling loss	
				meas.	calc.
Single-mode buried	8x8	0.25	<0.1	<0.05	0.01
				dB/point [*]	
Single-mode ridge	8x8	0.25	0.3	0.2	
				dB/point [*]	
Multimode ridge	40x40	1.0	0.2	1.8	1.7
				dB/(input+output) [**]	

[*]Single-mode fiber   :2a=8.9μm Δ=0.27%
[**]Multimode fiber(GI) :2a=50μm Δ=1.0% uniform mode excitation

intersecting angle is designed to be larger than 30 degrees. The waveguide gap, which is one of the features of the present technology, is formed simultaneously with the waveguide by using a single photomask. The gap can be used for hybridization of many optical chips such as wavelength filters.

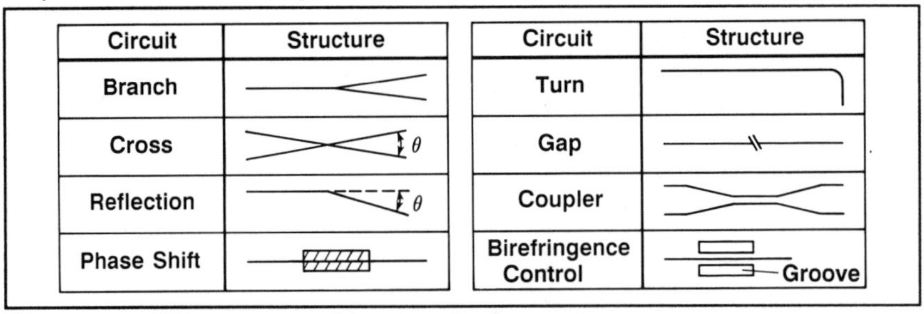

Circuit	Structure	Circuit	Structure
Branch		Turn	
Cross	θ	Gap	
Reflection	θ	Coupler	
Phase Shift		Birefringence Control	Groove

Figure 9.4 Silica waveguide circuit elements.

One of the most important elements is an optical directional coupler using single mode buried waveguides. Spacings of a few microns between the two waveguide cores are controlled precisely. It is confirmed that the experimental data on coupling efficiency of directional couplers agree with the calculated ones. The typical insertion loss is smaller than 0.5 dB, including input and output fiber butt joint losses.

A phase shifter utilizes the thermo-optic effect induced by a thin film heater deposited on the silica waveguide. The response time is about 1 msec. The phase shifter finds its use in affording optical switching/tuning functions to otherwise passive silica waveguides.

Silica waveguides exhibit stress-induced birefringence of an order of 10^{-4} due to the thermal expansion coefficient difference between the silica waveguides and the Si substrates. This birefringence can be controlled, if desired, by forming stress-release grooves along the waveguides.

9.4 Optical Chip Integration.

Micro-chips of passive and active devices such as fibers, wavelength filters, mirrors and light source/detectors are mounted onto the waveguide. Key points are a chip structure design suitable for hybrid integration and alignment-free assembling. The guiding block being fabricated using a single photomask provides the precise relative position between the chip and the waveguide.

It should be noted that the guiding groove structures allow optical device chips to be placed at any desired position on the substrate without being restricted to the substrate ends. This is a great advantage from the viewpoint of optical circuit design.

9.5 Integrated Optical Devices.

Various types of optical waveguide devices have been constructed and tested so far. These can be classified into four groups according to their functions:

[1]   branching devices
[2]   wavelength division multi/demultiplexing (WDM) devices
[3]   frequency control devices
[4]   switching and interconnection devices.

9.5.1 Branching device

The most simple devices are optical signal branching related ones such as the 1 x n splitter, the tap, the star coupler and the line monitor. There are two kinds of branching mechanisms: Y-branches and directional couplers. The Y-branch usually shows wavelength independent characteristics. The directional coupler can operate in a single mode system.

Figure 9.5 shows a Y-branch type 1x8 splitter configuration. The input optical signal is divided into 8 outputs through 3 step-cascaded Y-branches. This features compact size, stable operation and low insertion loss of 1-1.5 dB.

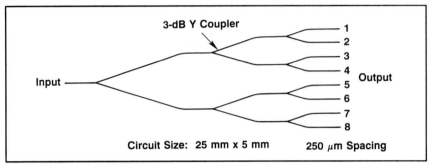

Figure 9.5 A 1x8 splitter configuration.

Figure 9.6 shows a multifunction single mode splitter configuration based on the directional coupler. This splitter can operate as a 1x2, 2x2, 1x4, 2x4, 1x8, or 2x8 splitter, according to the combinations of input and output ports.

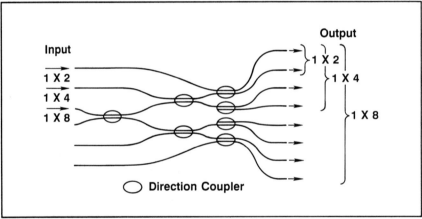

Figure 9.6 A multi-function splitter configuration.

9.5.2 WDM Devices.

Two types of WDM devices have been developed: the filter and the optical directional coupler. The filter type is constructed by inserting interference filter chips in the waveguide gaps as mentioned above. Figure 9.7 shows a 2-wavelength WDM device configuration. Based on this type, a WDM transmitting and receiving module was constructed by hybrid integration of laser diodes and APDs (avalanche photo-diodes) onto the WDM circuit for bidirectional transmission. Prior to mounting each packaged Si-APD, a small mirror chip with a

45-degree angle was placed at the waveguide end to reflect the light into the APD. Crosstalk of less than 30 dB is available from this configuration.

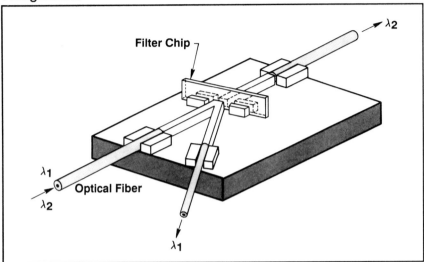

Figure 9.7 Filter type WDM device.

The optical directional coupler shows wavelength dependent coupling efficiency and can be used as a WDM with a broad wavelength bandwidth of 100-200nm.

Figure 9.8 shows the spectral loss curve for the directional coupler type WDM which operates as a multi/demultiplexer for 1.3/1.55 μm [5]. The insertion loss is 0.2dB and crosstalk is about 22 dB for both wavelengths. The pass-band width between 1-dB points is 170 nm and the stop-band width between 20-dB points is 25 nm.

9.5.3 Frequency Control Devices.

Two kinds of optical frequency control devices were constructed based on silica on silicon technology: frequency division multi/demultiplexers and ring resonators.

A Mach-Zehneder interferometer consisting of two 3dB-couplers and unequal arms operates as a multi/demultiplexer for optical frequency division multiplexing (FDM) systems.

Figure 9.9 shows the FDM configuration with waveguide arm length difference ΔL. The frequency spacing $\Delta f = f1 - f2 = c/2n\,\Delta L$ where c is the light velocity in a vacuum , n is the effective refractive index of the waveguide and $_\Delta L$ is the length difference between the two

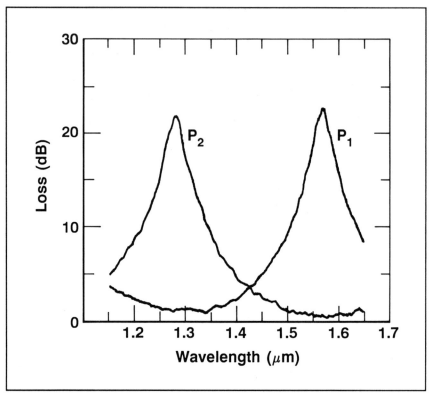

Figure 9.8 Spectral loss curve for a directional coupler.

waveguide arms.

Two light waves whose frequencies differ in a range of a few gigahertz to a few tenths of a terahertz can be demultiplexed.

A thin film heater, which acts as a phase shifter, is formed on one of the arms for precise frequency tuning.

Several kinds of FDM were fabricated and tested. The typical insertion loss and crosstalk were 2-5 dB and 15-25 dB, respectively.

A multi-channel FDM can be constructed by connecting multiple interferometers in a cascade configuration.

Figure 9.10 shows the ring resonator consisting of one ring and two directional couplers with weak coupling ratios [6].

The ring radius was designed to be 6.5 mm for a free spectrum range of 5 Ghz. High Δ waveguides were used in this resonator for small bending radius. A thin film heater was formed onto the ring for tuning the resonant frequency. Resonant characteristics were tested. The finesse was 15 and the insertion loss including input and output

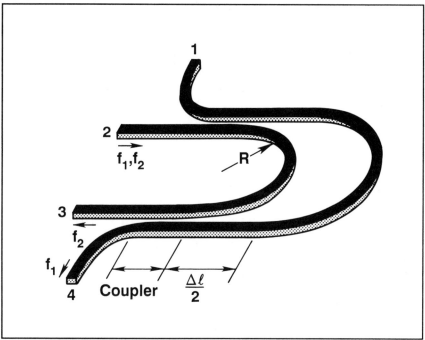

Figure 9.9 FDM configuration.

fiber coupling losses was 16 dB at the resonant frequency. By applying an electrical voltage to the thin film heater, the resonant frequency was changed. This result indicates that the ring resonator can be used as a tunable frequency filter.

9.5.4 Switching and Interconnection Devices.

The silica glass waveguide has passive and limited functions in itself. However, these functions can be extended by hybridizing suitable optical chips. Several kinds of optical signal switching and processing-related devices have been researched based on silica glass waveguide technology.

Figure 9.11 shows an optical switch fabricated by using the thermo-optic effect instead of the electro-optic effect for an LiNbO$_3$ waveguide. A thin film heater on one of the two single mode waveguide arms can act as a phase shifter. A chromium heater 5 mm long and 12 μm wide was formed just above one waveguide arm. The transmittance of the 1.3 μm LD (laser diode) light was measured while the voltage was applied to the heater. The extinction ratio was 17 dB

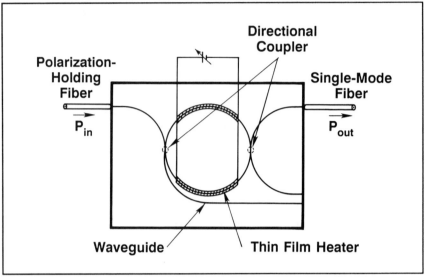

Figure 9.10 Ring resonator configuration.

and the excess loss was 0.8 dB, including fiber coupling losses.

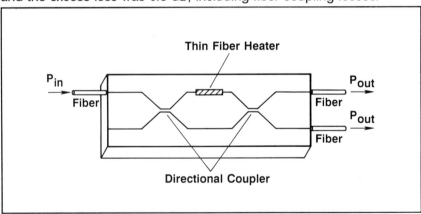

Figure 9.11 Thermo-optic optical switch configuration.

An optical matrix switch was demonstrated by combining silica glass waveguides and semiconductor LD gates. A preliminary experiment for a 4x4 matrix switch shows a size of 10 x 25 mm, a 23 dB insertion loss and -13 dB crosstalk. This type of optical gate matrix switch is expected to have many applications such as optical videosignal exchange and dynamic optical interconnections between computer circuit boards.

Optical interconnection is a potential solution to signal transmission capacity limitation and interference in LSI technology. An interconnection circuit was fabricated based on silica glass waveguide technology. The circuit is composed of a light source detector and a guided wave circuit which has four-channel waveguides for transmitting optical signals, and one mixer for distributing the signal. This circuit is used for broadcast (one-to-many chips) interconnections. An experimental 4-chip interconnection circuit had an optical signal transmission capacity of 1 Gbit/s. Large scale interconnections will be possible by improving the circuit performance.

9.6 Packaging.

Packaging techniques are being developed for silica waveguide on silicon devices. Special attention is paid to the optical performance, temperature stability, and long-term reliability. High precision alignment and fixing of optical chips and fibers to submicron tolerances are essential.

Figure 9.12 shows the bonding process of optical fibers to a 1x8 single mode splitter. Prior to the bonding, the 8 output fibers are arranged in 250 μm spaced grooves on the block which is mounted with the metal holder. These arrayed fibers are simultaneously aligned to the 8 output ports of the 1x8 splitter chip. Then, the metal holder is welded to the holder of the splitter chip.

Figure 9.13 shows the optical insertion loss characteristics of the packaged 1x8 splitter. The loss increase during the packaging process is 0.5-1.5 dB. Reliability tests are underway.

9.7 Conclusion.

Silica planar waveguide on silicon technology has been developed based on a combination of flame hydrolysis deposition with photolithographic and reactive ion etching processes.

The technology is flexible enough to provide a wide variety of high performance and cost effective optical devices, including branching, WDM, frequency control and switching, and interconnection devices.

9.8 References.

1   H. Nishihara, M. Haruna and T. Suhara, Hikari Shusekikairo, Ohm Shya, Tokyo (1985).

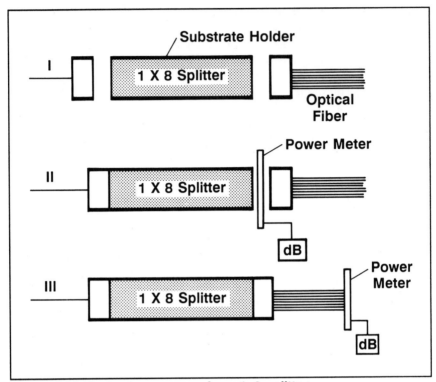

Figure 9.12 Packaging process for a 1x8 splitter.

2   N. Takato, K. Jinguji, M. Yasu, H. Toba and M. Kawachi, "Silica-based single-mode waveguides on silicon and their application to guided-wave optical interferometers", J. Light Technol. 6(6), 1003-1010 (1988).

3   Y. Yamada and M. Kobayashi, "Single-mode optical fiber connection to high-silica waveguide with fiber guiding-groove", J. Light Technol. 5(12), 1716-1720 (1987).

4   T. Miyashita, S. Sumida and S. Kakaguchi, "Integrated optical devices on silica waveguide technologies", SPIE Vol. 933, Integrated Optical Circuit Engineering VI, pp. 288-294 (1988).

5   N. Takato, M. Kawachi, M. Nakahara and T. Miyashita, "Silica-based single-mode guided-wave devices", SPIE Vol. 1177, Integrated Optics and Optoelectronics, pp. 92-100 (1989).

6   N. Takato, T. Kominato, K. Jinguji and M. Kawachi, "Performance and use of silica-based integrated-optic ring resonators", CLEO '88 Technical Digest, pp. 154-155 (1988).

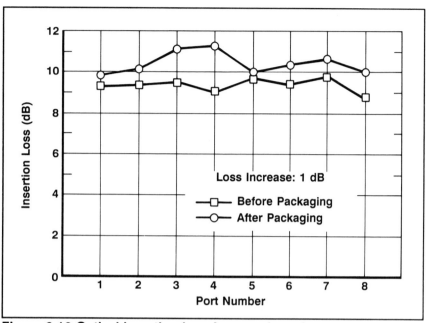

Figure 9.13 Optical insertion loss for a packaged 1x8 splitter.

10

ELECTROSTATIC AND ELECTRICAL OVERSTRESS DAMAGE IN SILICON MOSFET DEVICES AND GaAs MESFET STRUCTURES

DAVID S. CAMPBELL

VINCENT M. DWYER

10.1. Introduction

Perhaps the most remarkable achievement of the semiconductor industry, in recent years, has been the spectacular reduction in component size. The (basically financial) forces which continue to fuel this process have also brought about several technological advances. The reduction in transit time which results from reduced component size has led to improved gains and switching times [e.g. 1]. Smaller components require less external circuitry and, because internal circuitry is more reliable, more components may be interconnected with an associated increase in the complexity of the tasks performable by an integrated circuit. Larger memories are available for the same size of chip, thus manufacturers of, for example, Workstations and Minicomputers are able to increase their built-in RAM capacity with no loss of space. Of course, miniaturisation has also compounded some long standing problems. Some of the failure mechanisms which beset large scale integrated (LSI) circuits are magnified in the very large scale integrated (VLSI) circuits available today.

The metal-oxide-semiconductor field effect transistor (MOSFET) provides the single most important device for VLSI technology. A cross section of a typical device is shown in Figure 10.1. When active, the device carries a current I_{ds} between the source and the drain. A potential applied to the gate terminal, electrically isolated from the current in the source-drain channel by a thin layer of the insulator SiO_2, is then used to control I_{ds}. As the device area is decreased, the controlling voltages and the oxide thickness must also be decreased, so called device scaling. If this is not done the many undesirable "short channel" effects occur (velocity saturation, punch-through, etc.) [e.g. 1]. These very thin gate oxides must, of course, remain electrically robust or else transistor action will be degraded.

In a VLSI circuit the cost of the silicon makes up a substantial portion of a manufacturer's financial burden. It therefore stands to reason that, by increasing the number of transistors per unit area of silicon, the cost of each component is effectively reduced. The first MOSFETs were fabricated in 1960 [2] using a thermally grown oxide. These devices had a channel length of around $20\mu m$ and a gate oxide thickness of over $1000Å$. The continued pursuit of device size reduction has driven the gate oxide below $100Å$ and the channel length into the sub-micron region. Commercially available VLSI circuits can now have over 800 devices per mm^2 [3].

A byproduct of the reduction in size has been an increase in the susceptibility of devices to various overstress failure mechanisms.

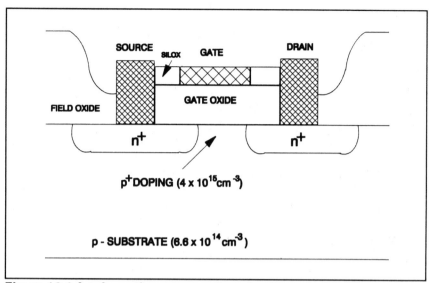

Figure 10.1 A schematic-cross section of an E-mode NMOS transistor.

Overstressing may arise in a variety of ways, from poor circuit design to improper handling by operators. In the former case this may lead to an applied voltage which is far above the normal operating range of the device, whilst in the latter case static charge accumulated over the body of a human operator may be discharged through the device on contact, thus presenting the device with a high voltage transient pulse. The failure mechanisms associated with such events are commonly referred to as electrical overstress (EOS) and electrostatic discharge (ESD) respectively. (It is perhaps worth noting that electromagnetic pulses (EMPs) also produce high energy transients). In addition, during fabrication, a device may accumulate static charge which then may be discharged once an earth path is opened (also a form of ESD) [4].

EOS and ESD account for sufficiently many failures (manufacturers estimate that EOS and ESD account for around 60% of all in-circuit failures [3]) - that the most sensitive devices are provided with "protection" circuitry, in the main consisting of diodes and resistors, which provide an earthing path for any voltage outside some specified safe limits. Thus, the reduction in size has led to an increase in the percentage of the total silicon wafer which is occupied by protection circuitry. Ways need to be found to reduce this protection circuitry and this in turn requires a knowledge of the true overstress hardness of each of these devices. Of course these failure mechanisms are not specific to FET structures, but occur also in many passive and other types of device [e.g. 5].

Recent work at Loughborough has attempted to characterize EOS and ESD failure of MOS and metal-semiconductor (MES) devices, particularly FETs. Naturally, the mechanism by which a particular FET fails depends on a variety of factors including which of the terminals is overstressed and where the FET resides in relation to the rest of the circuit. In practice, MOS gate oxides are only connected to the input address pins of an IC. The output pins have large CMOS (NMOS) transistors attached, with the drain connected to the outside world. These output pins tend to fail as a result of junction damage at one of the ends of the channel. In Figure 10.1 the source drain current sees an n^+-p-n^+ structure and either of the n^+-p diodes may fail due to junction burnout. In general, the gate terminal will be protected, but if, as often happens, the protection is insufficient or has been poorly designed, the overstress voltage will be felt directly by the device.

If the substrate is earthed, an EOS or ESD pulse applied to the gate places an enormous electrical stress on the thin insulating oxide which may, in some circumstances, suffer complete electrical breakdown. The (gate) oxide must be as defect and impurity free as possible so that it may display close to its real intrinsic breakdown strength (8-30MV/cm depending on oxide thickness [6]). Although the breakdown field E_{bd} increases with decreasing thickness, a given overstress or ESD voltage V will generally damage a thin oxide more readily than a thick one [6]. Consequently sensitivity to EOS and ESD is magnified as devices shrink.

While silicon is well established as the premier semiconductor, gallium arsenide (GaAs) has also received a great deal of attention in recent years. The higher mobility and lower effective mass of its carriers make GaAs more suitable for many, particularly high frequency, purposes. The GaAs counterpart of the Si MOSFET is the metal-semiconductor FET (MESFET). The structure of such a device is shown in Figure 10.2. Its operation is similar to that of the MOSFET except that, since GaAs has no natural insulating oxide, the electrical modulation of the channel current is performed by the biasing of a metal-semiconductor junction (Schottky barrier).

The failure mechanisms of a particular device will depend on both the physics of its operation and the level of protection provided by its protection circuitry. For unprotected MOS devices, the most sensitive part of the structure is the thin oxide. Consequently, high voltages applied to the gate of the device are likely to produce the most damage. In contrast, in MESFETs, failure is most likely to arise from contact metal migration, from degradation of the interface forming the Schottky barrier or from changes in the properties of the conducting layer. The diodes of the protection circuitry, and any others on an IC, are not immune from

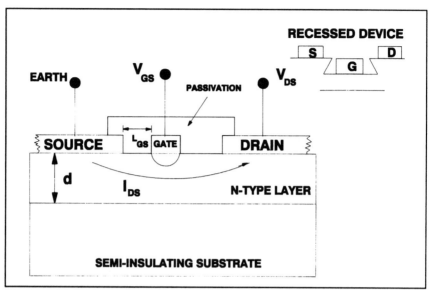

Figure 10.2 A schematic cross-section of an n-channel MES transistor.

overstress damage and junction burnout of these diodes may occur in the depletion region, effectively shorting its large resistance.

    The remainder of this paper is largely based on the studies carried out at Loughborough University of Technology into EOS and ESD failure in MOS, MES and diode structures, although some comments are also made regarding silicon-on-sapphire (SOS) devices. The structures of the devices are described first and the important features which determine the failure modes are identified. This is followed by a simple analysis of the structures in terms of their responses to a series of relevant routine experiments. The effects of variations in ambient temperature and overstress polarity were monitored as were the effects of single and multiple pulses. From these observations it is possible to build models of the failure mechanisms which can hopefully predict the true integrity of these devices. The failure mechanisms introduced above (gate oxide breakdown and junction breakdown) are discussed at length.

10.2. Structures

We begin our discussion with the structures of the devices. The MOSFETs used in these experiments were manufactured by Plessey Research (Caswell) Ltd., U.K. as part of a small geometry process characterisation. The devices were resident on two types of three-inch wafer providing both NMOS and PMOS devices. The NMOS devices consisted of thermally grown 425Å gate oxides with gate dimensions ranging from 1μm x 1μm to 100μm x 100μm on p-type wafers and a substrate dopant concentration of $6.6 \times 10^{14} \mathrm{cm}^{-3}$. In addition, the channel was preferentially doped either with boron to $4 \times 10^{15} \mathrm{cm}^{-3}$, forming enhancement (E-mode) devices, or with phosphorus to give a similar electron carrier density, forming depletion (D-mode) devices. The PMOS devices had 320Å gate oxides and an n-type substrate doped to a concentration of $2 \times 10^{14} \mathrm{cm}^{-3}$. E-mode devices were preferentially doped with arsenic to $1 \times 10^{16} \mathrm{cm}^{-3}$, while D-mode devices were formed by preferentially doping the channel with boron to the same concentration. Such devices might often be found in low noise, digital circuits. All devices were unprotected, allowing a direct access to the gate terminal.

The gate oxide is likely to be the most sensitive part of the structure to both ESD and EOS, consequently oxide integrity is of primary interest in the MOS investigations. Some capacitors, made with the same fabrication process and of the same oxide thickness, were also available on the test wafers. The capacitors enabled the integrity of the oxide to be tested more directly and, because of their relatively large surface area $(49 \times 10^3 \mu m^2)$, the electrical measurements were easier to perform and are expected to be more accurate. Before testing, all devices were given high temperature reverse bias (HTRB) of -12V at 150°C for approximately five minutes to minimise the effects of ionic impurities. All the MOS devices had highly doped polysilicon gates $(\approx 10^{21} \mathrm{cm}^{-3})$.

The MESFET structures used in the experimental study were resident on three two-inch wafers with substrate resistivities of $\approx 150\Omega/\square$. Each wafer contained 1500 FET structures with AuGeNi source/drain contacts and CrAu gates. The wafers (which we label I69X, I69W and E552), supplied by STC Technology Ltd., (Harlow) U.K., were, again, part of a process characterisation. A study of the degradation of MESFETs under ESD was carried out on the wafer E552, while all other MES studies were carried out on the I69X and I69W wafers. The I69X and I69W wafers were fabricated with the same mask set and may be regarded as identical except for the variations in n-type layer depth. The E552 wafer used a similar mask but had larger dimensions.

On the I69X and I69W wafers (Figure 10.2), the gate-source and gate-drain distances were 1μm and 2μm respectively and the gate length and width were 1μm and 150μm respectively. The channel was doped to $1 \times 10^{17} \mathrm{cm}^{-3}$ to a depth of $\approx 0.16\mu m$. All devices were proton isolated, passivated with polyimide and annealed at 100°C for 30 minutes. The E552 wafer had gate-source and gate-drain distances of 1.5μm. Throughout the experiments, only those devices with pinch-off voltages within ±0.2V of the mean value for the wafer were selected. In order to eliminate contamination effects the wafers were stressed with a 100°C temperature screen for 15 minutes before any experiments were carried out.

Under ESD or EOS, the device carries a relatively large current density which, combined with the high overstress voltage, will cause a considerable Joule heating effect. Thermal breakdown is, therefore, a prime candidate for overstress failure of MESFETs. Although an ESD pulse applied to any terminal may cause breakdown, it has been demonstrated that the gate terminal is again the most sensitive contact [7]. Consequently, in all experiments, the ESD/EOS pulses were applied to the gate. The MESFET may be thought to consist of two MES diodes, one between the gate and source, and the other between the gate and drain. If both the ohmic contacts are earthed and a pulse applied to the gate, current flow will be between the gate and both the ohmic contacts. In order to study the breakdown properties within a single well defined region of a device, our study has largely been confined to the breakdown characteristics of the gate-source MES diode alone. Of the two diodes within the MESFET structure, the gate-source diode was considered to be more appropriate for a detailed degradation analysis because, being physically smaller, it is more likely to breakdown than the gate-drain diode due to the greater field that exists in this region during overstress. Although we did not have the range of gate dimensions in the MES devices that we had in the MOSFET structures, a degree of geometry variation was afforded by the inclusion in the study of some recessed gate devices, also shown in Figure 10.2. These devices were also supplied by Plessey Research (Caswell). The n-type layer depth in the recessed devices was similar to that of the non-recessed devices, ≈0.16μm and the recess was approximately 0.15μm.

In the main, the input address pins of silicon ICs are provided with some degree of protection circuitry which can range from a simple diode and resistor combination to much more complex circuits. In the simplest case, in the event of an ESD pulse forcing the gate voltage outside the range of operating specification of the device, the protection diode switches on and an earthing path is opened. With a view to reducing the size of the protection circuitry, the hardness of the protection diode to ESD/EOS stress also becomes important. However, any models designed to predict MES diode breakdown will also be applicable to protection diode breakdown. As a result no special experiments were performed on protection diodes. In addition, n^+-p junction breakdown of the MOSFET due to source or drain overstressing will be covered by the same type of model.

Finally a test of whether we were indeed seeing the true integrity of the MOS gate oxide in the MOS tests was provided by a study of some silicon-on-sapphire (SOS) devices donated by GEC-Marconi Ltd. U.K. These devices also had an oxide thickness of 400Å and consequently should display roughly the same oxide breakdown voltage as the FETs under identical charge injection conditions.

10.3 Experiments

Several series of experiments were performed on each of the device types (MOSFETs, MOS capacitors, MESFETs, MES diodes and SOS MOSFETs). These include constant voltage overstressing (EOS) and single and multiple ESD pulses. As the operation of each of these devices depends on the polarity of the applied voltage, polarity variations were also included in tests.

The test circuit used for the ESD stressing attempts to model the effect of a human operator discharging previously acquired static charge through the device under test (DUT) to earth. There are several test circuits designed to study ESD effects [4,8,9]; we have used a circuit based on the American MIL-STD-883C "human body model" test system (Figure 10.3) [8]. The human body capacitor C_B is charged up to the ESD voltage V_o as the operator acquires the static charge. On contact with the DUT, i.e. on closing the switch S, C_B is discharged through the human body resistor R_B and the DUT to earth. In our electrical equivalent, the effects of the pulse transmission system must also be included. The MIL-STD model has become industry standard, although there are others. It has been suggested that the body resistance R_B should be absent from the test circuit [9], the so called "machine model" originating from Japan. Other ESD test circuits exist which mimic different physical discharge conditions. Of these probably the best known is the "charged device model" [4], which describes the possibility of a charged device suddenly finding an earthing path.

A commercially available instrument called the Hartley Autozap was used to create reproducible waveforms. However, our test circuit cannot properly be called MIL-STD since the rise time is greater than the specified MIL-STD figure of 10ns. A typical ESD waveform from our system would have a rise time of around 100ns. The relatively slow rise is caused by transmission line effects as the ESD pulse was carried to the device via a length of coaxial cable. In the MIL-STD the values of C_B and R_B are 100pF and 1500Ω respectively. The transmission line introduces extra inductive and capacitive elements, but is necessary in order to probe the devices on wafer.

It is clear from Figure 10.3 that the available charge in the system

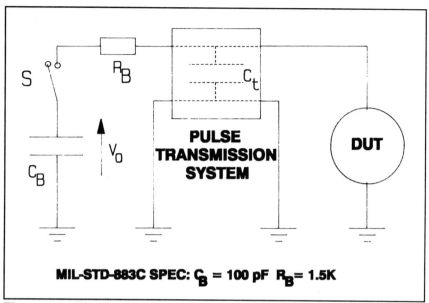

MIL-STD-883C SPEC: C_B = 100 pF R_B= 1.5K

Figure 10.3 The "human body model", including transmission system.

is $C_B V_o$ while the available energy is $\frac{1}{2}C_B V_o^2$. In order to separate the effects of charge and energy C_B and V_o were both varied.

For the EOS experiments, constant voltage stressing was created by an E-H Research Labs 132-L pulse generator, able to deliver constant voltage pulses of up to 50V with a duration of between 20ns and 100ms and a rise time of around 25ns. It was possible to capture any of the current and voltage waveforms on a 1GHz digital oscilloscope using a Hall probe placed around the input cable and a 10MΩ, 3pF voltage probe connected directly across the device. All devices were probed on wafer, using a manual Wentworth prober kit whose chuck had been adapted to allow wafers to be heated from room temperature to around 200°C, thus varying the local ambient temperature.

In all experiments, unless part of a multiple pulse experiment, each device was pulsed only once. The degradation exercise (Section 10.3.1) was statistical in nature, while the other experiments used only enough devices to establish what appeared to be the correct trend.

10.3.1 MOSFET degradation (ESD) [10-13]

A number of E-mode NMOS transistors (Figure 10.1) were subjected to a single voltage pulse of between 50V and 3kV to find a ball-park figure for the breakdown threshold. Voltages between 50V and 300V were found to be the most suitable. Beyond 300V, most devices showed catastrophic failure. For the main experiment 176 E-mode devices were stressed with a single pulse at each of five voltages (50V, 100V, 150V, 200V, 250V) and at each of five temperatures (25°C, 70°C, 110°C, 150°C, 200°C) making a total of 176x5x5=4400 transistors in all [10]. The selected temperature range encompassed the standard maximum operating temperatures for both commercial devices (70°C) and military devices (125°C). In addition, a much more limited series of experiments was performed on the D-mode NMOS and E-mode PMOS transistors [10].

The device I_{DS}-V_{DS} characteristics were recorded before and after stressing to test for degradation. A sensible classification of the degree of degradation arose naturally from the results, Table 10.1.

Table 10.1 MOSFET failure categories.

Category 1	No change	All parameters unaffected
Category 2	Degradation	Walking wounded
Category 3	Degradation with -ve g_m at high V_{GS}	I_{DS} decreases beyond V_{GS} = 3V
Category 4	Resistive characteristics	Linear I-V dependence
Category 5	Catastrophic failure	$I_{DS} \approx 0$ up to avalanche

Examples of categories 1, 4 and 5 are shown in Figure 10.4.

The most important category from a reliability stand-point is category 2, shown in Figure 10.5. These "walking wounded" devices show significant degradation while still exhibiting transistor action. They, therefore, may pass a simple quality control. The gate-drain and gate-source resistances indicate that a small leakage current is passing through the oxide, this prevents the n-channel from forming properly at the semiconductor-oxide interface and degrades the gate control. Further stressing or even normal use is likely to cause these devices to fail prematurely and this observation is supported by the results of experiments with sequential pulses. As the stress voltage is increased the percentage of the devices in this category rises rapidly from 20% at 50V to around 80% at 250V while the percentage of the devices unaffected by the pulse decreases from 70% to around 7%. Similar results were obtained for the D-mode NMOS and E-mode PMOS devices. Clearly ESD sensitivity is voltage dependent, Figure 10.6.

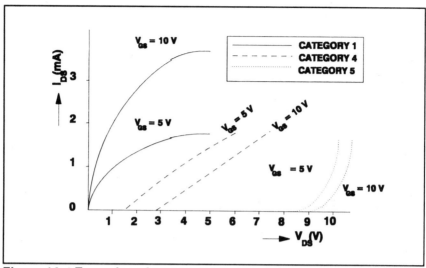

Figure 10.4 Examples of categories 1, 4 and 5 MOSFET degradations.

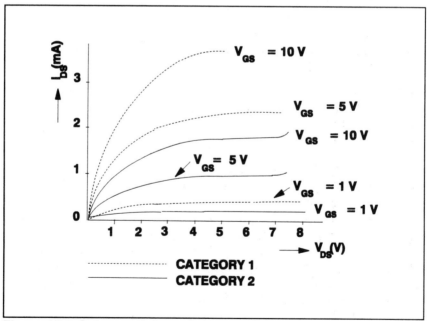

Figure 10.5 Category 2 degradation - walking wounded.

Variations in the ambient temperature of the device caused no noticeable changes in the numbers of devices in each category. One is

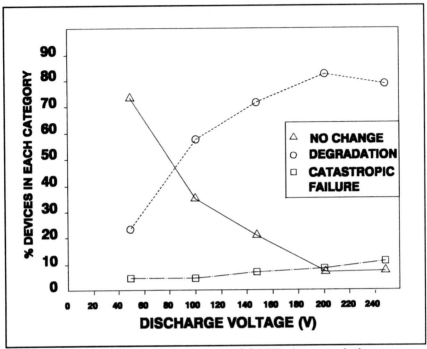

Figure 10.6 Voltage dependence of MOSFET characteristics.

forced to conclude that whatever is responsible for electrostatic discharge damage is unaffected by temperature. Other workers have observed an ESD sensitivity for NMOS LSI devices which is affected by temperature [14]. The cause of failure was traced to a diode in the protection circuitry [14]. As we shall see later it is possible for diodes in the input protection circuitry to introduce a temperature dependence.

The dimensional dependence of the no-change category is shown in Figure 10.7 at 150 and 200V. The increase in the percentage of the devices showing degradation, as the gate area is increased, supports the results of Wolters et al [6], for continuous voltage breakdown. The defects in the oxide, in the form of impurities or broken bonds, weaken the oxide, reducing its dielectric strength. These intrinsic defects may be assumed to occur randomly across the oxide area, consequently large area oxides will have more weak (defective) regions than smaller ones.

In the multiple pulse experiments the extent of degradation increases with the number of pulses. However, devices which had not shown degradation after a single voltage, did not appear to be affected by further pulses at the same voltage. This is in agreement with the

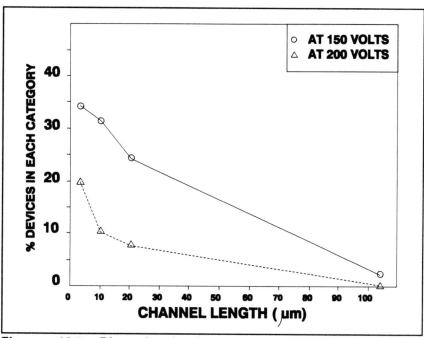

Figure 10.7 Dimensional dependence of MOSFET device characteristics.

assumption of Wolters et al [6] that the weakest defect determines the intrinsic breakdown strength of the oxide.

It is clear that an understanding of oxide integrity is vitally important and consequently, in an attempt to strip the problem to its minimum, capacitors fabricated by the same process were used to assess the oxide strength.

10.3.2 ESD thresholding in MOS capacitors [10,15-17]

The threshold voltages for breakdown under pulsed conditions were determined for various values of the (Autozap) capacitance of the discharge or human body capacitor C_B, as shown in Figure 10.8 for both positive and negative polarity pulses. Two striking features emerge from these relationships. First, to good approximation

$$V_{bd}^+ = |V_{bd}^-| + V_{av} \qquad\qquad [10.1]$$

where V_{av} is a constant approximately equal to 110V and $V_{bd}^{\pm}$ are the positive and negative polarity thresholds. Second, the $|V_{bd}^{-}|$ vs C_B curve represents an approximately constant energy contour. The reasons for this are outlined in Section 10.4.

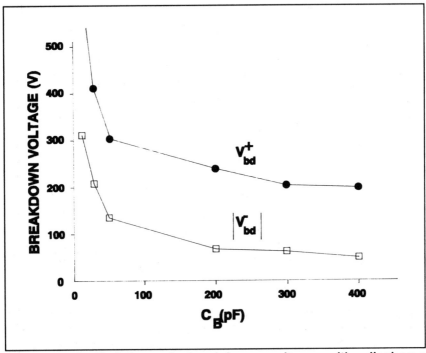

Figure 10.8 Variation of breakdown voltage with discharge capacitance.

10.3.3 Continuous Voltage Stressing of p-Si MOS capacitors [10,17]

Under continuous stress the three types of MOS capacitor showed breakdown thresholds of around ±40V, i.e. independent of the polarity of the applied voltage. However, for the p-MOS capacitors the breakdown time for positive (gate) stressing was significantly longer than for negative stress. In addition the continuous voltage threshold under positive stress was found to decrease as the temperature was raised [10,17] while negative polarity stress was found to be independent of temperature. The n-MOS capacitors showed no temperature dependence. This is significantly different from the pulsed voltage breakdown thresholds, implying that there are (at least) two different breakdown mechanisms involved.

10.3.4 Slow transient sensitivity of p-MOS capacitors [18]

A natural extension of the work reported in Sections 10.3.1-10.3.3 is an attempt to obtain a picture of the breakdown of MOS oxides over the complete range of stressing, from the fast transient ESD to the constant EOS voltages. This was done by gradually increasing the ramp speed of transient pulses applied to the structures. In all these experiments the polysilicon gates of the capacitors were negatively stressed, i.e. **negative** with respect to the substrate. A large number of capacitors were tested at ambient temperatures of around $\approx 25°C$ and $140°C$ with ramp speeds of between 10^4 and $10^7 Vs^{-1}$. By comparison a 1kV "human body" ESD transient, with a rise time of 100ns, produces a ramp of around $10^{10} Vs^{-1}$. An important feature of the slow transient results was that, as with ESD, no temperature dependence was observed [18]. The EOS temperature dependence seems only to occur for very long time stressing (>10ms).

The current and voltage waveforms were down-loaded from the oscilloscope on to a desk-top PC to enable some simple circuit analysis (this is developed further in Section 10.4). In this way we were able to extract the stress voltage felt by the oxide $V_{ox}(t)$. A typical $V_{ox}(t)$ profile, corresponding to a ramped voltage of $10^5 Vs^{-1}$, is shown in Figure 10.9. The crucial point is that, for <u>each</u> of the ramp speeds, the oxide voltage becomes pinned at between -35V and -40V. This is extremely close both to the breakdown voltage of the capacitor under continuous voltage stress and also to the voltage, predicted by the Fowler-Nordheim equation (Section 10.4), at which tunnelling of electrons into the oxide begins.

Although there is some disagreement in the literature about the final cause of oxide breakdown (Section 10.4), there is no doubt that it is initiated by electron injection. The amount of injected charge $Q_{ox}(t)$ may also be obtained from circuit theory. Its value, $Q_{bd} \equiv Q_{ox}(t_{bd})$, at the breakdown time, t_{bd}, is discussed in Section 10.4.

10.3.5 MESFET ESD degradation [13,19]

A series of experiments, similar in form to those conducted on the MOSFETs, was performed on the MESFETs from the E552 wafer. The same extrinsic variables of voltage (including polarity), charge and temperature were also used here. The three important DC electrical parameters of the MESFET, the pinch-off voltage ($V_p \approx -1.25V$), the saturated drain current ($I_{dss} \approx 10.2mA$) and the transconductance

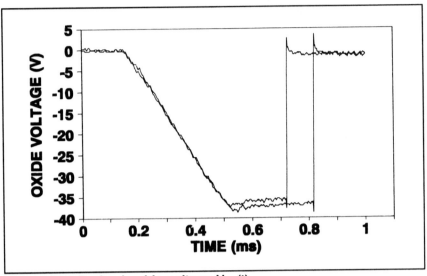

Figure 10.9 Derived oxide voltage $V_{ox}(t)$.

$(g_m \approx 9mS)$ were measured before testing began.

ESD pulses ranging from +350V to +900V, and from -50V to -250V were applied to the E552 devices. As with the MOS devices it was possible to classify the results in terms of a number of failure categories, Table 10.2.

Table 10.2 MESFET failure categories.

Category 1	No change	All parameters unaffected
Category 2	Degradation with >5% increase in I_{dss} or g_m	Some electrical improvement
Category 3	Degradation with >5% decrease in I_{dss} or g_m	Walking wounded
Category 4	Severe degradation in g_m, with I_{dss} constant to 5%	Gate-source burnout
Category 5	Resistive characteristics	Gate-drain burnout with resistive path

Category 2 devices seem to display some improvement in the quality of the Schottky barrier, while category 3 devices again join the ranks of the walking wounded. Devices in categories 4 and 5 show more serious damage. An SEM photograph of a category 4 (black) failure is shown in Figure 10.10.

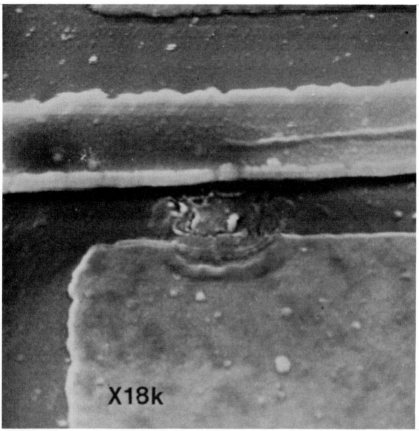

Figure 10.10 SEM photograph showing an MES category 4 black failure.

     The probability of a device falling into a given failure category follows a similar pattern to the MOS devices with the "no change" percentage falling from 90% at 350V to 15% at 900V. The thresholds for degradation or burnout clearly depend dramatically on the polarity of the pulse. The greatest degradation effects, under positive polarity pulses, occurred over a voltage range between +300V and +900V, while, under negative polarity, visible burn-out occurred above -50V and showed little change below this value. Categories 4 and 5 showed visible damage under an SEM, Figure 10.10. Negative polarity pulses caused severe damage in the channel region, which melted part of the ohmic contact, whereas positive polarity pulses only caused minor damage near the edge of the gate.

The multiple pulse experiments confirmed expectations. Pulsing at +300V caused no significant change in the I_{ds} vs V_{ds} characteristics, while pulsing at +500V showed only a gradual degradation. Visible damage could be observed after around eight pulses. No significant degradation in the I_{ds} vs V_{ds} characteristics was observed for pulses at -50V, while pulsing at -100V caused degradation effects similar to those observed for positive polarity pulsing at +500V. After the first pulse, damage occurred between the gate and source, with I_{dss} and g_m severely degraded (Category 2). Subsequent pulses caused further degradation in g_m alone eventually leading to a complete loss of gate control. It is clear that MESFETs are sensitive to ESD and also exhibit a "walking wounded" degradation of which device quality managers should be aware. The visible failure site was, in 95% of recorded cases, between the gate and source contacts and, as a result, most of the remaining experiments (and all of the modelling) were conducted on the gate source MES diode, with the drain and substrate floating.

10.3.6 Temperature effects [13,19]

A detailed study of the degradation of MESFET parameters was undertaken; devices were stressed over range of voltages (-150V, -100V, 0V (i.e. the control group), 100V, 300V, 500V, 900V) at 25°C and 100°C. In addition to I_{dss}, g_m and V_p, several other parameters were tested for signs of degradation. These included the source resistance R_s, the ideality factor η_f, the built-in voltage V_{bi} and the channel resistance R_{on}. At the lower positive voltages at 25°C, small changes appeared in many of the parameters, but at the higher voltages a definite decrease in the source resistance R_s is observed. At 100°C a very large increase in the channel resistance R_{on} occurs, much greater than observed at 25°C. This leads to significant decreases in I_{dss} and g_m.

Negative polarity pulses at -100V degraded the majority of devices out of specification, but the devices which remained measurable had much lower values of R_s. At 100°C more devices remained in specification after a negative polarity pulse of -100V than at the lower temperature. In general, these results show that an increase in the ambient temperature increases the likelihood of degradation during a positive polarity pulse and decreases the burnout probability during a negative polarity pulse.

Changes also occurred in the gate-source diode characteristics of the FET. For a working FET, in forward-bias, the current increases exponentially above 0.7V. This is followed by a linear region until ≈2V when the current begins to saturate until the device finally does breakdown at voltages >4.5V. In reverse-bias a small reverse current flows until the device enters the avalanche region at $V=-V_{br}$, as evidenced by white light emission from the edge of the gate. The device will finally exhibit irreversible breakdown at a reverse current of between $\approx$-1mA, at high values of $-V_{br}$, and $\approx$-5mA, at low values. As the ambient temperature was increased from 25°C to 150°C, the magnitude of the forward bias current limiting region in the I_{gs} vs V_{gs} characteristic decreased by $\approx$10mA at V_{ds}=5V. This decrease is qualitatively consistent with the changes in saturation velocity as a function of temperature [20]. The source resistance also increased with the temperature probably caused by the decrease in the electron mobility at the higher temperatures [1].

The reverse avalanche voltage $-V_{br}$ also increased with temperature and the change is also more pronounced at 150°C. This change is qualitatively consistent with the decrease in impact ionisation coefficients with increasing temperature [21]. In addition, the reverse leakage current increases with temperature. The effect can be understood by considering the effects of Schottky barrier lowering on the leakage current through a reverse bias depletion layer [1].

10.3.7 ESD thresholding in MES diodes [19,22,23]

The relatively large currents flowing under these high voltage transients pushes a substantial power through the small devices. The resultant Joule heating effect is likely to be the cause of most of the observed failures. This is backed up by the temperature results of the preceding section. As a result some ESD thresholding experiments were carried out on the MES gate-source diodes with a view to finding a link between the power input to the device, due to the ESD stressing voltage, and the time to failure. This link is developed further in the modelling Section 10.4. An important precursor to an understanding of how the diode deals with a (transient) ESD pulse is the constant voltage EOS thresholding.

10.3.8 Constant overstressing (EOS) of MES diodes [19,24,25]

The devices were subject to a power pulse, constant to within ±0.1W, and of duration up to 25μs. Evidence of thermal failure appeared as a rapid rise in the current/power profile. The time at which this occurs is taken to be the thermal failure time t_f. To estimate the effects of temperature the experiment was carried out at 25°C, 100°C and 175°C. The <u>input power/time to failure</u> relationship for these temperatures is

shown in Figure 10.11. There appeared, when viewed through the microprober, to be two types of <u>visible</u> failure. These appear to be related to
the point in the power pulse at which failure occurs. If the device fails
sufficiently early in the pulse, the rapid power increase causes severe
damage and the failure site is black in colour, Figure 10.12. Conversely,
if the device fails well into the pulse little power remains to do damage
and the failure site is white in colour.

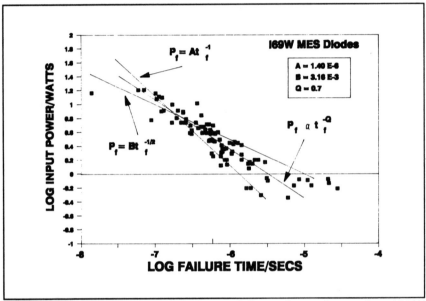

**Figure 10.11 Failure power/time to failure dependence for 169W
wafer.**

10.4 Modelling

10.4.1 Oxide breakdown [10,17,18,26]

We have observed previously that the thin gate oxide is likely to be the
most ESD sensitive part of an MOS device, and that dielectric breakdown
is initiated by the tunnelling of electrons into the oxide. While there is no
doubt that there are oxide leakage currents, it is still not clear exactly
what happens to the carriers once they are in the SiO_2 conduction band.
There is no doubt also that the oxide becomes positively charged
following electron injection [6,10] as is evidenced by the shift in the C-V
curve following injection. This has led to the idea that, once in the oxide

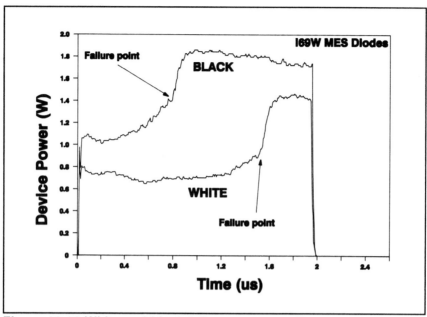

Figure 10.12 White and black failure power profiles.

conduction band, the electrons are accelerated in the stressing field, until they can generate electron-hole pairs via band-to-band impact-ionization (BBII). The holes drift back towards the injecting cathode under the applied field where they are trapped. This creates a positive charge close to the cathode which lowers the barrier to incoming electrons. The resulting positive feed-back soon leads to breakdown [27]. There have been objections to this model in the literature [28,29], mainly based on the fact that, because SiO_2 is a wide gap insulator ($\approx 8.8eV$), BBII is a relatively unlikely process. An alternative description of oxide breakdown is provided by the cumulative erosion models in which electrons already trapped in the bulk oxide enhance the local electric field. New traps can be created by two mechanisms. For sufficiently strong fields, the Si-O bonds may be broken creating new traps or, alternatively, the energy released in trapping may be used to create more traps. These trap sites then link together, eventually forming a conducting path, in the form of a tree, from anode to cathode [6,28,29]. The objections to the BBII model are usually levelled at low voltage breakdown [29]. However, as the BBII model, used here for large voltage stress, is simple and explains most of the observed breakdown phenomena, it is quite adequate as a working model.

    In an MOS device, under negative polarity stress, the polysilicon gate represents the injecting electrode and, as it is heavily doped, there is a ready supply of free electrons at the oxide interface. The negative polarity ESD failure threshold voltage is therefore just that voltage required to cause oxide breakdown.

Under positive polarity, electrons enter the oxide only through the bulk silicon. For MOS capacitors with a p-type substrate there are few electrons at the injecting interface until an inversion layer has formed. Inversion will only occur if an electron avalanche takes place in the bulk as it is relatively straight forward to show that thermal electrons cannot be generated sufficiently quickly to replenish the tunnelling electrons, thus preventing a voltage build-up across the oxide. As a result, the positive polarity ESD failure threshold voltage is given by the sum the oxide breakdown voltage and the bulk avalanche voltage. This simple observation explains the success of Equation 10.1 in describing ESD breakdown. The above explanation is confirmed by the facts that n-substrate capacitors show no polarity dependence and that the capacitors with n-type bulk have lower breakdown thresholds than those with p-type. Figure 10.8 has the appearance of a constant energy contour, i.e. $\frac{1}{2}C_B V_o^2 \approx$ constant. This would seem to point to a cumulative (possibly thermal) failure mechanism, but, as we shall see, this interpretation is incorrect.

Within the BBII model substantial lowering of the potential barrier to tunnelling is the critical point in oxide breakdown. This occurs only when a certain positive (hole) charge density Q_p^* is trapped in the oxide. In the constant current stressing experiments of Wolters et al [6] a critical level of injected electron current J_{cr} was identified. Applying this to the ESD case, for injection currents below J_{cr} the charge required for breakdown Q_{bd} (i.e. the injected charge required to produce a total trapped hole charge of Q_p^*) generally exceeds the limited charge available in the ESD system and breakdown cannot occur. For injection currents above J_{cr}, however, the charge required for breakdown is far smaller and breakdown is easily possible.

It is possible to see how this relates to our devices as follows. The tunnelling current density J_{ox} is given by the Fowler-Nordheim equation

$$ J_{ox} = kF^2 e^{-\frac{\beta}{F}} \propto J_o e^{-\frac{\beta}{F}} \qquad\qquad [10.2] $$

where F is the electric field at the cathode, and J_o can be regarded as constant. For an oxide of thickness T_{ox}, the impact ionization coefficient (i.e. the probability that an electron, having entered the oxide, causes BBII) is given by [30]

$$\alpha = \alpha_0 \, T_{ox} \exp\left(\frac{-H}{F}\right)$$ [10.3]

where α_o and H are constants. The total injected charge Q_{bd} required to support breakdown is given by

$$Q_{ox} = A \int_0^{t_{bd}} J_{ox} \, dt$$ [10.4]

where A is the oxide area and t_{bd} is the time to breakdown. Consequently the total trapped positive charge in the oxide at time t is given by [27]

$$Q_p(t) = \int_0^t \eta \, \alpha \, J_{ox} dt = \eta \, \alpha_o T_{ox} \exp\left(\frac{-H}{F}\right) \int_0^t J_{ox} dt$$

[10.5]

where η is the trapping efficiency. We have assumed here that the field F remains essentially constant across the oxide. This can be justified a posterei in that Fowler-Nordheim tunnelling only begins at around 37V and constant voltage breakdown occurs at around 40V, and also from the voltage pinning at around 40V seen in the capacitor waveforms. The final integral in Equation 10.5 represents the amount of charge which has been injected into the oxide by time t. At breakdown, $Q_p(t_{bd})=Q_p^*$ and $Q_{ox}=Q_{bd}$. Thus rearranging and solving for Q_{bd} we obtain [18]

$$Q_{bd} = \frac{AK}{T_{ox}} \exp\left(\frac{H}{F}\right) = \frac{AK}{T_{ox}} \exp\left(\frac{HT_{ox}}{V_{ox}}\right)$$ [10.6]

where V_{ox} is the voltage across the oxide. The oxide voltage may be obtained from

$$V_{ox}(t) = V - R_b\left(I - C_{sys}\frac{dV}{dt}\right) \qquad [10.7]$$

where C_{sys} represents the combined capacitance of the transmission line the probes and the microprober obtained using a Wayne-Kerr 4210 LCR bridge and V and I are the voltage and current measured by the probes.

Equation 10.6 shows how an oxide, under a relatively low stress voltage, can withstand considerable charge injection prior to breakdown. As the voltage is increased, the injected charge required to support breakdown rapidly rises. This is crucial for ESD stressing since the charge in the pulse is relatively small even for fairly high voltages. Consider an ESD pulse at a voltage V_o. Simple circuit theory gives an effective pulse voltage V_{eff} felt by the device to be

$$V_{eff} = \frac{V_o C_1}{C_1 + C_{ox} + C_t} \qquad [10.8]$$

where C_{ox} and C_t are the capacitances of the oxide and the pulse transmission system. Breakdown can only occur if the available charge $C_B V_o$ exceeds Q_{bd}. Figure 10.13 shows a plot of both the available charge and the charge required for breakdown against V_{eff} for two realistic values of C_B (100pF and 200pF). It is clear that, almost independent of the value of the stressing conditions, the breakdown threshold is around 40V. Consequently, although it appears that Figure 10.8 represents a constant energy contour, in each case only around 40V is dropped across the oxide, the remainder falling across the discharge capacitor and the transmission system capacitance.

During constant stressing of an n-MOS capacitor there are plenty of electrons ready for injection into the oxide, whether from the highly n-doped gate (-ve gate stress) or the n-type substrate (+ve gate stress). As a consequence the same breakdown mechanism described above for ESD breakdown will apply here and consequently no temperature dependence is observed. This will also be the case for negative gate stressing of p-MOS capacitors. In the final case, positive stressing of the p-MOS capacitors, a different, temperature dependent, mechanism enters. There is now plenty of time for the usual mechanism of thermal carrier generation to invert the substrate-oxide surface and sustain a voltage across the oxide. Consequently an electron avalanche is not

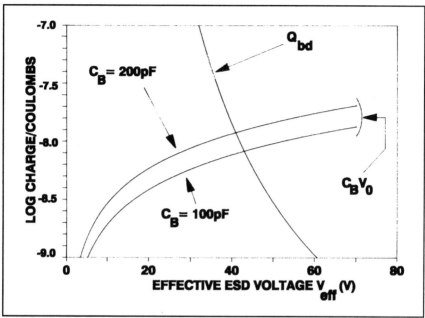

Figure 10.13 Negative polarity ESD breakdown criterion for MOS capacitors.

required. It appears that a second, temperature dependent, breakdown mechanism which also becomes activated at around 40V, but which occurs on a much longer time scale is causing breakdown here.

The above results were confirmed by a similar series of experiments on silicon-on-sapphire FETs of the same oxide thickness. Again an oxide failure voltage of 40V was obtained.

Further work on positive polarity stressing is continuing at Loughborough.

10.4.2 MES diode burnout [19,22-25]

We have identified the problem of a thermal runaway, caused by Joule heating, as the most likely form of breakdown in the (gate/source) MES diodes. Consider first the case of constant (EOS) stressing of such a device at an ambient temperature T_a. It is likely that the current distribution in the diode is non-uniform, so that some portion of the channel will heat up more quickly than the rest. One assumes that when the "defect" reaches some critical temperature T_c the device fails. T_c may be a melt, dissociation or other critical temperature.

Several analytic models have been developed in the past [31-33] to describe overheating of a reverse biased silicon p-n diode. In these models the defects have always been assumed to be high symmetry shapes. Wunsch and Bell [31] took the defect, the region dissipating heat, to be a infinite two dimensional plane, representing the depletion junction of the diode. In this case, as we shall see below, it is possible to show that the relationship between the power required to cause damage $P_f(t_f)$ and the time to failure t_f is

$$I \times V \equiv P_f(t_f) = \frac{B}{\sqrt{t_f}} \qquad [10.9]$$

where B is known as the Wunsch-Bell coefficient.

Tasca et al [32] assumed that the device defect was spherically symmetric, with the resulting relation

$$P_f = \left(\frac{A_1}{t_f} + \frac{B_1}{\sqrt{t_f}} + C_1 \right). \qquad [10.10]$$

Although strictly not an accurate interpretation of Equation 10.10, it has been customary to divide the time domain into three regions. For small failure times, typically below 10ns, the surface of the defect loses little heat, so that an adiabatic (t^{-1}) term dominates. For very large failure times, typically above 100μs, and after thermal equilibrium is established, the constant or steady state term dominates. During the intermediate region a Wunsch-Bell dependence is observed.

Arkhipov et al [33] tested this model against a "needle shaped" defect, in the form of an infinite cylinder. In this case one obtains

$$P_f = \frac{A_2}{\log_e (t_f / t_o)} \qquad [10.11]$$

for some characteristic time t_o. In practice this logarithmic dependence is seldom reported. Usually something of the form

$$P_f = const. \times t^{-Q} \qquad [10.12]$$

is fitted, where Q is generally close to 0.5. The constants in these equations depend on the thermal parameters of the material (thermal conductivity - K, density γ and the specific heat capacity C_p) and on the increase in temperature required to cause failure, i.e. T_c-T_a. It is clear that, under these assumptions, thermal failure of a protection diode will introduce an ambient temperature (T_a) dependence into the observed failure pattern of protected MOSFET devices which is absent from the unprotected failure reported above.

At Loughborough we have been working on a more general model of diode failure [19,22-25], aimed particularly at our MES diodes, but equally applicable to other devices. This model attempts to explain both pulsed (especially ESD) and constant stress (EOS) failure. There have been some attempts in the past to translate the constant power results to ESD stress [34-36] but these exclusively assume a Wunsch-Bell dependence for the constant power stress. Speakman [34] replaces the ESD pulse by an equivalent constant power pulse. However, because there is no unique equivalent, the method gives no unique solution. As below, Tasca et al [35] and Pierce et al [34] use the Duhamel theorem to supply the translation from EOS to ESD. These methods are numerically rather complex whereas we have tried to retain the simplicity of the Speakman model. Our model is also more general geometrically in that it does not rely on the high symmetry shapes (plane, sphere, cylinder) used before [31-33].

We begin with a defect volume Δ through which the majority of the gate-source current is being channelled. The resultant Joule heating is assumed to be generated uniformly within Δ, this assumption is convenient but not essential. Thus a localised source $P(t)=V(t)I(t)$ generates heat within Δ. When the centre of the defect (the hottest point) reaches the critical value T_c breakdown occurs. It is convenient to define the defect volume in terms of three perpendicular dimensions taken in the order $c \leq b \leq a$, these may be thought of as being the three principal axes of the defect. The defect in the MES diode will be that part of the channel which is carrying current [24]. Thus, if some substrate conduction is taking place, c will be slightly greater than the channel depth ($\approx 0.16\mu m$ in this case); if we allow for current flowing underneath the ohmic contacts, b will be slightly greater than the gate-source distance ($1\mu m$); and a will be some portion of the ($150\mu m$) gate width. A schematic of the defect in the MES diode is shown in Figure 10.14.

The temperature inside the defect will be given by the solution of the heat transfer equation [37]

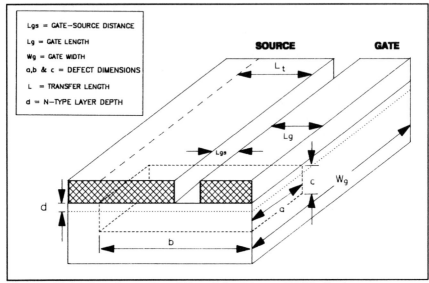

Figure 10.14 A schematic of the defect in an MES diode.

$$\frac{\partial T}{\partial t} - D\nabla^2 T = \left\{ \begin{array}{ll} \left(\dfrac{q(t)}{\rho C_p}\right) & \text{if } \underline{r} \in \Delta \\ \\ 0 & \text{if } \underline{r} \notin \Delta \end{array} \right\} \qquad [10.13]$$

where D ($=K/\gamma C_p$) is the thermal diffusivity and q(t) is the rate of heating per unit volume ($=V(t)I(t)/\Delta$). For constant power P, q(t) is also independent of time ($=P/\Delta$). By specifying the appropriate geometry and employing the Green's function method, Equation 10.13 is readily solved to obtain Equations 10.9-10.11 [24]. To simplify the analysis here we shall assume that the defect volume is a rectangular parallelepiped. In this case Equation 10.13 becomes separable for the three defect dimensions so that, setting T($\underline{0}$,t) to the critical temperature T_c, we obtain the required relation between P_f (the value of the input power at which the device fails) and the failure time t_f [24]

$$\frac{1}{P_f} = \frac{1}{\rho C_p \Delta (T_c - T_a)} \int_0^{t_f} erf\left(\frac{a}{4\sqrt{D\tau}}\right) erf\left(\frac{b}{4\sqrt{D\tau}}\right) erf\left(\frac{c}{4\sqrt{D\tau}}\right) d\tau.$$

[10.14]

At this point it is convenient to define three diffusion times associated with the lengths a, b and c as follows

$$t_a = \frac{a^2}{4\pi D}, \quad t_b = \frac{b^2}{4\pi D}, \quad t_c = \frac{c^2}{4\pi D}.$$ [10.15]

These represent approximate times for thermal equilibrium to be established in the directions of the respective principal axes. The three times t_c, t_b and t_a divide the time domain into four regions I-IV. Inside region I (i.e. $t \leq t_c$), to a good approximation, all the error functions are approximately equal to one and Equation 10.14 easily integrates to give an adiabatic dependence [24]

$$P_f = \frac{A_2}{t_f}.$$ [10.16]

In region II (i.e. $t_c \leq t \leq t_b$) the error function in c departs from one, becoming approximately $(t_c/t)^{1/2}$. Integrating Equation 10.14 yields [24]

$$P_f = \frac{B}{\sqrt{t_f} - \sqrt{t_c}/2}.$$ [10.17]

Note that as c tends to 0, the Wunsch and Bell dependence (Equation 10.9) is regained. At still larger times, in the range $t_b \leq t \leq t_a$ (region III), the error function in b becomes approximately $(t_b/t)^{1/2}$. In this case Equation 10.14 integrates to a form which is similar to Equation 10.11

$$P_f = \frac{A_2}{\log_e(t_f/t_b) + const.}.$$ [10.18]

Finally in region IV, for times greater than t_a, the power required to cause failure rises to an asymptotic steady state value

$$P_f = \frac{D}{const. - 2\sqrt{\dfrac{t_a}{t}}}. \qquad\qquad [10.19]$$

The point here is that, in general, any defect displays all four different types of time dependence. The analysis of previous workers, where only one or two time dependences are shown, is the (high symmetry) exception rather than the rule [24]. The times at which the failure-time dependence of the input power changes may be used to estimate the size of the defect dimensions c, b and a. In the case of the MES diodes, using the melting point of GaAs (1511K) as the critical temperature, one obtains $c \approx 1.5\mu m$, $b \approx 6.9\mu m$ and $a \approx 30.7\mu m$. Of these values the least realistic, and the one most probably in error, is c. The value given would imply significant substrate conduction (the channel is only $0.16\mu m$ deep), and, although this has been suggested in the past, the value still appears a little high. Unfortunately the breakdown times which determine c, less than 100ns, are very difficult to obtain accurately [25].

The analysis presented above can be readily extended to the ESD situation. During an ESD event the current waveform is exponential to a very good approximation. However, due to parasitic elements in the discharge circuit the time constant associated with the decay, τ, is closer to 260ns than the expected 150ns ($100pF \times 1500\Omega$). Under positive polarity ESD stress, reconstructing the I-V characteristic from the current and voltage waveforms I(t) and V(t), it is clear that the device behaves somewhat like a diode with junction voltage $V_j = 5V$ with a series resistor of around $R_D = 5\Omega$, Figure 10.15. This can also be seen from the voltage waveforms, Figure 10.16. The unfailed device voltage pins at around 5V for some time. It is interesting to note that, while the current is decaying exponentially, the voltage often shows a recovery. This rising voltage with decaying current indicates that (under this positive polarity stressing) the device enters a region of negative differential resistance (NDR) [20].

A complete analysis based on the above is possible [24], however, for simplicity, let us assume that the failures occur within the Wunsch-Bell region, Equation 10.8. The power in the pulse will be given by

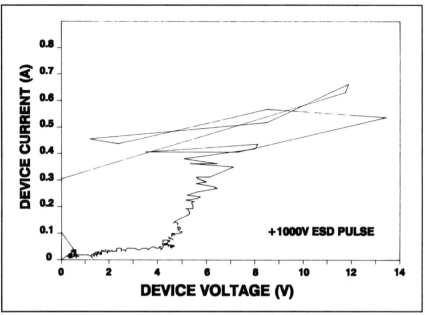

Figure 10.15 Reconstructed gate-source MES diode I-V characteristic.

$$P(t) = V_j I(t) + R_D I^2(t)$$

$$= V_j I_0 \exp\left(\frac{-t}{\tau}\right) + R_D I_0^2 \exp\left(\frac{-2t}{\tau}\right)$$

[10.20]

where I_0 is the peak current

$$I_o = \frac{V_o - V_j}{R_D - R_B} ,$$
[10.21]

$$\tau = (R_D + R_B) C_B$$
[10.22]

and V_o is the ESD voltage.

It can be shown [22] that I_o is only physically meaningful in a small time window between 0.427τ and 0.854τ, where τ is the pulse decay time. This is in contrast to the assumed failure time of 5τ set by Speakman [34] and encompasses the numerical value of Pierce et al

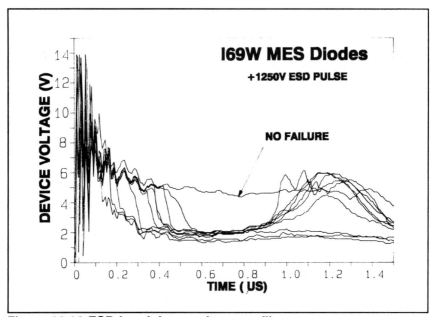

Figure 10.16 ESD breakdown voltage profile.

[36]. As there is little variation in the power required to cause breakdown within this window, it is possible to obtain approximate values for the threshold failure current and threshold failure time as [22]

$$I_{th} = \frac{V_J}{\sqrt{2}R_D}\left[\sqrt{1 + \frac{2\sqrt{2}\dfrac{B}{\sqrt{\tau}}}{0.521\dfrac{V_J^2}{R_D}}} - 1\right] \qquad [10.23]$$

and

$$\sqrt{\frac{t_f}{\tau}} = \frac{1}{2\,(0.521)}\left[\frac{V_J + I_{th}R_D}{V_J + \sqrt{2}I_{th}R_D}\right] \qquad [10.24]$$

respectively.

Using the value of the Wunsch-Bell parameter calculated from the EOS experiments, it is possible to estimate theoretical values for the threshold current and failure time. For our MES diodes the appropriate values are I_{th}=1.3A and t_f=200ns [23]. (Note these values are, in fact, obtained with the more accurate version of this analysis, given in reference [22], rather than the simple Wunsch-Bell formula).

An experiment involving 40 devices was performed (10 devices at 1000V, 20 devices at 1250V and 10 devices at 1500V) using the "human body model" circuit. At 1000V no devices failed catastrophically. At 1250V, 9 out of the 20 devices failed (six showed visible white failures, the remaining three displayed black failures). At 1500V all the devices failed in less than 100ns indicating a pulse above threshold. The threshold voltage may be assumed to be around 1250V. According to Equation 10.21 this corresponds to a peak current of 0.83A. The observed times to failure were all between 200 and 350 ns [23]. The theoretical values are in still in significant error of the experimental values, there are many reasons why this might be so. ESD is a current controlled phenomenon whereas EOS is voltage controlled and the negative differential resistance behaviours seen in the I-V plot, Figure 10.15, of these are different. Thus, it is not immediately obvious that, for example, the Wunsch-Bell coefficient derived from voltage controlled EOS will be the most appropriate value for current controlled ESD.

Finally, the voltage profiles for ten of the devices pulsed at 1250V are shown in Figure 10.16. These include all nine which failed and one which did not (for reference). The six devices which failed with a visible white burnout site all show a recovery in their voltage waveform, while the three which failed with a black burnout site showed no such recovery. In the EOS experiments white failures were associated with failure towards the end of the pulse, while the black failures occurred relatively early on, Figure 10.12. In addition, white failures did not show serious I-V degradation when tested subsequent to the pulse and also if pulsed a second time at the same voltage a second white region generally appeared. These observations strongly suggest that a white failure is associated with the development of an open circuit path between the gate and source through the defect, while a black failure corresponds to short circuit path through the defect.

10.5 Conclusions

(1)   MOSFETs and MESFETs are highly sensitive to EOS and ESD stressing, with commercial MOSFETs showing lower breakdown voltages (50V - 300V) than their MESFET counterparts (350V - 900V). It is also clear that the FETs show an ESD voltage and polarity dependence.

(2)   Both devices show a systematic degradation of their I-V characteristics prior to breakdown, this is confirmed by multiple pulsing experiments. It is quite possible that devices which have suffered a relatively low voltage ESD pulse could pass a quality control while at the same time being sufficiently damaged to fail within their working life. The existence of "walking wounded" devices and the possibility of latent failures are important conclusions for quality managers.

(3)   There appear to be two mechanisms which causes the oxide breakdown under EOS and ESD both of which become important at an oxide stress voltage of around 40V. The particular mechanism which causes a breakdown depends only on the supply of electrons at the injecting surface. When the injecting electrode is n-doped the relatively fast acting, temperature independent Fowler-Nordheim tunnelling causes breakdown. If the injecting electrode is p-doped, under constant EOS stressing, the oxide surface is inverted thermally before electrons may enter the oxide. This process appears to be relatively slow and the breakdown process which follows is temperature dependent. In the ESD case the pulse is two short lived for the injecting electrode to become inverted by thermal electrons and an avalanche in the bulk is required before Fowler-Nordheim tunnelling may begin.   The temperature dependent EOS breakdown mechanism does not become important until the pulse length exceeds around 10ms.

(4)   A clear gate dimension dependence is seen in the ESD results which arises primarily because larger area oxides have more defects to trap holes, this results in an increased trapping efficiency.

(5)   In the case of p-substrate capacitors, positive and negative polarity ESD breakdown thresholds only differ by the constant avalanche voltage required to invert the p-Si substrate surface. The negative polarity "constant energy contour" is coincidental. For realistic ESD breakdown, a constant voltage of around 40V must fall across the oxide.

(6)   The systematic degradation of the MESFET structures shows a strong temperature dependence. The reduction in I_{dss} and g_m caused by an ESD pulse is enhanced as the temperature is increased. The breakdown mechanism is likely to be Joule heating failure of the gate-source diode.

(7)   MES diodes and protection circuit diodes show ESD and EOS temperature sensitivity. A thermal model has been developed to predict the failure of such devices. The model provides a natural framework for

both constant and pulsed power stressing. It also provides the correct interpretation for previous work in this area and yields simple analytical expressions for the ESD breakdown thresholds. Two types of failure mode are established (white and black) which are seen to arise due to the creation of open circuit and short circuit paths (respectively) through the defect.

(8)   Under positive gate stressing the MES diodes also exhibit near classical current controlled negative differential resistance. This phenomenon gives rise to the current filaments (lumped to form the defect) which heat the device.

10.6 Acknowledgements

The authors are happy to acknowledge and thank Drs. E. A. Amerasekera and A. J. Franklin, and M. J. Tunnicliffe for their invaluable contributions to this work. The work was carried out with the support of the Procurement Executive (ECG), Ministry of Defence U.K. We would like also to thank Dr. J. Woodward (of RSRE, Malvern) and Mr. D. Shore (of ECG) for many valuable discussions on these subjects, and STC Technology Ltd. (Harlow) U.K., Plessey Research Ltd. (Caswell) U.K. and GEC-Marconi Ltd. U.K. for supplying us with wafers.

10.7 References

1   S.M. Sze, Physics of semiconductor devices, John Wiley, New York 1969.

2   R.S.C. Cobbold, Theory and applications of field effect transistors, John Wiley and sons, New York 1970.

3   E.A. Amerasekera and D.S. Campbell, Failure mechanisms in semiconductor devices, John Wiley and sons, London 1987.

4   P.R. Bossard, R.G. Chemelli and B.A. Unger, Proc. EOS/ESD 2, 17 (1980).

5   T.T. Lai, IEEE Trans. Comp. Hyb. and Man. Tech. CHMT-12, 627 (1989).

6   D.R. Wolters and J.J. van der Schoot, Phillips J. of Res., 40, 115 (1985); 40, 137 (1985); 40 164 (1985).

7   C.L. Huang, F. Kwan, S. Y. Wang, P. Galle and J. Barrera, Proc. 17th. Ann. Rel. Phys. Symp., paper 5.1, 1979; R.E. Lundgren and G.O. Ladd, Proc. 16th Ann. Rel. Phys. Symp., pp. 255-259, 1978.

8   MIL-STD-833C, Method 3015.2, Electrostatic Discharge Sensitivity Test, 1983, p. 1.

9   G. Kosonocky, H. Pon and T. Maloney, Proc. EOS/ESD 11, 78 (1989); L. van Roozendaal, E.A. Amerasekera, P. Bos, P.Ashby, W. Baelde, F. Bontekoe, P.kersten, E. Korma, P.Rommers, P. Krys and U. Weber, EOS/ESD 12, (1990) to be published.

10   E.A. Amerasekera, PhD thesis, Loughborough University of Technology, 1986.

11   E.A. Amerasekera and D.S. Campbell, QRE Inter. 2, 107 (1986).

12   E.A. Amerasekera and D.S. Campbell, Solid. St. Electr. 32, 199 (1989).

13   A.J. Franklin, E.A. Amerasekera and D.S. Campbell, Proc. ERA Seminar on ESD in electronics, London, 1986, pp. 4.3.1 -4.3.14.

14   A. Hart, T.T. Teng and A. McKenna, Proc. IRPS 18, 190 (1980).

15   E.A. Amerasekera and D.S. Campbell, Proc. RELCON 1986, 325.

16   E.A. Amerasekera and D.S. Campbell, Proc. ESSDRC 17, 733 (1987).

17   E.A. Amerasekera and D.S. Campbell, EOS/ESD Proc. 8, 208 (1986).

18   M.J. Tunnicliffe, V.M. Dwyer and D.S. Campbell, Proc. CERT conf. Gatwick U.K. (1990).

19   A.J. Franklin, PhD thesis, Loughborough University of Technology, 1990.

20   J.P.R David, J.E. Sitch and M.S. Stern, IEEE Trans. ED-29, 1548 (1982).

21   G.A. Baraff, Phy. Rev. 128, 2507 (1962).

22   V.M. Dwyer, A.J. Franklin and D.S. Campbell, IEEE Trans. ED. Submitted for publication.

23   A.J. Franklin and V.M. Dwyer, Proc. CERT conf. Gatwick U.K. (1990).

24   V.M. Dwyer, A.J. Franklin and D.S. Campbell Solid. St. Electr. 33, (1990) 533.

25   A.J. Franklin, V.M. Dwyer and D.S. Campbell Solid. St. Electr. In press.

26   M.J. Tunnicliffe, V.M. Dwyer and D.S. Campbell, EOS/ESD Proc. 12 (1990), to be published.

27   I-C. Chen, S.E. Holland and C. Hu, IEEE Trans. ED-32, 413 (1985).

28   P.B. Budenstein, IEEE Trans. EI-15, 225, 1980.

29   E. Avni and J. Shapir, J. Appl. Phys. 64, 743, 1988.

30   M. Knoll, D. Braunig and W.R. Fahrner, J. Appl. Phys. 53, 6946 (1982).

31 D.C. Wunsch and R.R. Bell, IEEE Trans. NS-15, 244 (1968).

32 D.M. Tasca, IEEE Trans. NS-17, 364 (1970).

33 V.I. Arkhipov, E.R. Astvatsaturyan, V.I. Godovitsyn and A.I. Rudenko, Int. J. Electronics. 55, 395 (1983).

34 T.S. Speakman, Proc. IRPS 12, 60 (1974).

35 D.M. Tasca, J.C. Peden and J.L. Andrews, GEC Missile and Space Division, Doc. No. 735D4289, 1973.

36 D.G. Pierce, EOS/ESD Proc. 9, 51 (1987);D.G. Pierce, W. Shiley, B.D. Mulcahy, K.E. Wagner and M. Wunder, EOS/ESD Proc. 10, 137 (1988).

37 H.S. Carslaw and J.C. Jaeger, Conduction of Heat in Solids, 2nd ed. (Oxford University Press, London, 1959).

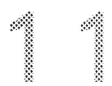

CLEANING SURFACE MOUNT ASSEMBLIES: THE CHALLENGE OF FINDING A SUBSTITUTE FOR CFC-113

J. KIRK BONNER

11.1 Introduction

Up to now, the majority of cleaning solvents used in the electronics industry have been formulated using CFC-113 (1,1,2-trichloro-1,2,2-trifluoroethane: CCl_2FCClF_2). This is because CFC-113 has many truly excellent properties. Chief among these are its outstanding stability and compatibility with almost all materials used in the electronics industry, low toxicity (i.e., high threshold limit value or TLV, for example, $TLV_{CFC-113}$ = 1000), nonflammability, and ability to azeotrope (or boil as a compound at a single temperature) with other ingredients -- especially lower molecular weight alcohols and other select ingredients -- thus enhancing its solvency power for contaminant residues.

One particular azeotrope consisting of CFC-113, methanol, and nitromethane (b.p. 104°F) has been employed in the electronics industry for over sixteen years. Another azeotrope based on CFC-113, methanol, acetone, isohexane, and nitromethane (b.p. 104°F) has exhibited outstanding cleaning performance for cleaning printed wiring assemblies (PWAs) and surface mount assemblies (SMAs) [1,2].

11.2 Ozone Depletion - Greenhouse Warming Problem

There is, however, accumulating scientific evidence that chlorofluorocarbons (CFCs) are potentially damaging to the layer of ozone (O_3) in the stratosphere [3,4 p. 235]. The stratosphere is, of course, the ozone-rich layer of the earth's atmosphere beginning roughly six to ten miles above the earth's surface and extending to about a height of thirty miles. In March of 1988, a distinguished panel of world atmospheric scientists concluded that the amount of total column ozone in the northern hemisphere had been depleted by 1.7 to 3.0% since 1969, and that man-made chlorine species, most likely from CFCs, were strongly suggested to be a chief contributor to this decline. Undoubtedly the excellent stability of CFCs contributes to this problem since they are not subject to breakdown until they reach the stratosphere.

Atmospheric chemistry processes are rather involved. There is some evidence that heterogeneous chemical reactions may be responsible for ozone depletion problems over the polar regions (primarily Antarctica) during the unique conditions of the polar winter. This means that the chiefly destructive reactions do not occur in one phase, for example, the gaseous phase, but in several phases. In fact, particles in the stratosphere are believed to provide surfaces on which these chemical reactions involving the destruction of ozone take place [5,6].

It is almost certain that active chlorine is the main culprit. It propagates a free radical chain reaction whereby one active chlorine-containing fragment catalyzes the destruction of about one hundred thousand ozone molecules before the reaction is finally terminated. In the case of the Antarctic ozone hole, it appears that the chlorine, which may come initially from man-made chlorine-containing species, such as CFCs and also 1,1,1-trichloroethane (CH_3CCl_3), is decomposed into inactive reservoir species such as hydrochloric acid (HCl), and chlorine nitrate ($ClONO_2$). Laboratory studies have shown that polar stratospheric clouds (PSCs) convert photochemically inactive species such as HCl and $ClONO_2$ into active forms such as chlorine monoxide (ClO) which are capable of destroying ozone molecules. When the temperature drops below a certain point, these reservoir species are entrapped in frozen cloud particles. In the Antarctic spring the reservoir compounds start vigorously releasing active chlorine which then attacks the ozone [7]. The impact of the heterogeneous processes on both polar region and global ozone diminution is not fully understood at this time. There is much that must be learned regarding this critical aspect of atmospheric chemistry.

CFCs have been suggested as a source of chlorine in the stratosphere because of their excellent stability. It takes years for them to diffuse to the stratosphere. Other than defluxing printed circuit boards and degreasing applications, CFCs are also used in refrigeration and air conditioning applications as sterilant gases and as foam blowing agents. These findings in stratospheric ozone depletion have prompted the search for new molecules to replace today's CFCs in all of these applications. In this paper several new molecules will be discussed which have excellent potential for defluxing applications.

Another phenomenon, the greenhouse effect, is also very important and makes the search for new molecules more difficult. Trace gases, such as carbon dioxide, strongly absorb infrared radiation and trap the emitted energy from the earth's surface. Consequently, the earth's surface temperature is raised over what it would be if no trace gases are present. Climate modelers claim to have detected the signs of global warming and a major contributor to this are these trace gases. Along with carbon dioxide, methane and water vapor, CFCs are also considered to be contributors to the greenhouse effect. The quest for new molecules has to include their contribution to the greenhouse effect also.

11.2.1 The Montreal Protocol

Recently the ozone problem has been broached under the auspices of the United Nations.   In 1987, the so-called Montreal Protocol was proposed by the United Nations Environment Program (UNEP) to regulate CFCs (and halons) [8].   The Montreal Protocol has been ratified by a majority of signatory nations.   Basically, it says that CFC consumption (where consumption = production + imports - exports) will be fixed at 1986 levels by 1989. In 1993 the consumption of CFCs will be phased down to 80% of the 1986 level and by 1998 consumption will be cut back to 50% of the 1986 level.   However, based on further scientific findings, which will be the basis of "advanced notice of proposed rule making" (ANPRM) by the U.S. EPA, the consumption levels may be cut back even more. In addition, more severe tightening of the current regulations may be incorporated into the Montreal Protocol.

The CFC materials covered by the Protocol are: CFC-11, -12, -113, -114, and -115.   In addition, Halon-1211, -1301 and -2402 are covered.   The halon molecules contain bromine in addition to chlorine and fluorine.   The bromine radical is even more destructive towards ozone than chlorine. For example, the ozone depletion potential (ODP) of Halon-1301 is 10.0 as opposed to that of CFC-11 which is taken as a reference standard at an ODP of 1.0. The ODP of CFC-113 is 0.8. CFC-11 and CFC-12 make up about 75% of the CFCs regulated by the Protocol. For the most part CFC-11 finds use as a foam blowing agent and CFC-12 as a refrigerant.

11.3 Stratospherically Safe Fluorocarbons

New materials are being actively investigated to replace CFCs. Among these are materials which, although they have many of the properties of conventional solvents, have substantially lower ozone depletion potentials (ODPs) and lower greenhouse warming potentials (GWPs).   Because these new solvents are much less detrimental to stratospheric ozone, they are known as stratospherically safe fluorocarbons (SSFs). All of these materials contain the elements carbon (C) and fluorine (F) in the molecule. Many also contain the elements chlorine (Cl) and hydrogen (H).   Those that contain hydrogen (H), chlorine (Cl), fluorine (F), and carbon (C) combined together are known as hydrochlorofluorocarbons, or HCFCs.   Among the most significant of these HCFCs are two in particular: HCFC-141b (1,1-dichloro-1-fluoroethane) and HCFC-123 (1,1-dichloro-2,2,2-trifluoroethane). Figure 11.1 depicts the two dimensional structural formulas of these two compounds and also depicts the two dimensional structural formulas of CFC-113 (1,1,2-trichloro -1,2,2-trifluoroethane) and 1,1,1-trichloroethane.   An example of an SSF not containing chlorine (Cl) is HCFC-134a. The boiling points (°F), ozone depletion potentials (ODPs) and greenhouse warming potentials (GWPs) of the five CFCs covered by the Montreal Protocol are given in Table 11.1 in contrast with the boiling points, ODPs and GWPs of the three SSFs mentioned above and also tabulated.

Table 11.1 Environmental properties of 5 Montreal Protocol CFCs and 3 SSFs.

Solvents	B.P. °F	ODP	GWP
CFC-11	75	1.0	1.0
CFC-12	-22	1.0	2.4
CFC-113	118	0.8	1.3
CFC-114	39	0.8	4.5
CFC-115	-38	0.4	9.2
HCFC-123	82	0.02	0.02
HCFC-141b	89	0.08	0.12
HCFC-134a	-15	0.0	0.14

HCFC-141b and HCFC-123, with boiling points between 80° and 90°F, are being considered in solvent applications. HCFC-134a is presently being pursued as a prime alternate fluid for refrigeration applications in place of CFC-12.

11.4 Solubility and Cleanliness - Some Theoretical Remarks

Cleanliness of printed wiring assemblies (PWAs) and surface mount assemblies (SMAs) after defluxing depends on a number of factors. The most important of these is probably the solubility of fluxes and pastes in the solvents used. Cleanliness also depends on the penetration power of the solvent under the components, especially when the stand-off is very low, such as with leadless chip carriers (LCCs). This second effect is very important in the case of cleaning under surface mount components (SMCs). The interaction between the surface of the printed wiring assemblies having rosin and activator residues in combination with solvents also plays a part in the cleaning.

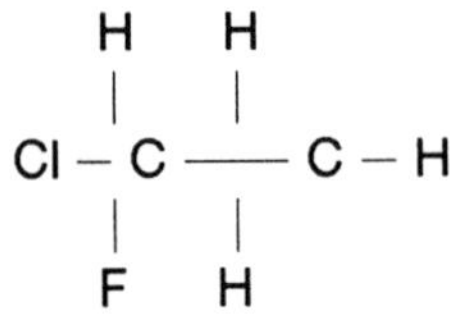

(a) 1,1-dichloro-1-fluoroethane (HCFC - 141b)

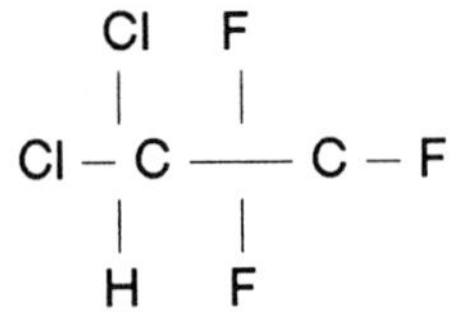

(b) 1,1-dichloro-2,2,2-trifluoroethane (HCFC - 123)

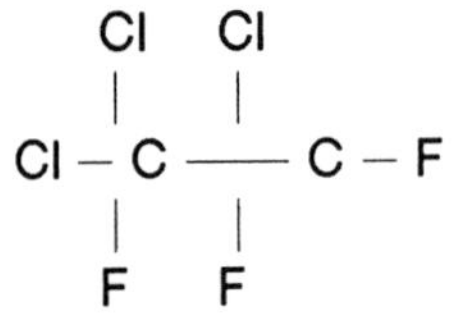

(c) 1,1,2-trichloro-1,2,2-trifluoroethane (CFC-113)

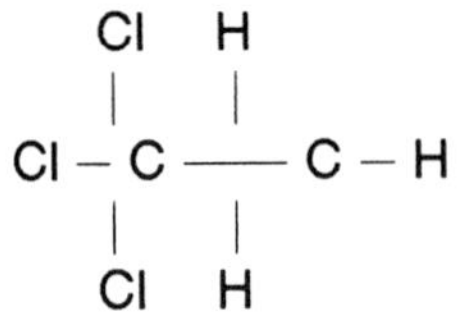

(d) 1,1,1-trichloroethane (methyl chloroform)

Figure 11.1 Selected structural formulas.

The solubility of rosin-based fluxes and pastes in the solvent will be briefly touched upon. In the case of these fluxes, it is well known that they contain primarily two distinct varieties of material. The first one is an organic component, rosin, or abietic acid,* which is soluble in a large number of organic solvents, and the second one is an ionic activator which is primarily water soluble and therefore has reasonably high solubility in lower molecular weight primary alcohols such as methanol (CH_3OH) or ethanol (CH_3CH_2OH).

Pastes are more complex because of the addition of thixotropic agents. However, in the solubility investigations the solubility of abietic acid and activators was considered primary with the assumption that the removal of these two components in the paste will assure the cleanliness of the SMAs to pass required specifications.

In industry today, cleanliness of PWAs and SMAs is judged chiefly on the basis of ionic material left on the board via the determination of ionic conductivity of the residue left on the board after defluxing. Cleanliness depends on the solubility of both rosin and activators since in many cases the activators are trapped under the rosin and are not removed unless the rosin is well dissolved. This is why the dissolution of both rosin and activators is considered here.

In order to try to understand the solubility behavior of flux components, especially the abietic acid in the new solvent blends based on HCFC chemistry, solubility parameter theory will be used. The solubility parameter model was originally developed by Hildebrand [9]. He developed what is commonly known as the theory of regular solutions and described the solubility behavior of a large number of solute-solvent systems in terms of parameters generally known as solubility parameters. The relation between solubility and solubility parameters can be rigorously defined in terms of intermolecular forces and is valid for a large number of non-electrolyte systems where the intermolecular forces are short-ranged.

A crystalline solid dissolving in a liquid may be described in the following way. Both the solid and the liquid are heated to the melting temperature of the solid, then they are mixed and cooled to room temperature. It is assumed that they remain in one phase at room temperature. In a solution formed in that manner, solubility can be expressed as (Basu et al[10])

---

*The principal rosin acid is abietic acid, but there are a number of others. Almost all rosin acids have the same chemical formula, namely $C_{19}H_{29}COOH$, but with different structural formulas. These are termed isomers.

$$-\ln x = \ln \gamma + \frac{\Delta H_f}{R}(1 - \frac{T}{T_m})$$ [11.1]

where γ is called the activity coefficient, ΔH_f is the heat of fusion, T_m is the melting point of the substance in °K (degrees Kelvin), x is the mole fraction of the solid dissolved, R is the gas constant and T is the temperature at which the solubility is desired, also expressed in °K.

In Equation 11.1, the most difficult quantity to estimate is γ, the activity coefficient of the solute in the solution. It depends on the actual interactions between molecules. In the case of regular solutions the activity coefficient γ is given by

$$RT\ln\gamma_2 = V_2^L \phi_1^2 (\delta_1 - \delta_2)^2$$ [11.2]

where V_2^L is the molar volume of the subcooled liquid, ρ_1 is the volume fraction of the solvent in the solution and the δ's are the well known Hildebrand solubility parameters given by

$$\delta = \sqrt{\frac{(\Delta H_v - RT)}{V^L}}$$ [11.3]

In Equation 11.3, ΔH_v is the heat of vaporization of the solvent or the subcooled liquid. In these equations the indices 1 and 2 stand for solvent and solute respectively. Hansen [11] found that the use of δ alone did not give satisfactory results. He found splitting δ into three parts, non-polar, polar, and hydrogen bonding, makes this parameter more usable in solubility estimations. The various contributions can be estimated and detailed lists may be found in the literature [12].

Using Hansen's parameters Equation 11.2 may be written as

$$RT\ln\gamma_2 = V_2^L \phi_1^2 [(\delta_{np}^{(1)} - \delta_{np}^{(2)})^2 + (\delta_p^{(1)} - \delta_p^{(2)})^2 + (\delta_{HB}^{(1)} - \delta_{HB}^{(2)})^2]$$ [11.4]

Here δ_{np}, δ_p, δ_{HB} stand for the non-polar, polar and hydrogen bonding parts of the solubility parameter. These are known as 3-D solubility parameters. The indices 1 and 2 again stand for the solvent and solute. For systems where the 3-D solubility parameters are not known there are methods of estimating these values. For abietic acid the 3-D solubility parameters are estimated and given in Table 11.2. Solubility parameters can also be plotted in a triangular plot where each of the axes represent one of the three fractional parts of the solubility parameter. One can define an area in the 3-D plot such that substances whose solubility parameters reside in that area can dissolve one another. One can also use Equation 11.4 to quantitatively determine the solubility. This method is widely used in the paint industry to formulate solvent blends.

Table 11.2 Solubility parameters and liquid molar volumes.

Solvent	Molar Vol.	Solubility Parameter (cal/cc)$^{1/2}$			
	(g-mols/cc)	Non-Polar	Polar	Hydrogen	Total
		(δ_{np})	(δ_p)	(δ_{HB})	(δ_t)
CFC-113	119.2	7.2	0.8	0.0	7.3
HCFC-123	117.0	6.5	2.2	1.3	7.0
HCFC-141b	116.0	6.8	2.7	2.9	7.9
Methanol	40.7	7.4	6.0	10.9	14.4
Ethanol	58.6	7.7	4.4	9.5	12.9
Acetone	74.0	7.6	5.1	3.4	9.8
Abietic Acid	213.2	10.3	2.2	5.0	11.6

11.4.1 Solubility Measurements

Solubility measurements of flux components were first performed in the laboratory. In this section some of these results and their comparisons to theoretical calculations will be discussed. In particular, the solubility results of two of the major ingredients of flux, namely, abietic acid and dimethylamine hydrochloride, will be dealt with. Dimethylamine hydrochloride is a common activator.

The solubility of both dimethylamine hydrochloride and abietic acid in various solvents was measured by gravimetric titration. Weighed amounts of solutes were introduced in a test tube (screw-on cap). Solvent was gradually added to the test tube and stirred continuously in a thermostatically controlled bath with the temperature kept at 25°C

(78°F) ± 0.02°C. To avoid overshooting, the titration was stopped when a piece of single crystal was found left in the solution after continuous stirring for 24 hours. The solvent was then added in very small increments, and the solution was stirred continuously until this crystal disappeared.

This was repeated several times and the solution was weighed carefully to determine the weight percent of the solute in the solution. The screw-on cap type of test tube assures no loss of solvent during the experiment. The reproducibility of the solubility was found to be within ± 0.1% and an overall accuracy of ± 0.2% is estimated in the final values.

The measured solubility of dimethylamine hydrochloride has shown that among pure components methanol is the best solvent for activators, such as dimethylamine hydrochloride, as expected. HCFC-123 has significantly higher solubility for the activator than CFC-113. The data indicate that the azeotropic blend of HCFC-141b/HCFC-123/methanol has higher solubility for dimethylamine hydrochloride than stabilized CFC-113/alcohol based blends.

Theoretical solubility estimations for abietic acid were done using Equation 11.1 with the activity coefficient calculated by the solubility parameter model described above (Equation 11.4). Equation 11.1 uses three unknown quantities, of which, two are physical properties of the solute, ΔH_f the heat of fusion and T_m the melting temperature of the solute. These two quantities were measured for both dimethylamine hydrochloride and abietic acid by means of thermogravimetric analysis (TGA). The heat of fusion of dimethylamine hydrochloride was found to be 26.65 cal/gm and the melting point was found to be 175.2°C (347.4°F). The abietic acid showed a more complex behavior and an effective heat of fusion of 19.84 cal/gm and an effective melting point of 160°C (320°F) were used. The other unknown quantity in Equation 11.1 is the activity coefficient which depends on the interaction between molecules and is calculated by using the solubility parameters of solutes and solvents. The abietic acid used showed a glass transition and therefore, as mentioned before, an effective heat of fusion and melting temperature were used to estimate its solubility in various solvents. The simple solubility parameter theory cannot explain the solubility of the activator, which is ionic in nature; therefore, no attempt was made to explain its solubility using this theory. The solubility parameters and the molar volumes of the compounds are shown in Table 11.2.

Abietic acid solubility data agree reasonably well with solubility parameter theory. In the case of mixtures a mole fraction average of the solubility parameters of the individual components in the solvents was used. The solubility data have shown that the HCFC-141b/HCFC-

123/methanol azeotrope has a higher solubility for abietic acid than CFC-113/alcohol based azeotropes.

Solubility data for an RMA paste are presented in Table 11.3. In this case a measured amount of RMA paste was put in a pre-cleaned glass slide with a dip in the middle to contain the paste. This was reflowed in an oven at 200°F and then defluxed in a vapor degreaser using vapor rinse for 1 minute. The removal of paste has been determined by accurately determining the weight change of the glass slides. The solubility results have shown that the azeotropic blend of HCFC-141b/HCFC-123/methanol has a better solubility for this paste than the stabilized solvent azeotropes using CFC-113 which are currently used in defluxing SMAs.

Table 11.3 RMA paste dissolution study.

Solvent	% Paste Removal
HCFC-141b/HCFC-123/Methanol/MeNO$_2$	48
CFC-113/MeOH/Acetone/Isohexane/MeNO$_2$	38

11.5 Performance Data with HCFC Solvents

A new family of solvents has been introduced which show a significant reduction in ozone depletion potential (ODP). These solvents are based on the chemistry of two specific hydrochlorofluorocarbons (HCFCs), substances made up of the elements hydrogen (H), chlorine (Cl), fluorine (F), and carbon (C). These two HCFCs are HCFC-141b and HCFC-123 [13-16].

To replace defluxing solvents based on CFC-113, preliminary tests indicate that HCFC-141b-based solvents are good defluxing materials. It was discovered also that the addition of a small amount of HCFC-123 suppressed the flame limits of HCFC-141b-based solvents. Hence, the most promising solvent of the HCFC-141b family of solvents is one containing both HCFC-141b and HCFC-123 with the addition of methanol. The ODP of this solvent is 0.07 to 0.06 depending on the amount of HCFC-123 in the formulation. However, an azeotropic solvent based only on HCFC-141b and methanol also shows promising results. The physical properties of HCFC-141b and HCFC-123 are given in Table 11.4 along with those of CFC-113. The properties of the two stabilized solvents based on HCFC compounds plus methanol is shown in Table 11.5.

Table 11.4 Physical properties of HCFCs contrasted with CFC-113.

	HCFC-141b	HCFC-123	CFC-113
Molecular weight	116.9	152.9	187.4
Appearance	Colorless liquid	Colorless liquid	Colorless liquid
Boiling point:			
°F	89.7	82.2	117.6
°C	32.0	27.9	46.6
Vapor pressure Psia @ 77°F (25°C)	11.6	13.2	5.5 *
Liquid density @ 77°F (25°C):			
g/cc	1.236	1.463	1.565
lbs/gal	10.28	12.20	13.06
Latent heat of vaporiz- ation at NBP, Btu/lb	74.0	95.4	63.1
Solubility of water (ppm) @ 77°F (25°C)	420	660	90 *
Solubility in water (ppm) @ 77°F (25°C)	660	2100	270 *
Liquid viscosity (cP) @ 77°F (25°C)	0.43	0.48	0.69
Vapor flammability @ 77°F (25°C):			
Lower limit (vol.%)	7.6	None	None
Upper limit (vol.%)	17.7	None	None
Flash point	None	None	None
Ozone depletion potential	0.08	0.02	0.8
Greenhouse warming potential	0.12	0.02	1.3
* @ 68°F (20°C)			

11.5.1 Defluxing Results with Stabilized HCFC-141b/HCFC-123/Methanol

Defluxing results based on a solvent composed of HCFC-141b, HCFC-123 and methanol (CH_3OH) show equal, if not improved, ability to CFC-113-based solvents for removing both flux and paste residues. For the formulation of the solvent based on HCFC-141b, HCFC-123, and methanol, see Table 11.5. Performance evaluations of this solvent were conducted in both a batch defluxer and a conveyorized in-line defluxer. For example, the results (averages for four boards per run) for defluxing conventional PWAs and SMAs in a batch defluxer are presented in Table 11.6. Figure 11.2 depicts the conventional PWA used in the investigation, and Figure 11.3 shows the SMA used.

Figure 11.2 Printed wiring assembly (PWA) with conventional through-hole components (THCs).

The fluxes and pastes used in the experiments described herein are classified as follows:

Flux #1	RA-MIL
Flux #2	RA

Table 11.5 Compositions of two HCFC solvents.

(a) Composition of HCFC-141b/HCFC-123/Methanol Azeotrope	
Note: This particular composition is known as Genesolv®2010	
Ingredient	Wt. Percent
HCFC-141b	86.1
HCFC-123	10.0
Methanol	3.6
Nitromethane	0.3
Boiling point (°F)	85.2°F
Ozone depletion potential	0.07

(b) Composition of HCFC-141b/Methanol Azeotrope	
Note: This particular composition is known as Genesolv®2004	
Ingredient	Wt. Percent
HCFC-141b	96.0
Methanol	3.9
Nitromethane	0.1
Boiling point (°F)	84.9°F
Ozone depletion potential	0.08

Flux #3	RA
Flux #4	RA
Flux #5	SA
Flux #6	RA
Paste #1	RA
Paste #2	RA
Paste #3	RMA

where RA stands for rosin-activated, (RMA mildly activated, RA-MIL-spec)

Table 11.6 Defluxing results of the stabilized azeotrope of HCFC-141b/HCFC-123/methanol vs. a CFC-113 based solvent in a two-sump batch degreaser.

	Defluxing Results (μgNaCl/in^2)						
	Ionic Contamination Residue						
	(PWAs) Flux					(SMAs) Paste	
Solvent	#1	#2	#3	#4	#5	#1	#2
HCFC-141b/HCFC-123/ MeOH/MeNO$_2$	6.7	6.3	9.7	12.9	9.9	6.7	11.0
CFC-113/ MeOH/Acetone/ Isohexane/MeNO$_2$	9.1	7.0	13.9	9.9	14.2	12.3	9.6

	Defluxing Results(μgCont./in^2)						
	Organic Contamination Residue						
	(PWAs) Flux					(SMAs) Paste	
Solvent	#1	#2	#3	#4		#1	#2
HCFC-141b/HCFC-123/ MeOH/MeNO$_2$	139	109	130	147		283	38
CFC-113/ MeOH/Acetone/ Isohexane/MeNO$_2$	123	162	138	154		404	88

and SA for synthetic-activated.

Two solvents were used for comparison of defluxing ability. The first solvent, based on HCFC-141b, was the stabilized azeotrope containing HCFC-141b, HCFC-123, and methanol. The other solvent was a stabilized azeotrope of CFC-113, methanol, and several other ingredients. The cleaning cycle for the PWAs was 1 minute vapor dwell, 3 minutes rinse sump immersion followed by 1 minute vapor dwell. For the SMAs the cleaning cycle was 30 minutes direct boil immersion, 30 seconds rinse sump immersion followed by 30 seconds vapor dwell. The results include (1) ionic results using a conventional ionic contamination tester for the PWAs and the new type of ionic contamination tester having

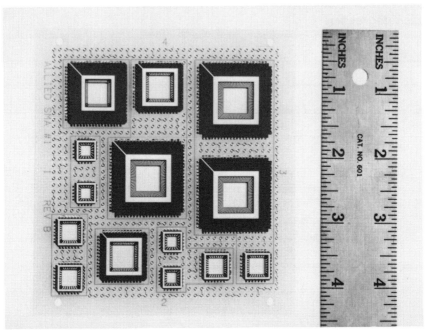

Figure 11.3 Surface mount assembly (SMA) with leadless ceramic chip carriers (LCCCs) (dummy packages).

spray headers in the test tank for the SMAs, and (2) organic (rosin residue) results using a UV/VIS-spectrophotometer which are given in micrograms of rosin or equivalent per square inch (μg Rosin/in^2). The ionic results are expressed in micrograms of sodium chloride or equivalent per square inch (μGnacl/in^2). Flux #5, the SA flux, did not absorb in the UV/VIS region; hence, it does not appear under the results for residual rosin. It can be seen that the solvent based on HCFC-141b performs as well as, if not better, than the conventional CFC-113-based solvent.

Table 11.7's results (averages of four boards per run) are given for cleaning PWAs and SMAs run in a conveyorized, in-line defluxer at 3 feet per minute. The same two solvents as described in the paragraph above were used. Three different fluxes and two different pastes were employed. The first two recirculating sumps had top spray pressures of 54 psi and bottom spray pressures of 30 psi. The last recirculating sump had a top spray pressure of 195 psi; the bottom spray pressure was 125 psi. The results are expressed in micrograms of sodium chloride or equivalent per square inch (μGnacl/in^2). The PWAs were tested in the conventional ionic contamination tester; the SMAs in the new type of ionic contamination tester containing spray headers in the test tank and heated test solution. Again it can be seen that the solvent based on HCFC-141b performs as well as, if not better, than the conventional CFC-113-based solvent.

Table 11.7 Defluxing results of the stabilized azeotrope of HCFC-141b/HCFC-123/methanol vs. a CFC-113 based solvent in an in-line liquid seal defluxer (3 ft/min).

	Defluxing Results (μGnacl/in^2)				
	Ionic Contamination Residue				
	(PWAs) Flux			(SMAs) Paste	
Solvent	#1	#2	#3	#1	#2
HCFC-141b/HCFC-123/MeOH/MeNO$_2$	4.1	8.2	9.7	6.1	5.8
CFC-113/MeOH/Acetone/Isohexane/MeNO$_2$	5.4	8.3	7.9	8.9	5.9

11.5.2 Defluxing Results with Stabilized HCFC-141b/Methanol

In addition to examining the stabilized HCFC azeotrope solvent containing HCFC-141b, HCFC-123, and methanol (CH_3OH), the stabilized azeotrope based solely on HCFC-141b and methanol was also examined. The HCFC-141b/methanol/nitromethane azeotrope was used to deflux both conventional PWAs and SMAs. The composition of this solvent is given in Table 11.5. Performance evaluations of this solvent were conducted in a batch defluxer using the same cleaning cycle specified above. As before, the CFC-113 solvent used in the investigation was the stabilized azeotrope of CFC-113, methanol, acetone, and isohexane. The fluxes and pastes used for the investigation were:

Flux #2	RA
Flux #3	RA
Flux #6	RA
Paste #3	RMA

Defluxing results based on the HCFC-141b and methanol also show equal, if not improved, ability to that of CFC-113-based solvents for removing both flux and paste residues. Table 11.8 gives the defluxing results for ionic residues for the conventional PWAs, and also the defluxing results for the SMAs for the RMA paste. For these results high performance liquid chromatography (HPLC) was used.

Table 11.8 Defluxing results of the stabilized azeotrope of HCFC-141b/methanol vs. a CFC-113 based solvent in a two-sump batch degreaser.(*Flux applied by wave).

	Defluxing Results (μgNaCl/in^2)		
	Ionic Contamination Residue		
	(PCAs) Flux		
Solvent	#2	#3 *	#6
HCFC-141b/HCFC-123/MeOH/MeNO$_2$	5.2	17.4	6.7
CFC-113/MeOH/Acetone/Isohexane/MeNO$_2$	7.2	21.2	7.0

	Defluxing Results(μgCont./in^2)	
	Organic Contamination Residue	
	(SMAs) Paste	
Solvent	#3	% Left on Components
HCFC-141b/HCFC-123/MeOH/MeNO$_2$	110	76
CFC-113/MeOH/Acetone/Isohexane/MeNO$_2$	120	76

It is evident that both HCFC-based solvents perform as well as, if not somewhat better, than solvents based on CFC-113. It is expected that the new HCFC solvents will prove entirely adequate for defluxing electronics assemblies. Since the solvents behave similarly to conventional solvents regarding the kinds of operations on the manufacturing floor, it is anticipated that the electronics manufacturing engineer will not have to address himself or herself to learning an entirely different process technology.

11.5.3 Equipment To Hold the New HCFC Solvents

For using the new HCFC-based solvents in an in-line defluxer, it would be best to incorporate liquid seals since the solvent boils at 85°F. If this is not feasible, the machine should be equipped with finned-tubing for both its primary condensing coil and entrance and exit end vapor traps. A batch cleaner should have a minimum of 100% freeboard to contain the new solvent. Suitable low emissions equipment, both batch and in-line, is available to adequately contain the new solvents based on HCFC-141b.

11.6 Alternative Technologies

In response to the threat of CFCs to stratospheric ozone, a number of alternative technologies have arisen as possible substitutes for defluxing applications. Some of these technologies have been around a long time, such as water soluble fluxes. Some of them are more recent, such as the hydrochlorofluorocarbons (HCFCs) and "no clean" fluxes. In order to facilitate acceptance of these alternative technologies by the military, the EPA arranged for a consortium of EPA, military, and industry to tackle the problem of military acceptability of new technologies. This consortium is known as the EPA/DOD/IPC (Environmental Protection Agency - Department of Defense - Institute of Interconnecting and Packaging of Electronic Circuits) Ad Hoc Solvents Working Group. This group of people, drawn from the EPA, the military, and industry, got together to define and specify both a suitable printed wiring assembly test board and a detailed test procedure for cleaning performance acceptance of alternative technologies.

11.7 Phase 1 and Phase 2 Testing

The EPA/DOD/IPC Ad Hoc Solvents Working Group elaborated a standard test procedure to be used for cleanliness performance acceptance of alternative defluxing technologies for the purpose of replacing CFC-113 in defluxing and cleaning operations [17]. This document outlines in detail the procedures that were followed in the Phase 1 test; these same procedures must be followed by anyone performing a Phase 2 test. The Phase 1, or Benchmark, test was performed at two government facilities, EMPF in Ridgecrest CA, and NAC in Indianapolis IN. The results of the Phase 1 test are available from the IPC [18]. All manufacturers of alternative technologies who wish to have their material accepted by the military must undergo a Phase 2 test with their alternative technology.

11.7.1 Purpose

The Phase 2 test was designed by the EPA/DOD/IPC Ad Hoc Solvents Working Group as a test for evaluating alternative cleaning technologies for reducing the level of CFCs (chlorofluorocarbons) used in electronics manufacturing cleaning processes. The Phase 1, or Benchmark, test defined by the IPC Cleaning and Cleanliness Testing Program specified the use of a batch cleaning operation and employment of the nitromethane-stabilized azeotrope of CFC-113 and methanol as the cleaning agent. (Note: commercially available as Freon®TMS).

11.7.2 Process Sequences

Five process sequences were run: A boards, B1 boards, B2 boards, C assemblies and D assemblies. The definition of these are:

A boards are precleaned and tested. No other processing operation is performed on them. See Figure 11.4.

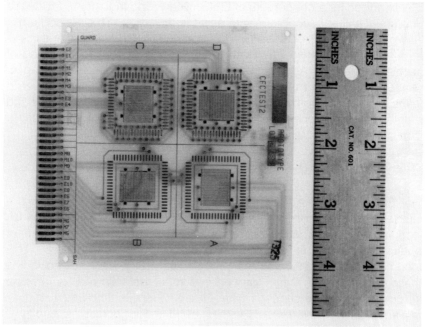

Figure 11.4 Benchmark/Phase 2 printed wiring board (A board).

B1 boards have the specified solder paste screened on the bare board substrate and subsequently vapor phase reflowed. They are tested, but they are not cleaned.

B2 boards have the specified solder paste screened on the bare board substrate and subsequently reflowed. They are also exposed to the wave soldering operation using the specified wave solder flux. They are tested, but they are not cleaned.

C Assemblies have components attached to the bare board substrate using the specified solder paste and vapor phase reflow. They are then cleaned and tested. See Figure 11.5.

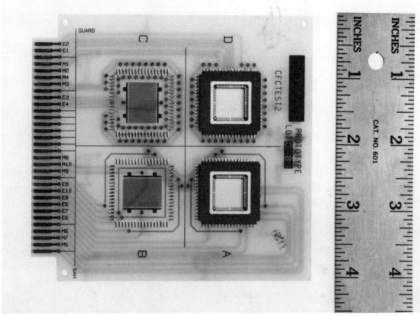

Figure 11.5 Benchmark/Phase 2 surface mount assembly (C assembly) with two LCCCs.

D Assemblies have components attached to the bare board substrate using the specified solder paste and vapor phase reflow. They are also exposed to the wave soldering operation using the specified wave solder flux. They are cleaned twice - once after vapor phase soldering and once after wave soldering. They are tested after the two cleaning operations. The only boards used for the official criteria were the A, C, and D boards. The B1 and B2 boards were used for reference only.

11.7.3 Cleaning Process

The cleaning process followed at Allied-Signal Inc. during their Phase 2 test was identical to the batch process used in Phase 1 (Benchmark) with the exception of the solvent. In place of the stabilized azeotrope of CFC-113 and methanol, the stabilized azeotrope of HCFC-141b, HCFC-123, and methanol was used instead. No other process change was made. Essentially, the cleaning cycle consisted of (1) 30 seconds vapor equilibration, (2) 3 minutes direct boil sump immersion, (3) 1 minute rinse sump immersion, (4) 30 seconds vapor equilibration.

11.7.4 Phase 2 test results

Three official cleanliness test methods were used to ascertain cleanliness. These were:

1. ionic contamination test
2. residual rosin test
3. surface insulation resistance (SIR).

For the ionic contamination test, a tester having spray headers in the test tank and heated test solution was used (Omega Meter 600 SMD). For the residual rosin test a UV/VIS spectrophotometer was employed. The SIR data are for reference only. They are presented in the complete Allied-Signal Phase 2 Report [19] but not here.

The results for the official Allied-Signal Phase 2 test are given in Table 11.9. Again, it must be stressed that the only difference between the Benchmark, or Phase 1, test and the Allied-Signal Phase 2 test was the solvents used. The Benchmark test employed the stabilized azeotrope of CFC-113 and methanol and the Allied-Signal Phase 2 test employed the stabilized azeotrope of HCFC-141b, HCFC-123, and methanol. The entire phase 2 test was repeated with the stabilized azeotrope of HCFC-141b, HCFC-123, and methanol. However, batch cleaning was not used. Rather, the material was run in an in-line conveyorized defluxer with the same pressure parameters as previously used. The conveyor speed was 4 ft/min. The results are presented in Table 11.10.

11.8 Safety, Handling and Containment Aspects

For the new molecules such as HCFC-123 and HCFC-141b, there are a few issues to be addressed in terms of safety, handling and containment

Table 11.9 Allied-Signal Phase 2 test results contrasted with the benchmark results.

Benchmark (Phase 1)			Phase 2* (Allied-Signal)	
CFC-113/Methanol Azeotrope			HCFC-141b/HCFC-123/ Methanol Azeotrope**	
Assembly Type	Ionic Residue	Rosin Residue	Ionic Residue	Rosin Residue
	$(\mu g/in^2)$	(μg)	$(\mu g/in^2)$	(μg)
A	2.0	301	0.6	47
C	3.8	3,135	2.5	3,128
D	10.7	3,945	8.1	1,536
* Test officially monitored by TMVT members on site.				
** Genesolv®2010 = HCFC-141b/HCFC-123/Methanol Azeotrope				

because these are lower boiling than CFC-113. The hydrolytic stability of the azeotropic blend in the presence of metals will be addressed. In terms of safety, two aspects are vital, namely (1) exposure levels to people working around the solvents and (2) the flammability hazard, if any, under various circumstances.

Any new molecule must be subjected to extensive toxicological studies before it can be used commercially. Both HCFC-123 and HCFC-141b are undergoing these tests. A Program for Alternative Fluorocarbon Toxicity (PAFT) as been set up by a consortium of a large number of current CFC manufacturers all over the world. It is divided into two parts, PAFT I and PAFT II, to complete the long term toxicological studies (two year toxicological studies) for stratospherically safe fluorocarbons (SSFs). Fourteen fluorocarbon manufacturers from all over the world have joined PAFT I in which HCFC-123 is included for toxicological studies and nine fluorocarbon manufacturers have joined PAFT II which includes HCFC-141b for similar studies. After the long time toxicological studies are completed the threshold limit values (TLV) for these molecules will be assigned. CFC-113 has a TLV of 1000. At this point both HCFC-141b and HCFC-123 are expected to have lower TLVs than CFC-113 and permissible exposure limits of 100 for HCFC-123 and 500 for HCFC-141b have been used. This indicates, however, that these materials have to

Table 11.10 Allied-Signal Phase 2 test results contrasted with in-line results.

Phase 2* (Allied-Signal) Batch Unit			Phase 2 (Allied-Signal) In-Line (Liquid Seal) Conveyorized Defluxer		
HCFC-141b/HCFC-123/Methanol Azeotrope			HCFC-141b/HCFC-123/ Methanol Azeotrope**		
Assembly Type	Ionic Residue	Rosin Residue	Ionic Residue	Rosin Residue	
	$(\mu g/in^2)$	(μg)	$(\mu g/in^2)$	(μg)	
A	0.6	47	0.7	2	
C	2.5	3,128	1.6	155	
D	8.1	1,536	4.1	237	
* Test officially monitored by TMVT members on site.					
** Genesolv®2010 = HCFC-141b/HCFC-123/Methanol Azeotrope					

be handled more cautiously in terms of exposure than CFC-113.

The HCFC-141b/HCFC-123/methanol stabilized azeotrope does not exhibit any flash point according to the tag open-cup test (ASTM D 1310-86). Hence it is classified as a non-flammable fluid. HCFC-141b alone, however, does exhibit flame limits when mixed with certain compositions of air. The flame limits of HCFC-141b are very similar to those shown by commercially available defluxing solvents based on 1,1,1-trichloroethane. HCFC-123 alone does not exhibit any flame limit when mixed with air. In case of the azeotropic blend of HCFC-141b/HCFC-123/methanol, HCFC-123 helps to suppress the flame limit of HCFC-141b.

Due to the lower boiling point of the HCFC solvent blends (~85°F), some special considerations are needed for this material. The material will be packaged in 55-gallon drums and 5-gallon pails but it should not be exposed to temperatures greater than 90°F.

The lower boiling point affects the machine design also. The optimal design will be to use machines with liquid seals. Allied-Signal/Baron-Blakeslee has an active machine design program in place to use this solvent in both in-line and batch operations. At present it has been seen that the use of finned-coiled condensers is adequate to minimize vapor loss from the machines. With these types of condensers the solvent loss for HCFC-141b/HCFC-123/methanol has been found to be comparable to that of CFC-113. A retrofit program for existing equipment in the field is also in existence.

A few words about the hydrolytic stability of the molecule will be added. The stability of the HCFC-141b/HCFC-123/methanol azeotrope was tested by refluxing it in the presence of various metals and water for a two week period. The solvent showed excellent stability even without the stabilizer. Its stability compares very well with CFC-113, and it seems far superior to 1,1,1-trichloroethane in hydrolytic stability. It seems that the presence of a fluorine in the molecule ensures better stability in the case of HCFC-141b over 1,1,1-trichloroethane.

11.9 Conclusion

A suitable solvent substitute for CFC-113 has been found, namely, HCFC-141b. This material azeotropes well with methanol and HCFC-123. Either the stabilized azeotrope of HCFC-141b with HCFC-123 and methanol or the stabilized azeotrope of HCFC-141b and methanol can be used for cleaning electronics assemblies, both PWAs and SMAs. Performance studies show either solvent removes solder flux and paste residues as well as, if not somewhat better than, commonly used cleaning azeotropes based on CFC-113. In addition, the stabilized azeotrope of HCFC-141b, HCFC-123, and methanol has successfully passed the Phase 2 test. Therefore, it will receive recognition by DOD for incorporation into military specifications.

11.10 Acknowledgements

The following individuals made substantial contributions to this paper. Any errors that may remain are solely due to me. My colleague, Dr. Rajat Basu, at Allied-Signal's Buffalo Research Laboratory (BRL) contributed substantially to Sections 11.4 and 11.4.1. Indeed, the data presented in Section 11.4.1 and Table 11.3 was made at the BRL by Ms. E.L. Swan. I would also like to extend my appreciation to Mr. Jerry Gozner, an applications engineer working at Allied-Signal's Melrose Park Applications Laboratory, for the data presented in Tables 11.6-11.10. Jerry also made a substantial contribution to Allied-Signal's new HCFC-based solvent passing the Phase 2 test.

11.11 References

1   John K. Bonner, "A New Solvent For Post Solder Cleaning of Printed Wiring Assemblies", Proc. Nepcon West '86 (1986), 763-777.

2   John K. Bonner, "A Comparison of four Solvents for Defluxing Printer Wiring Assemblies", Circuit Expo '86 Proc. (1986), 94-103.

3   Richard Monastersky, "Antarctic Ozone Reaches Lowest Levels", Science News, 132 (10/10/87) 230.

4   Richard Monastersky, "Decline of the CFC Empire", Science News, 133 (4/9/88) 234-236.

5   Richard Monastersky, "Arctic Ozone: Signs of Chemical Destruction", Science News, 133 (11/11/88) 383.

6   Richard Monastersky, "Clouds Without a Silver Lining", Science News, 134 (10/15/88) 249-251.

7   Ibid.

8   "Montreal Protocol On Substances That Deplete The Ozone Layer (Final Act)", UNEP 1987.

9   J.H. Hildebrand, J.M. Prausnitz and R. L. Scott, Regular and Related Solutions, (VanNostrand Reinhold, New York, 1970).

10   R.S. Basu, H. Pham and D.P. Wilson, "Prediction of solubility of long-chain hydrocarbons using UNIFAC", Internat. J. Thermophysics, 7, 319-330, (1986).

11   C.M. Hansen, "The three dimensional solubility parameter - key to paint component affinities", J. Paint Technol., 39, 104-117, and 505-514, (1967).

12   A.F.M. Barton, CRC Handbook of Solubility Parameters and Other Cohesion Parameters, (CRC Press, Boca Raton FL, 1983).

13   J.K. "Kirk" Bonner, "Searching for a substitute: alternatives to CFC-113" Circuits Mfg., 29 (7), 42-45, (July 1989).

14   David W. Bergman, "CFCs: the search for alternatives," Printed Circuit Assembly, 3 (9), 40-44, (Sept 1989).

15   J.K. "Kirk" Bonner, "New solvent alternatives," Printed Circuit Assembly, 3 (9), 36-38, (Sept 1989).

16   R.S. Basu and J.K. Bonner, "Alternative to CFCs: new solvents for the electronics industry, "Surface Mount Technol., 3 (8), 34-37, (Dec 1989).

17   "Cleaning and Cleanliness Testing Program: A Joint Industry/Military/EPA Program to evaluate Alternatives to Chlorofluorocarbons (CFCs) for Printed Board Assembly Cleaning," March 30, 1989 (available from the IPC).

18   "Cleaning and Cleanliness Test Program: Phase 1 Test Results," IPC-TR-580, October 1989.

19   "Allied-Signal Phase 2 Final Report" (available from J.K. Bonner, Allied-Signal
        Gensolv/Baron Blakeslee, 2001 N. Janice Avenue, Melrose Park, IL 60160.

12

ELECTRICAL BONDING

OF CONNECTORS ON

JET ENGINE ELECTRONICS

STEPHAN J. MESCHTER

12.1 Introduction

External electrical connectors that interface with the jet engine cabling are required to be electrically conductive for EMI and lightning requirements. The DC resistance between these connectors and their mounting surfaces must be less than 2.5 milliohms. Electronic control systems have experienced difficulty in meeting this bonding specification after environmental stress screening. The particular interface in question is between stainless steel MIL-STD-38999 connectors and chromate conversion (Alodine) coated aluminum. To improve the electrical bonding the addition of nickel plating on the aluminum plate and the connector has been evaluated. The configurations were evaluated in thermal cycling, humidity and salt spray. Bonding resistance versus normal force was determined for the various configurations. The results show that electroplated nickel stainless steel connectors will meet the connector to engine control bonding resistance specification of 2.5 milliohms.

12.2 Background

Mission critical electronic systems in aerospace are being used in increasingly harsh environments. These systems are sensitive to electromagnetic interference (EMI) and lightning. They are also significant electromagnetic energy generators. Effective EMI and lightning protection is critical to the reliable operation of these systems.

Electrical bonding is a key parameter for effective EMI shielding and lightning protection. A typical control system that is installed on a jet engine is shown in Figure 12.1. A low resistance path must be established through the metal-to-metal interfaces in question for good bonding. The level of metal-to-metal bonding is dependent upon the amount of current expected to pass though the joint. Table 12.1 quantifies the general level of bonding required for various applications.

There is no military specification requirement for the bonding of the external interface connectors/cables to the enclosure. The fundamental requirement for the engine cabling is that the system works and passes EMI testing. The specific bonding resistance for the engine cabling interface is usually specified in the customer's equipment specification. Presently, the bulk-head connector to electronic enclosure bonding resistance is specified as 2.5 milliohms. On the electronic control, stainless steel flange mount and jam nut mount connectors per MIL-C-38999 and MIL-C-83723 are secured to enclosures which are 6061 aluminum with a chromate conversion coating per MIL-C-5541.

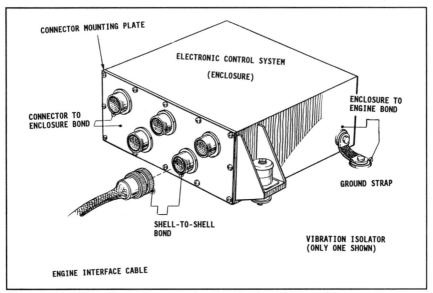

Figure 12.1 Bonding of an engine control and its cables.

Table 12.1 Typical bonding resistance magnitudes from MIL-B-5087.

Application	Bond Resistance
Static charge	1 ohm
Shock Hazard	0.1 ohm (current dependent)
RF Interference	0.0025 ohm

The control systems are having difficulty meeting the 2.5 milliohm bonding resistance from the connector to the enclosure. Upon initial assembly of the connector to the enclosure the bonding resistance is well within the 2.5 milliohm specification. After thermal environmental stress screening, some of the connectors would fail the 2.5 milliohm specification. The variability of the bonding resistance is very high. Values range from 0.5 milliohm to 90 milliohms in the present configuration.

The electrical bonding requirements for aerospace systems are defined in the military specification MIL-B-5087. Class R of this specification is for applications with radio frequency (RF) potential. Class R is chosen because the EMI potentials are in the RF frequency range.

Per MIL-B-5087, the bonding method must result in a direct current impedance of less than 2.5 milliohms from enclosure to structure. As shown in Figure 12.1, the bond is typically made with a ground strap around the vibration isolator to the engine structure. The DC bonding resistance of 2.5 milliohms is used to assure an adequate RF bond. The correlation between a 2.5 milliohm bonding resistance and EMI shielding effectiveness is not fully understood. Bonding resistance alone cannot be used to establish effective EMI shielding. Dr. Jan Gooth of the Georgia Institute of Technology [1] has shown that the 2.5 milliohm bonding resistance requirement should not be relaxed at this time. He investigated a stainless steel joint with a 0.030 inch gap filled with conductive sealant and determined that for resistances above 2.5 milliohms the shielding effectiveness of the joint rapidly decreases. Dr. Gooth is continuing shielding effectiveness testing.

With regard to lightning protection, according to Ohm's law a 2.5 milliohm bond resistance can pass 200,000 amps at 500 volts. The simple application of Ohm's law does not consider the frequency characteristics of the sharp spiked nature of the lightning wave form.

12.3 Analysis

The purpose of electrical bonding is to allow electrons to flow freely from the atomic lattice of one contact member to the atomic lattice of another member. Therefore, topography and film resistivity of the mating surfaces become factors in determining the contact resistance of an electrically conductive couple. The majority of the information in this section is obtained from conversations with John Salvaggi, the resident metallurgist at GE [2], and information from the Ney contact manual [3].

Practical engineering surfaces produced by machine tools are quite rough when compared to atomic or molecular dimensions. A magnified metallic surface, ideally clean, will display many asperities. The probability of a large number of such asperities on two mating surfaces coming into exact alignment and contact is quite remote. Relatively few will be perfectly aligned and will touch. The result is that a large proportion of the space at the contact interface is occupied by air.

All metallic electrical contact is achieved via the minute metal-to-metal contact sites that are formed when two deformable metallic bodies are brought together. These minute asperities, described above, are the locations where the current will flow. The number and location of these asperities are dependent on surface finish, contact geometry, metallurgy and the force used to press the two members together.

Resistivity across the contact area can be defined as the property of a material that impedes electrical current when a sample of specified unit dimensions is considered. It is characterized by the following equation:

$$\rho = \frac{RA}{L}$$

[12.1]

where:

ρ = resistivity (ohm-cm)
L = length (cm)
R = resistance (ohm)
A = contact area (sq cm)

Contact resistance consists of three terms as shown Figure 12.2. The following equation describes the resistance components.

$$R_t = R_b + R_c + R_f$$

[12.2]

where:

R_t = Total Resistance,
R_b = Bulk Resistance,
R_c = Constriction Resistance, and
R_f = Film Resistance.

Bulk resistance is a function of the cross-sectional area of the column of metal leading up to and going away from the area of contact (where the asperities are aligned). It can be expressed as

$$R_b = \frac{\rho L}{A}$$

[12.3]

where:

ρ = resistivity, ohm cm
L = length, centimeters
A = cross-sectional area, sq cm.

Constriction resistance, R_c is a function of asperity size, quantity and distribution across the contact area. The size and number of asperities is a function of applied force and the distribution is a function of contact geometry. The resistance to current flow develops because the current must take divergent paths through the small asperity contact points rather than through one wide path.

The constriction resistance can vary by factors of two or three at a given pressure, depending on the contact geometry. Surfaces of the

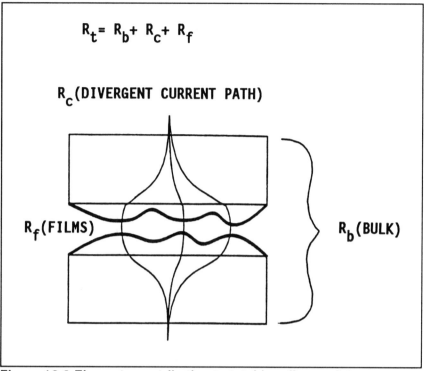

Figure 12.2 Elements contributing to total bonding resistance.

softer contact member will deform only enough to support the applied load. With larger contact surface areas, there are more sites available to support this load. The constriction resistance, like the bulk resistance, is also a relatively small resistance.

The most significant variable in the R_t equation is the effect of films, R_f. All metals, even noble metals used in contact metallurgy, have surface films. Base metals such as copper, aluminum, iron and nickel form oxides from pollutants almost immediately upon exposure to air. These are defined as normal or thin films. Their thickness varies from 10 to 30 Å. Films produced by heat aging or chemical passivation are usually referred to as thick films and are measured as 100 Å or greater. The film resistance for thin films is negligible as shown in Table 12.2.

It is apparent that resistivity increases rapidly as a function of film thickness. Studies have shown that in static contact situations, with no wiping action present, no reasonable amount of force will penetrate 100+ Å films (aging films) to provide stable contact resistance.

Table 12.2 Typical tunnel resistivities for various film thicknesses.

Film thickness (Angstroms)	Tunnel resistivity (Ohms/cm)
5	10^{-8}
15	10^{-7} TO 7 X 10^{-7}
20	2 X 10^{-7} TO 10^{-6}
100	10^{-8} TO 10^{-2}

12.3.1. Hypotheses

The following hypotheses are proposed for the bonding resistance degradation which occurs during environmental stress screening:

1. The chromate conversion coating on the aluminum breaks down and forms a high resistance film above its rated temperature of 60°Celsius.

2. Pressure relaxation is occurring with the jam nut connector resulting in a lower interfacial pressure and a high bonding resistance.

3. Through environmental exposure a high resistivity oxide layer is building up on the stainless steel.

4. The soft aluminum cannot readily break through the stainless steel oxide, so good contact is not maintained through the life of the connector assembly.

12.4 Variables

An in-depth characterization of oxide resistance was not the objcct of this evaluation. The approach focused on various coatings/platings and the assembly procedures used in connector installation. The bonding resistance is monitored through temperature cycling, salt spray, and humidity. Long term temperature cycling is the most severe engine environment.

There are many variables in the electrical bonding resistance determination. Two stainless steel jam nut connectors assembled on the same chromate conversion coated aluminum plate to the same torque can yield completely different resistances after temperature cycling. A summary of the basic variables is given in Figure 12.3. A detailed discussion on each of the variable groups is presented next.

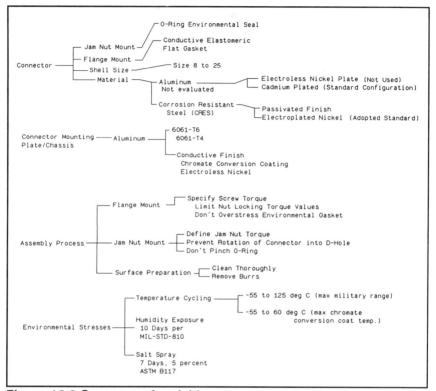

Figure 12.3 Summary of variables.

12.4.1. Baseline Configuration

The engine control systems have typically used a stainless steel circular connector on a chromate conversion coated aluminum mounting plate. The jam nut connector uses an o-ring for an environmental seal as shown in Figure 12.4. The flange mount connector shown in Figure 12.5 does not have a built-in environmental seal, so a flat gasket is used for sealing. This flat gasket must be conductive to maintain a connector to mounting panel EMI bond.

12.4.2. Connector Variables

The two primary connector materials that are used in aircraft applications are aluminum and corrosion resisting steel (stainless). Aluminum connector finishes are olive drab cadmium plate and electroless nickel

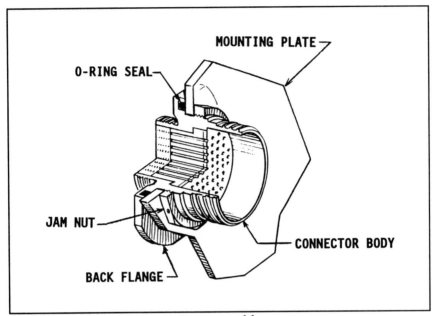

Figure 12.4 Jam nut connector assembly.

plate. The stainless steel connector is passivated or nickel electroplated. There is a significant difference between electroless and electroplated nickel. Electroplated nickel is nearly pure nickel which is quite soft. Electroless nickel plate is a nickel-phosphorous alloy that can be as hard as 55 Rockwell C. The pure nickel will tend to conform to the surface of the aluminum mounting plate and increase the bonding area. Table 12.3 gives the hardness comparisons for the base metals and the platings.

An important link in the shielding effectiveness of a cabling system is the bonding across the mated connector pair. This is defined in the military connector specifications as shell-to-shell conductivity. A summary of the connector data is given in Table 12.4. The shell-to-shell bonding resistance and shielding effectiveness of the nickel plated connectors is superior to the other configurations.

12.4.3. Enclosure Variables

The electronic control unit is typically made of dip brazed aluminum. The connectors may be installed directly on the chassis or on a removable panel. The chassis is made of 6061-T4 aluminum and the removable panel is machined from a piece of 6061-T6 aluminum. There is a

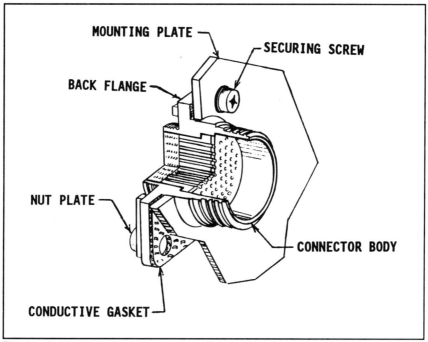

Figure 12.5 Flange mount connector assembly.

Table 12.3 Metal hardness.

Material	Alloy	Hardness
Aluminum	6061-T6 6061-T4	95 Brinell 65 Brinell
Nickel Electroplated	99% Ni	90-120 Brinell 140-250 DPH
Corrosion Resisting Steel	303	160 Brinell
Electroless Nickel	Ni-Ph	45-50 Rockwell C

significant strength difference between the two.

The exposed aluminum surfaces that are required to be electrically conductive have a chromate conversion coating applied. MIL-C-5541 defines this coating. Class 1A provides the maximum corrosion protection. The plate is immersed in the chromate conversion

Table 12.4 Connector shell-to-shell bonding information.

Connector	Shell-to-shell Bonding (milliohms)	Shielding Effectiveness at 10 Ghz
Corrosion Resistant Steel	10	-45 Db
Cadmium Plated Aluminum	2.5	-50 Db
Electrodeposited Nickel Plated Corrosion Resistant Steel	1	-65 Db
Electroless Plated Aluminum	1	-65 Db
Note: The shielding effectiveness information is obtained from Pyle-National Co., Chicago, IL		

bath for one to three minutes. Class 3 is intended for use in high conductivity areas and it has an immersion time of 15 to 45 seconds. The only difference between the two coatings is the exposure time in the coating bath. The dependence on bath exposure time makes the chromate conversion coating properties highly variable.

The maximum temperature for this coating is 60°C. Above this temperature a reduction in corrosion protection is seen. The corrosion resistance is time and temperature dependent. This is a significant concern because the maximum military temperature is 125°C.

Evaluations were made with 6061-T6 aluminum connector mounting plates that have an electroless nickel plating per MIL-C-26074 Class 4 Grade A. The nickel plating is a simple process for a removable mounting plate but becomes impractical for an entire chassis. Pinholes in the nickel plating over aluminum are unavoidable and result in significant aluminum corrosion.

In nonconductive areas many applications use paint. Proper masking of the paint is required so conductive areas are not contaminated. Also in some instances paint electrically insulates the jam nut from the mounting plate. This eliminates a conduction path through threaded jam nut interfaces. Conductive paths through the threaded fasteners should not be relied upon in the electrical bonding design.

12.4.4. Assembly Methods

The quality of the mechanical assembly defines the electrical bonding resistance. A clean joint with high interfacial pressure will yield a good low resistance bond.

<u>12.4.4(a) Jam Nut.</u> The jam nut connector torque must be specified by the designer to give the proper pressure. Figure 12.6 shows the jam nut torque used for the stainless steel connector assembly, (except that the shell size 25 connectors were evaluated at 300 in-lbs because the standard was not yet in place). Since there is only one and a half jam nut threads in the larger connector sizes, the torque is severely limited. This limited torque, the high stiffness of the connector and the low stiffness of the aluminum mounting plate result in a poorly designed bolted joint.

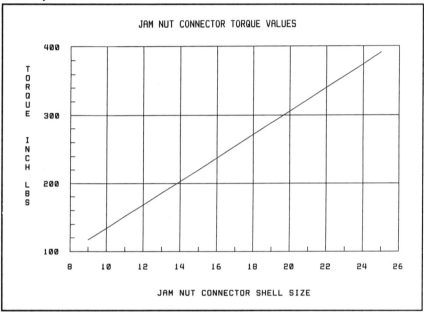

Figure 12.6 Jam nut torque values.

The biggest assembly issue with the jam nut connector is the improper restraint of the connector during tightening. This allows the connector to rotate into the D-hole as shown in Figure 12.7. Thread galling is a major issue in the stainless steel configuration because both the nut and the connector body are 303 steel. Thread galling permanently damages the connector and significantly changes the connector to enclosure pressure.

<u>12.4.4(b) Flange Mount.</u> Four screws in conjunction with the nut plate are used to secure the flange mount connector to the mounting plate. The screw torque must be specified in the assembly. The nut plate has nuts with a deformed thread that provide thread locking. Experience shows

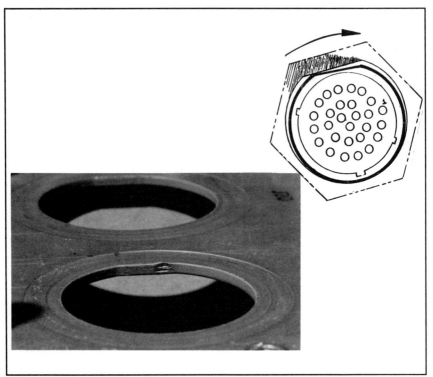

Figure 12.7 Jam nut connector rotation into the D-hole.

that the locking torque is highly variable and must be controlled to tight limits on the procurement document. High locking torque consumes all of the applied torque and no torque will be available to generate interfacial pressure.

<u>12.4.4(c) Environmental Stresses.</u> The electrical bonds are evaluated through temperature cycling, humidity and salt spray. The two thermal cycling temperature tests are from -55 to 125°C and from -55 to +60°C. The temperature cycles are shown in Figure 12.8. The 60°C temperature is the maximum rating for the chromate conversion coating per MIL-C-5541. Early in the evaluation it was thought that the chromate conversion coating was degrading at temperatures above 60°C, causing the increase in electrical bonding. The humidity exposure and the salt spray are tailored for the specific program requirements. The duration of these exposures is given in the test section.

<u>12.4.4(d) Measurement Methods.</u> A four wire precision ohm meter, model

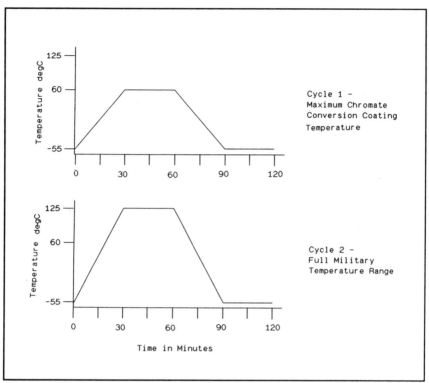

Figure 12.8 Thermal cycling profiles.

1705B made by Electro Scientific Industries (ESI) with digital converter model 1700, is used for the resistance measurements. The results were compared to a portable double Kelvin bridge made by General Electric. For resistances below one milliohm, readings were obtained within 0.2 milliohms. For higher resistances, the readings were within 0.4 milliohms.

Typically, there is no correlation between the bonding resistance values obtained just after assembly and the bonding resistance obtained after environmental exposure. The minimum environmental exposure prior to bonding measurements is program dependent. Caution must be used in the interpretation of the resistance data obtained. In configurations that exhibited poor electrical bonding, the securing hardware provides a significant alternate conduction path. For example, with the jam nut connector configuration, the lockwire should not be installed during the bonding resistance measurement.

The connector manufacturers measure the shell-to-shell bonding resistance of the mated connector pairs. Typically a maximum millivolt drop is allowed when one amp of current is passed through the mated connectors. For the MIL-STD-38999 connectors the bonding test is defined in MIL-STD-1344 Method 3007. This method is not used to check the bonding resistance of the connector to the mounting plate because of the risk to the control system electronics.

12.5 Test Procedure

The connectors are assembled onto the connector mounting plate. A sample assembly is shown in Figure 12.9. Initial bonding resistance measurements are made between a defined part of the connector and a defined location on the mounting plate. Initial measurements are taken. Then the samples are subjected to thermal cycles and the bonding resistance is measured periodically through the cycling and after each environment.

12.5.1 Constant Pressure Bonding Resistance of the Jam Nut Connector.

The constant pressure evaluation is designed to eliminate the variable of pressure from the bonding resistance of the jam nut connector. Due to the changes in temperature, it is possible that the interfacial pressure changes and/or stress relaxation occurs due to the difference in coefficients of thermal expansion. The evaluation is designed to determine if there is a basic material compatibility problem with the materials being used. An analysis of a D38999/24KJ35PN stainless steel jam nut connector in a 0.080in thick aluminum mounting plate shows that the mounting plate will contract 38 percent more than the connector at -55°C.

To eliminate potential jam nut relaxation, a clamping press was fabricated to evaluate the bonding resistance after temperature cycling exposure. The press is shown in Figure 12.10. The press is designed to apply 400lbs load to the sealing flange of the connector body. There is no jam nut in the assembly and the only conduction path is through the rear flange of the connector. A connector and mounting plate are loaded in the fixture. The fixture assembly is then put into an Instron Tension/Compression Tester and the springs are compressed to 400 lbs. Initial bonding resistance measurements are made and then the assembly is thermally cycled. The results for the various configurations are given in Section 12.6.1.

Figure 12.9 Typical test sample.

12.5.2 Interface Pressure and Bonding Resistance.

The bonding resistance of any interface is sensitive to the pressure at the material interface. The Instron tension/compression tester is used to evaluate the connector to mounting plate bonding resistance for interfacial pressures up to 1000 lbs. The connector body is placed in contact with the plate and resistance measurements are taken as the assembly is loaded. The results of the various configurations are given in Section 12.6.2.

12.5.3 Jam Nut Connector.

The bonding resistance of jam nut connectors secured to a mounting

Figure 12.10 Constant pressure clamping press.

plate is evaluated in thermal cycling and salt spray. Earlier jam nut connector configurations that this author tested were subject to humidity with no major increase in bonding resistance; therefore the humidity test was not performed. Various plating combinations of the stainless steel connector and the 6061 aluminum mounting plate are evaluated.

The connectors were evaluated with and without insulating epoxy paint under the jam nuts. Since a proper design does not allow conduction paths to go through threads, the data obtained with the insulated jam nut is preferred. A fungus inhibited epoxy coating used for the jam nut insulation is per MIL-C-22750, gray color No. 36231 of FED-STD-595.

The test matrix for the jam nut connector testing is shown in Table 12.5. Stainless steel jam nut connectors with the standard passivation per QQ-P-35 on the chromate conversion coated aluminum mounting plate are the baseline. The electroless nickel plate is evaluated as an alternative to the chromate conversion coating as a conductive enclosure coating.

All of the samples are exposed to a minimum of 175 thermal cycles from -55 to +125°C and 9 days of salt spray per ASTM 117B

Table 12.5 Jam nut connector test.

Connector Finish	Mounting Plate Finish	Insulated Jam Nut	Bonding Resistance (milliohms)	Thermal Cycles
Standard QQ-P-35 Passivation	MIL-C-5541 Class 1A Chromate Conversion	Yes	3.7 to 90 Failed	175
		No	Below 2.0 Passed	175
* Standard QQ-P-35 Passivation	MIL-C-26074 Class 4 Grade A Electroless Ni	No	0.43 to 1.3 Passed	126
** QQ-N-290 Electroplated Nickel	MIL-C-5541 Class 1A Chromate Conversion	Yes	0.3 to 0.53 Passed	400
QQ-N-290 Electroplated Nickel	MIL-C-26074 Class 4 Grade A Electroless Ni	Yes	0.2 to 0.33 Passed	175
MIL-C-26074 Electroless Nickel	MIL-C-5541 Class 1A Chromate Conversion	No	0.58 to 5.0 Failed	175
MIL-S-5002 Electropolish	MIL-C-5541 Class 1A Chromate Conversion	Yes	3.70 to 17 Failed	175

Notes:
Connector base material: 303 stainless steel
Mounting plate base material: 6061-T6 aluminum
Thermal cycles are from -55 to +125°C
Humidity exposure is per MIL-STD-810C Method 507.1
All samples exposed to 9 days of salt spray (MIL-STD-202)
 * 240 hours humidity
 ** 18 days of salt spray

(except the passivated stainless steel on electroless nickel which had 126 cycles). In order to fully understand the phenomenon behind oxide film development, one connector of each configuration is exposed to 18 hours of the humidity cycle shown in Figure 12.11 with 8 hours at 55°C, 4 hour ramp and 5 hours at 20°C.

Extended testing is performed on the electroplated nickel connector on the chromate conversion coated 6061-T6 aluminum mounting plate with the insulated jam nut is evaluated for 400 thermal cycles, and 18 days of salt spray. The test results are given in Section

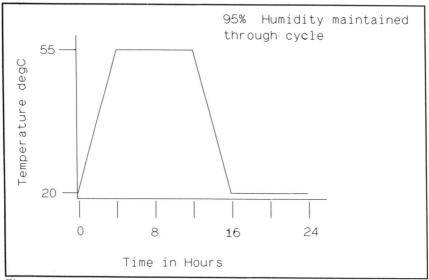

Figure 12.11 Flange mount connector humidity cycle.

12.6.3.

12.5.4 Flange Mount.

The flange mount connector configuration is tested in thermal cycling, humidity and salt spray. The stainless steel connector finishes are passivated per QQ-P-35 and electroplated nickel per QQ-N-290. The mounting plate material is a 6061 dip brazed aluminum with a T4 temper. The mounting plate is finished with a chromate conversion coating per MIL-C-5541 Class 1A or brushplated nickel per MIL-C-865 0.0002 inch thick. A 4-40 corrosion resisting steel screw torqued to 8 inch-lbs is used for the testing. The sealing gasket is a woven monel wire in a fluoro-silicone elastomer. A short ground strap is installed on one of the samples to see if the connector bonding is improved. The strap is made of 0.003 inch thick tin plated copper sheet 0.300 inch wide with a length of 0.650 inch. The ground strap is bent into a C shape and provides a bonding path around the gasket.

The test configurations are shown in Table 12.6. Each configuration is thermal cycled from -55 to +125°C 400 times, exposed to 15 days of humidity per Figure 12.11 and 96 hours of salt spray per MIL-STD-810. The results are summarized in Section 12.6.4.

Table 12.6 Flange mount test.

Connector Finish	Mounting Plate Finish	Monel Gasket /Ground Strap	Resistance (milliohms)			
			Initial	After Thermal Cycles	After Salt Spray	After Humidity
Standard QQ-P-35 Passivation	MIL-C-5541 Class 1A Chromate Conversion	Yes/No	0.59	5.24	6.91	8.15 Fail
		Yes/Yes	0.60	3.18	3.95	6.0 Fail
		No/No	0.88	1.19	1.78	3.4 Fail
Mechanically Polished		No/No	1.34	1.54	2.79	3.25 Fail
QQ-N-290 Electro- plated Nickel		Yes/No	0.21	0.42	0.74	1.32 Pass
		Yes/Yes	0.35	0.54	1.03	1.48 Pass
	(no salt spray) ----->	No/No	0.21	0.21	0.43 Pass	
Standard QQ-P-35 Passivation	QQ-P-290 Electro-plated Ni	Yes/No	0.49	1.72	2.25	2.25 Marginal Pass

Notes:
Connector base material: 303 stainless steel.
Mounting plate base material: 6061-T4 aluminum.
Thermal cycles are from -55 to +125°C.
Humidity exposure: 15 days (95% humidity through cycle, 55°C for 8 hours, 20°C 8 hours, 4 hour ramp between temperatures.)
Salt spray: 96 hours per MIL-STD-202.
All samples exposed to 9 days salt spray per MIL-STD-202.

12.5.5 Temperature and Bonding Resistance.

The bonding resistance of passivated stainless steel connectors on a chromate conversion coated mounting plate is evaluated over a -30 to +80°C temperature range. Three connectors are installed in one plate. An Invar washer was designed to match the thermal coefficient of expansion and installed under two of the connectors. The assembly is exposed to 266 thermal cycles from -55 to +125°C. The resistance is then checked over temperature.

Three additional connectors are assembled and cycled like the first assembly except the mounting plate is electroless nickel per MIL-C-26074 Class 4 Grade A. The tightening torque at installation and after cycling is closely monitored. The results are given in Section 12.6.5.

12.6 Results.

The electroplated nickel stainless steel connector on a chromate conversion coated aluminum mounting plate is the best configuration overall. This configuration yielded bonding below 0.5 milliohms for the jam nut configuration and below 1.5 milliohms for the flange mount connector with a monel gasket.

12.6.1 Constant Pressure Evaluation Results.

There is no real difference in electrical bonding between the -55 to +60°C thermal cycle and the -55 to +125°C thermal cycle. The bonding results for the chromate conversion coating per MIL-C-5541 Class 3 (MIL-STD-171 Paragraph 7.3.3) are shown in Figure 12.12. The chromate conversion coat bonding resistance behavior is highly variable, ranging from 30 to 2.5 milliohms.

12.6.2 Interface Pressure and Bonding Resistance Results.

The most pressure sensitive bond is the passivated stainless steel connector on the chromate conversion coated aluminum. The least pressure sensitive electrical bond is the stainless steel with electroplated nickel connector on a chromate conversion coated aluminum plate. With a pressure of 140 psi a bonding resistance of 0.4 milliohms is obtained. The results are shown in Figure 12.13.

12.6.3 Jam Nut Results.

The final bonding resistance values are given in Table 12.5 along with the configuration description. The best configuration is the electroplated nickel stainless steel connector on an electroless nickel plated mounting plate. The electroplated nickel connector on a chromate conversion coated aluminum mounting plate is the next best configuration and the

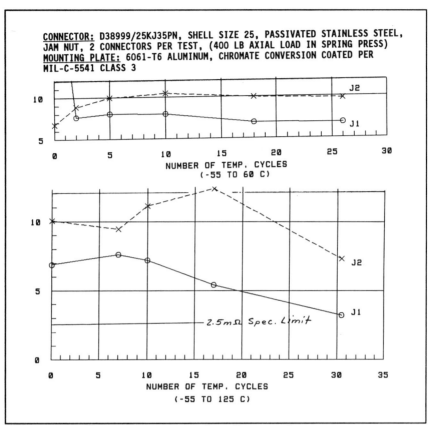

CONNECTOR: D38999/25KJ35PN, SHELL SIZE 25, PASSIVATED STAINLESS STEEL, JAM NUT, 2 CONNECTORS PER TEST, (400 LB AXIAL LOAD IN SPRING PRESS) MOUNTING PLATE: 6061-T6 ALUMINUM, CHROMATE CONVERSION COATED PER MIL-C-5541 CLASS 3

Figure 12.12 Constant pressure bonding results.

production costs are much less. Figure 12.14 shows plots of thermal cycles versus bond resistance for the best and the worst configurations.

12.6.4 Flange Mount Results.

The final bonding resistance values are given in Table 12.6. The best configuration with a monel wire gasket is the electroplated nickel connector on the chromate conversion coated aluminum mounting plate.

12.6.5 Temperature and Bonding Resistance Results.

The resistance measurements are taken after the samples complete 240 thermal cycles from -55 to +125°C. The resistances over the temperature range of -30 to +80°C are shown in Figure 12.15. The

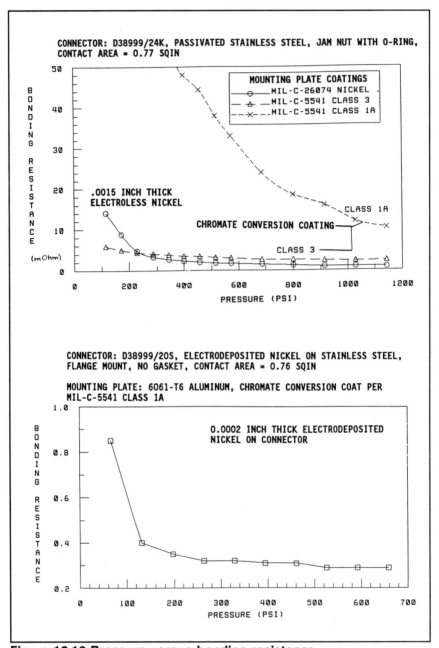

Figure 12.13 Pressure versus bonding resistance.

torque values of the six connectors was evaluated after the cycling and

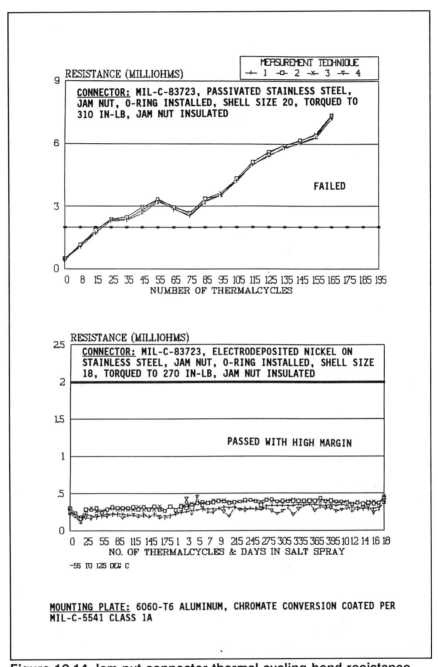

Figure 12.14 Jam nut connector thermal cycling bond resistance.

the results are shown in Table 12.7. The Invar washer designed to match thermal coefficients of expansion did not make a difference in torquing results.

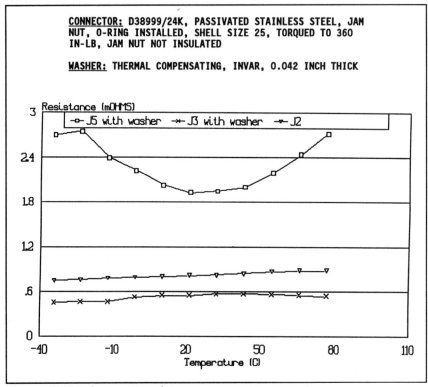

Figure 12.15 Jam nut resistance versus temperature.

One connector had a tightening torque increase of 61 percent after thermal cycling and failed the bonding resistance. Thread galling is likely.

12.7 Conclusion/Discussion.

12.7.1 Recommended Configuration.

The electroplated nickel (per QQ-P-290) connector on a chromate conversion coated (per MIL-C-5541 Class 1A) aluminum plate (6061 T4-T6) is the recommended configuration for future applications requiring stainless steel connector shells. This configuration assures a good bonding path that does not rely on a threaded interface for contact.

The pressure versus resistance data shows that this combination

Table 12.7 Post-thermal cycling torque values.

Mounting Plate	Connector	Tightening Torque (in-lb)	Percent change
Electroless Nickel	J2	220	-39
	J3 (washer)	360 (break torque)	0
	J5 (washer)	320	-11
Chromate Conversion	J2	280	-22
	J3 (washer)	260	-28
	J5 (washer)	580	+61

All connectors initially torqued to 360 in-lbs.
Connector - D38999/24KJ35PN (stainless steel jam nut).
All connectors subject to 266 cycles from -55 to +125 °C.
Chromate Conversion Coating is per MIL-C-5541 Class 1A.
Electroless Nickel is per MIL-C-26074 Class 4 Grade A.

has the least pressure sensitive bonding resistance. The reduced pressure sensitivity allows more pressure variation during installation and life. The additional cost of nickel plating is insignificant compared to the $125-$300 cost of the connectors.

12.7.2 Electroless Nickel Plating (MIL-C-26074)/Electroplated Nickel (QQ-P-290).

Electroless nickel plating the mounting plate provided a significant improvement when used with the passivated stainless steel jam nut connector. A hypothesis is that the hard electroless nickel on the mounting plate breaks through the oxide layer on the passivated stainless steel connector achieving a good metal contact.

The electroplated nickel connector achieved good bonding results because the connector surface is changed from the hard stainless steel to the malleable electrodeposited nickel. According to Dr. Vook of the Chemical Dept. at Syracuse University [4]:

"One should expect substantially lower contact resistances across a nickel surface over that of stainless steel. Even with comparable oxide film thicknesses the malleability of electrodeposited nickel would allow the asperities to give-way or crush under the securing pressure. This crushing of the asperities would create holes in the protective film where good metal-to-metal contact could be achieved. This good metal-to-metal contact will result in the lower resistance values."

Dr. Vook then provided further justification by comparing the protective film on nickel and stainless steel to ice on a sheet of rubber and on a slab of concrete. If one is to stand on the rubber configuration the ice would crack and expose portions of the rubber surface. Concrete, on the hand, is too rigid to yield under the applied pressure and the ice will not crack, thus not exposing any of the concrete surface.

In most cases, once a metal's protective oxide film is broken it will mend itself. However this will not happen if the electrodeposited nickel's film is broken as the flange is pressed up against the chassis or front panel. Oxidizing agents are unable to reach the exposed metal. This is due to the tight fit that results from the torquing of the jam nuts.

In addition to the protective oxide film, volume resistivity is still another significant difference between nickel and stainless steel. According to the second edition of Materials and Processes [5], the volume resistivity of stainless steel is approximately 11 times greater than that of nickel, or 33.1 to 3.08 microhm-in., respectively.

12.7.3 Jam Nut Torque.

The ability to translate the data determined in the pressure versus resistance section hinges on determining the interfacial pressure in the actual assembly. The torque applied to the nut/screw translates to interfacial pressure between the connector and the mounting plate. Calculating this pressure requires friction coefficient knowledge. The general equation is given in Shigley's machine design book [6] as

$$Torque = 0.2\, Force \times Nominal\, thread\, diameter$$

[12.4]

where the thread diameter, force and torque are conventionally in inches, lbs, and in-lbs, respectively, and 0.2 is the coefficient of friction for standard size nuts and bolts. This coefficient is valid for all standard threaded fasteners regardless of size or the pitch of the threads. Preliminary testing of a size 25 stainless steel jam nut connector shell on aluminum yielded friction coefficients higher than 0.2. Not enough data was obtained to statistically determine a coefficient because the data is highly variable.

The variation in interfacial pressure over the temperature range and the level of pressure after extended thermal cycling is not well understood. J.H. Bickford [7] states that "The thread engagement for steel fasteners should be at least 0.8 times the nominal thread diameter of the fastener. If the engagement length is too short (too few threads to support the load), thread contact areas are again smaller than those intended by the fastener manufacturer and excessive relaxation can result." The jam nut connectors are not tightened to the loads of a conventional bolted assembly, but the thread section is only 0.07 times the nominal diameter (0.125 inch thick/1.7 inch diameter). Analysis of this threaded interface is continuing but a torque reduction of up to 20 percent can be expected.

12.7.4 Resolution of Hypotheses.

The chromate conversion coating on the aluminum does not break down and form a high resistance film above its rated temperature of 60°C. If this were true the electrodeposited nickel on the connector would not have worked.

The pressure relaxation hypothesis is very difficult to test because interface pressure data cannot be directly obtained. This hypothesis is not fully tested. It is true that the electroplated nickel connector on the chromate conversion coating had the bonding resistance that was the least sensitive to pressure.

The effect of an oxide layer on the bonding resistance is also very difficult to determine. Oxide thickness in the 30 to 80Å thickness range are not readily measurable. The growth of the oxide layer along the surface would result in a reduction of bare metal contact area. This would increase the bonding resistance. The oxide progression on the surface is fostered by the differences in thermal expansion rates in the radial direction. A configuration with a limited number of contact sites will tend to degrade faster than one with many sites.

Adding a soft electroplated nickel to the surface of the stainless steel has the lowest bonding resistance through environmental exposure. Dr. Vook's rationale on the hardness compatibility between the stainless steel and aluminum seems to hold true. In thermal cycling, the malleable electrodeposited nickel continually forms clean metal sites promoting good bonding.

12.7.4 Other Benefits of Electroplated Nickel Connectors.

The metal oxide varistor (MOV) transient pins and the ground pins in the hermetic connectors will drastically improve with the addition of the nickel plating. It had been determined that many of the ground pins in the connectors have resistances between the ground pin and the connector shell (pin-to-shell resistances) as high as 250 milliohms. The internal configuration of a custom pin connector is represented in Figure 12.16, where the pin is press fitted into a nickel plated ceramic disk. The disk is then press-fitted into a copper ring that is pressed against the stainless steel shell. The oxide film on the stainless steel is the reason for the inconsistently large pin-to-shell ground path resistances, and the addition of the electrodeposited nickel plated shell will improve the designed path to ground. The connector manufacturers Pyle and Bendix have committed to meeting less than 5.0 milliohms consistently if the connector shell surface is changed from stainless steel to electrodeposited nickel.

The shell-to-shell engine cable mounting resistance between the electronic control system's receptacle and the plug on the cable will be dropped from the 200 milliohms max to 5.0 milliohms max if the plug is changed from class Y to class N (MIL-C-83723). A class N plug is a class S plug with electrodeposited nickel plating. Similarly, a change from the MIL-STD-38999 Stainless Steel firewall (Class K) to the Electrodeposited Nickel Plated Stainless Steel (Class N) will result in a shell-to-shell bonding resistance reduction from 10 milliohms to one milliohm.

12.8 Acknowledgments.

Thanks are due to Phil Kingsley and Ron Kurz Jr. for their help in the preparation of this document. Much of the test data and discussion points are obtained from earlier memos they prepared [8,9].

12.9 References.

1   J.W. Gooth, J.K.Daher, Electromagnetic Pulse and Electromagnetic Interference versus Corrosion Protection Methods, written for Warner-Robins Air Logistics Center MMEMC Robins AFB, Georgia, Georgia Institute of Technology Atlanta, Georgia, August 31, 1987

2   J. Salvaggi (GE Internal Memorandum) March 1990

3   K.E. Pitney, Mgr. Electronic R&D, NEY Contact Manual, The J.M. Ney Co., Bloomfield, CT

4   Dr. Vook, Chemical Dept. Syracuse University, (private conversation)

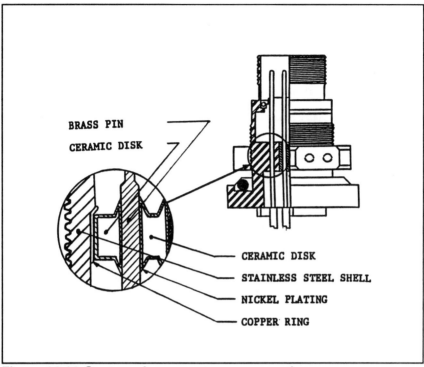

Figure 12.16 Custom pin connector cross-section.

5 J.F. Young, Materials and Processes 2nd Edition

6 J.E. Shigley and L.D. Mitchell, Mechanical Design Engineering, 4th Edition (McGraw-Hill, New York, NY) pp. 378

7 J.H. Bickford, An Introduction to the Design and Behavior of Bolted Joints, 2nd Edition (Marcel Decker, Inc, New York, NY)

8 P. Kingsley (GE Internal Memorandum) July 1990

9 R. Kurz Jr. (GE Internal Memorandum) March 1990

Appendix

MILITARY AND FEDERAL SPECIFICATIONS

MIL-C-5087   Bonding, electrical, and lightning protection

MIL-C-26074D   Military Specification requirements for coatings and electroless nickel

MIL-STD-171E   Military Specification requirements for finishing of metal and wood surfaces

MIL-C-83723D   Connectors, Electrical (Circular, Environment Resisting), Receptacles and Plugs

MIL-C-38999   Connectors, (Circular, Environment Resisting), Electrical

QQ-N-290   Nickel Plating (Electrodeposit)

MIL-C-5541   Chemical Conversion Coatings on Aluminum and Aluminum Alloys (Chromate Conversion Coatings)

PARAMETERIZATION OF

FINE PITCH PROCESSING

GEORGE R. WESTBY

MICHAEL D. SNYDER

13.1 Introduction

Assembly of large fine pitch gull wing leaded surface mount components has become a necessity in the printed circuit board assembly industry. The assembly process yields are limited by practical barriers originating in component and bare board construction constraints and material variations as well as the assembly process itself. It is the purpose of this paper to identify and quantify these areas and present assembly yield data obtained in a carefully controlled manufacturing environment. The variations observed in the materials is then summarized and a Monte Carlo technique used to simulate a pick-and-place process yield model.

13.2 Assembly Process Characterization

13.2.1 Solder reflow

The process flow for a mass reflow assembly process as described in Figure 13.1 was the model used for this paper. Solder paste is deposited on the fine pitch land areas using a solder paste stencil, after which components are picked and placed and attached using infrared or vapor phase mass reflow techniques. Components unsuitable for mass reflow would be picked, placed, and attached subsequent to mass reflow. Components addressed in this paper are limited to QFP, PQFP, and TapePak* devices at 0.025" and 0.5mm pitch. TQFP and TAB type packages must be assembled with pick, place, and attach techniques since mass reflow temperatures will cause deformation of the lead support polyimide and therefore the leads.

   The printed circuit board attributes listed below must be monitored to insure consistency of yields during the process:

   - Height of mask relative to pads
   - Coplanarity of pads
   - Quantity of solder on pads
   - Solderability
   - Intermetallic thickness
   - Feature size repeatability
   - Feature position repeatability and accuracy

   The height of the printed circuit board solder mask must be coplanar to or below the attachment pads to provide a sufficiently planar

---

*TapePak is a registered trademark of National Semiconductor Corporation.

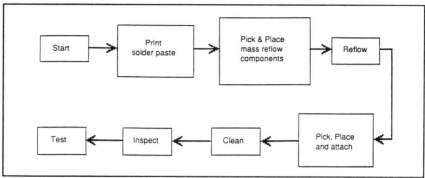

Figure 13.1 Process flow for combined mass reflow and pick, place and attach assembly.

surface onto which solder paste is deposited. Excessive solder mask thickness inhibits intimate contact of the stencil with the pads, significantly reducing print definition, and encouraging shorts between adjacent pads. Attachment pads must provide coplanarity within 0.001" to allow accurate dispensing of solder paste and contact to all component leads.

Hot air leveling of the attachment pads produces different results depending upon the direction of attack of the board to the solder wave. Pads parallel to the wave have been shown in Figure 13.2 to produce an average of 603 microinches of solder on the pad, whereas solder deposited on pads longitudinal to the wave produce an average of 165 microinches as shown in Figure 13.3. Large pads will have a tendency to have large thickness variations across the pad itself, whereas small pads will have smaller variations and higher thicknesses.

Solderability problems can occur with the hot air leveling process if the copper tin intermetallic layer is exposed when excessive air knife pressure is used. The intermetallic should be between 20 and 25 microinches for properly deposited solder layers [1].

Localized printed circuit board warpage (see Figure 13.4) that occurs from unbalanced ground plane inner layer constructions can result in substrates which present non-coplanar attachment sites exacerbating component coplanarity problems.

General printed circuit board warpage or localized board warpage can also create a non-coplanar condition that can affect soldering yields of the board. Warpage is generally limited to 0.008" per industry IPC standards, localized warpages should also remain less than or equal to 0.008"/inch.

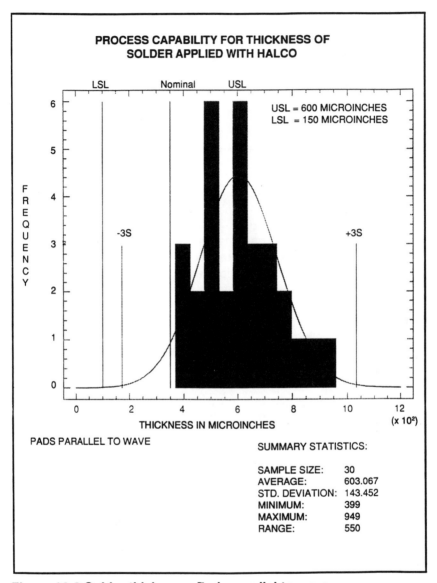

Figure 13.2 Solder thickness: Pads parallel to wave.

13.2.2 PCB characterization.

Figure 13.5 is a graphical representation of the technique that was used for measuring a group of printed circuit boards. Distances between global fiducials and the centers of component pads were measured using

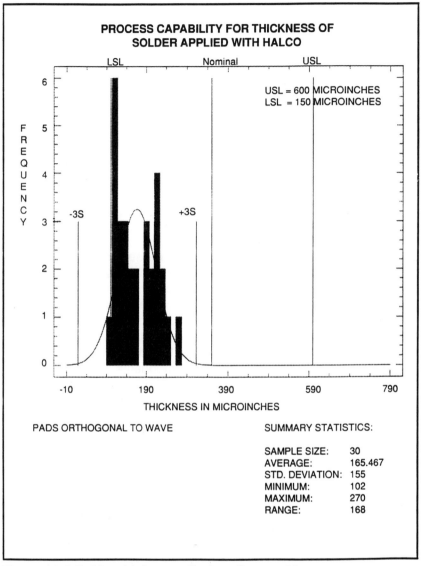

Figure 13.3 Solder thickness: Pads orthogonal to wave.

eight pads from the printed circuit board for each component, and the center of the component pads used to define the perimeter of a circle. The center of the circle defined the center of the component pad set.

Figure 13.6 shows a process capability study for the position of the component pads relative to global fiducials. The centroid of the

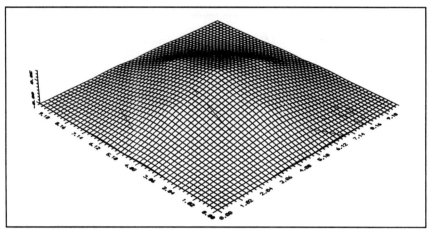

Figure 13.4 Printed circuit board warpage.

distribution is approximately 0.001" below the design nominal position. Figure 13.7 is the process capability study for the position of component pad relative to local fiducials. The position of the local fiducial to the component pads fell on the design nominal, however the three sigma standard deviation of the position exceeded the upper and lower specification limits of +/- 0.001". While these tolerances exceed those established for any class of printed circuit board in IPC-D-300G, they are required for the assembly of fine pitch components.

Component placement using global fiducials would result in a short placement. Local fiducials would reduce the consistently short placement, however, the high standard deviation of the pad center to fiducial distance would still produce high standard deviation placements.

Figure 13.8 is a process capability study for the feature size relative to the original master artwork. The feature sizes are significantly below the design specification. The size of this feature was 0.002" below the nominal design point. Additionally, the three sigma standard deviations fell beyond the upper and lower specifications limits of +/- 0.001". This data demonstrates a need to specify the end use of the product to the substrate manufacturer as well as specifying the acceptable tolerances required on the printed circuit board. The feature position and size affect the yields in the fine pitch process.

The effect of an "out of spec" feature size increases as component pitch is reduced and attachment pads become smaller. With the overall error unchanged, the percentage of attachment pad area affected increases. Alterations of 0.002" in pad width affect the overall process yield when pad widths extend below 0.015".

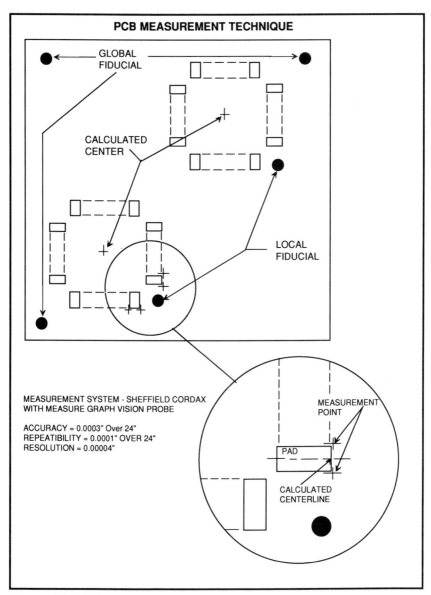

Figure 13.5 PCB measurements.

13.2.3 Components.

Components contribute significantly to the process variability through splayed tweezed, swept, non-coplanar and unsolderable leads.

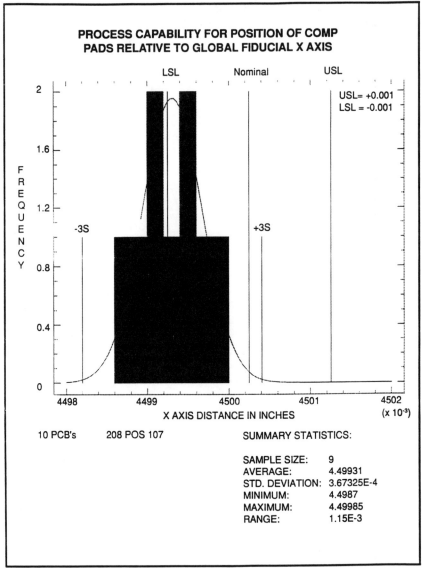

Figure 13.6 Process capability: Global positioning.

Variations in lead width as shown in Figure 13.9 suggest that either several lead frames were used in the manufacture of these 132 pin PQFP components or the process does not produce normally distributed product. Figure 13.10 demonstrates the pitch of these components to be centrally distributed around 0.025" with an overall range of 0.0083".

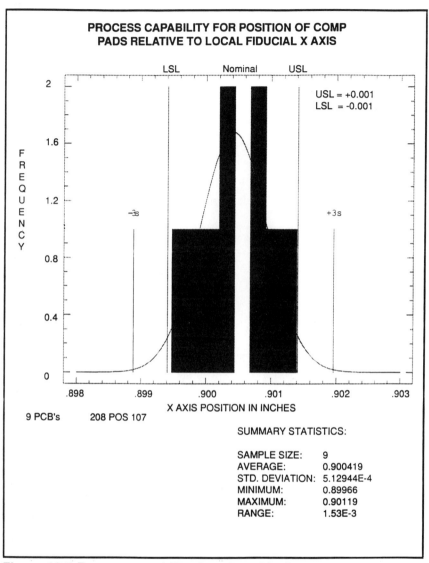

Figure 13.7 Process capability: Local positioning.

Component lead coplanarity can produce unsoldered leads when they are sufficiently deformed to inhibit intimate contact with the solder paste. A triangulation laser mounted on a Sheffield Cordax CMM was used to measure the relative lead heights to conduct analysis of open joints on assembled test panels.

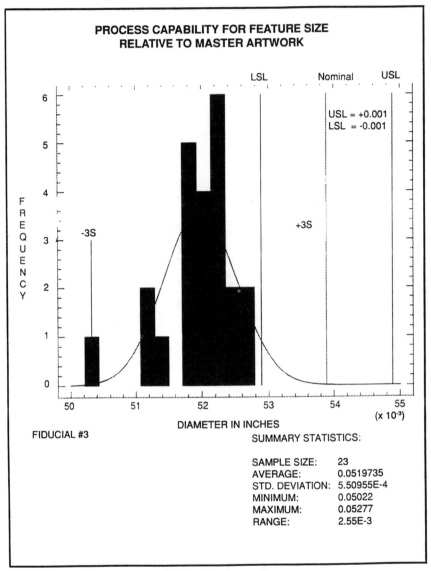

Figure 13.8 Process capability: Feature size.

13.2.4 Solder printing.

Solder paste viscosity, tackiness, metal content, particle size, and flux activity must be monitored to assure adequate process yields. Solder paste affects process steps beyond the solder paste print process as shown in Figure 13.11. The viscosity of solder paste used for fine pitch

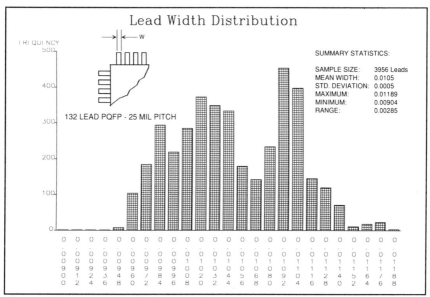

Figure 13.9 Lead width distribution.

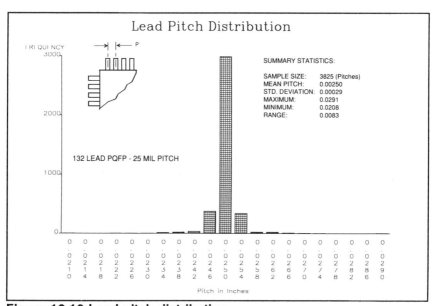

Figure 13.10 Lead pitch distribution.

is significantly higher than that used for 0.050" pitch components. Higher viscosities and metal content enhance print definition reducing shorting.

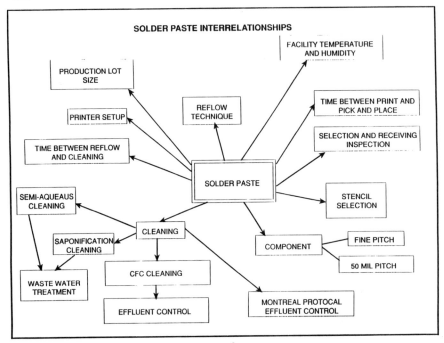

Figure 13.11 Solder paste process issues.

        The viscosity of solder paste used for fine pitch when measured
with a Brookfield RVTD Viscometer at 25 degrees centigrade and 5 rpm
ranges from 1 to 1.8 million centipoise.   This experiment utilized 1.8
million centipoise solder paste as shown in Figure 13.12.   Higher
tackiness solder pastes are typically used for fine pitch applications due
to the importance of retaining the placed location of the component.
Figure 13.13 shows the thickness results on the solder paste used in the
experiment. The tests were conducted utilizing an Instron machine with
a one square inch ram test fixture. The maximum tear resistance mean
was calculated to be 6.9 pounds per square inch.
        Solder paste disposition with paste stenciling equipment requires
proper stencil selection. The printed width to pad width ratio and the
stencil thickness to pad width ratio dictates process yields.
        Brass, beryllium copper, and stainless steel are used in the
construction of stencils. All can be successfully utilized providing the
substrate is sufficiently flat to provide adequate support for the stencil.
The print width or narrowest print dimension should not be less than the
thickness of the stencil. Stencil thickness of 0.0045" to 0.006" should be
used on 0.025" to 0.0196" (0.5mm) pitch devices to provide adequate
solder fillet volumes. Since fine pitch components typically reside on

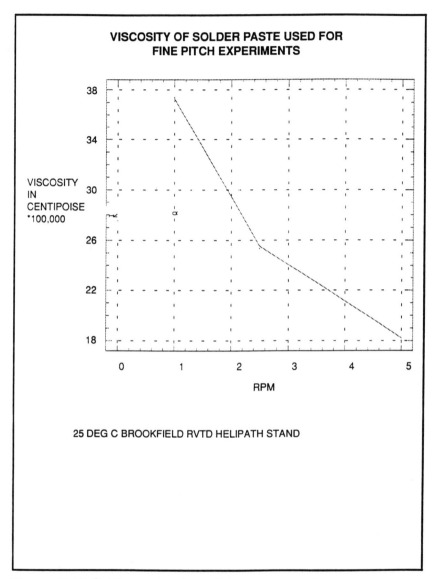

Figure 13.12 Solder paste viscosity measurement.

printed circuit boards with coarse pitch devices, the use of multi-thickness stencils is required.

Solder paste height and volume determinations must be conducted with non-contact sensors. Figure 13.14 shows a topographical projection of solder paste deposited with a 0.006" thick

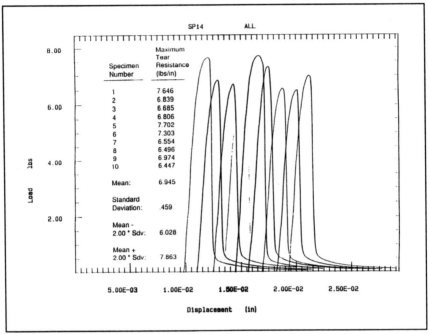

Figure 13.13 Solder paste tack test.

stencil and measured with a triangulation laser. The technique was used to measure a group of samples printed on both bare laminate and a patterned test board. The vertical pads, those normal to the direction of squeegee travel, were found to be significantly wider than the horizontal pads, those orthogonal to the direction of squeegee travel. (See Figure 13.15) Wider prints enhance shorting between adjacent pads during reflow, suggesting that the fine pitch stencil openings normal to the direction of squeegee travel should be made somewhat narrower.

An experiment was conducted on both 132 pin 0.025" pitch and 208 pin 0.5mm pitch devices to determine the overall yields expected and the effect of varying the print to pad ratio. The yields with a 0.010" print and pad ranges from 48 to 239ppm at 95% confidence interval. A 0.015" pad and 0.012" print produces 0ppm process problems. (See Figure 13.16) Failure analysis of the devices showed 0 to 85ppm open due to component coplanarity at 0.003". These analyses were conducted with the triangulation laser.

Experiments on the 208 pin device showed that the optimum pad to print width ratio was 1.22 as described in Figure 13.17. This corresponds to a 0.0109" wide pad and 0.0089" wide print. Opens in the range of 600ppm due to insufficient solder were found. No solder shorts

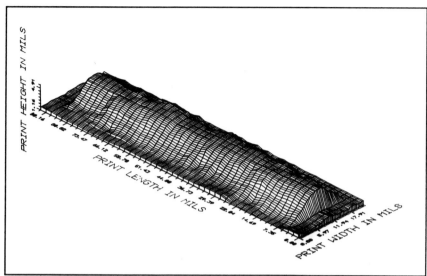

Figure 13.14 Solder paste test print (large) for 208L PQFP.

were observed, but severe coplanarity problems produced 26,000ppm opens. Failure analysis showed coplanarity opens occurred at 0.0025".

13.2.5 Assembly summary.

In summary, to obtain optimum yields in the assembly process, the following process features must be controlled.

-PWB fabrication process for flatness, solder leveling quantities, and feature location.

-Component coplanarity of 0.003" is required for 0.025" pitch and 0.002" for 0.0196 pitch components.

-Solder paste selection must be based upon component pitch and cleaning technique.

-Ratio of stenciled width to attachment pad width may be altered to improve soldering yield.

In order to properly understand the entire fine pitch component assembly process component, machine, and component handling issues must be considered.

The component handling and placement portion of fine pitch assembly typically encompasses process steps shown in Figures 13.18 and 13.19. A quantification of errors associated with these process steps will allow prediction of the likelihood of the assembly process to generate particular yield results.

SOLDER PASTE STENCILING RESULTS

6 MIL THICK STENCIL ON BARE LAMINATE

11. 0 MIL WIDE PRINT

VERTICAL PADS=   4.35 MIL HIGH   X 13.07 WIDE   X 72 LONG   =2698 X 10^{-7} IN^{-3}

HORIZONTAL PADS= 5.30 MIL HIGH   X 11.24 WIDE   X 74 LONG   =2939 X 10^{-7} IN^{-3}

8.8 MIL WIDE STENCIL

VERTICAL PADS=   4.51 MIL HIGH   X 11.94 WIDE   X 71 LONG   =2549 X 10^{-7}IN^{-3}

HORIZONTAL PADS= 5.51 MIL HIGH   X 9.66 WIDE   X 72 LONG   =2555 X 10^{-7} IN^{-3}

6 MIL THICK STENCIL ON TEST PCB

11.0 MIL WIDE PRINT

VERTICAL PADS=   4.59 MIL HIGH   X 13.2 WIDE   X 72 LONG   =2908 X 10^{-7} IN^{-3}

HORIZONTAL PADS= 5.74 MIL HIGH   X 11.6 WIDE   X 74 LONG   =3285 X 10^{-7} IN^{-3}

8.8 MIL WIDE STENCIL

VERTICAL PADS=   4.47 MIL HIGH   X 11.55 WIDE   X 70 LONG   =2409 X 10^{-7} IN^{-3}

HORIZONTAL PADS= 5.63 MIL HIGH   X 10.51 WIDE   X 72 LONG   =2804 X 10^{-7} IN^{-3}

VOLUME CALCULATIONS BASED ON PARABALOID EXTRAPOLATION.

Figure 13.15 Solder paste stenciling measurements.

13.3 Placement Yield Model.

Once the process steps have been identified, and the ways in which they interact understood, a placement yield model can be constructed. For

ASSEMBLY YIELD
132 LEAD 25 MIL PITCH - PQFP
10 MIL WIDE PAD- 10 MIL WIDE PRINT

PROCESS PROBLEMS:

INSUFFICIENT SOLDER	48 TO 188 PPM
EXCESSIVE SOLDER / SHORT	0 TO 51 PPM

TOTAL PROCESS DEFECT RATE   **48 TO 239PPM**

COMPONENT VENDOR PROBLEMS:

LEAD SOLDERABILITY	13 TO 116 PPM
DAMAGED PIN (PLCC)	0 TO 32 PPM
COPLANARITY	0 TO 85 PPM

TOTAL DEFECT RATE   **61 TO 472 PPM AT 95% CONFIDENCE**

ASSEMBLY YIELD
132 LEAD 25 MIL PITCH - PQFP
15 MIL WIDE PAD- 12 MIL WIDE

PROCESS PROBLEMS:

NONE	0 PPM

COMPONENT VENDOR PROBLEMS:

SOLDER BRIDGE ON LEAD SHOULDER	0 TO 301 PPM
SOLDERABILITY	0 TO 403 PPM
COPLANARITY	0 TO 87 PPM

TOTAL DEFECT RATE   **0 TO 891 PPM AT 95% CONFIDENCE**

NOTES: DEFECTS CALCULATED ON A SOLDER JOINT BASIS,
FAILURE DATA ASSUMED TO BE NORMALLY DISTRIBUTED.

Figure 13.16 Process yield with varying print to pad ratio.

one typical machine structure, the error model is as pictured in Figure 13.20. Although several metrics may be used to assess placement yield, the one selected for the model to be discussed measures lead to pad

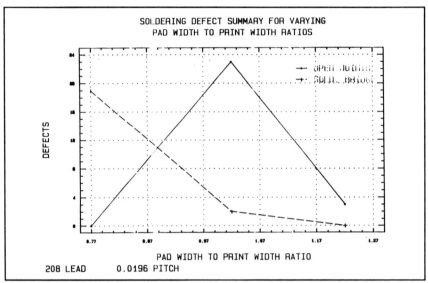

Figure 13.17 Soldering defect summary for varying pad width to print width ratios.

percentage coverage. The use of this metric makes the assumption that if all leads achieve a sufficient degree of coverage on the respective pads, good quality solder joints will be reliably created.

Referring to Figure 13.18, the first portion of the placement process deals with computing the location and orientation of the final resting place of the component to be placed. This information is customarily determined through an application of vision and X - Y positioning system technologies.

Typically, circuit boards associated with fine pitch assembly have vision identifiable features or fiducials. These fiducials have a known location relationship to the placement sites of the fine pitch components. Usually, fiducials located at the corners of the circuit board are used to determine the gross location and angular registration of the board relative to the placement machine X, Y, Theta coordinate system. These "global" fiducials allow the placement machine to automatically compensate for variations in registration of the board within the placement machine. In addition, if CAD data is available for the board and assembly pattern, compensation can be made for board stretch and shrink geometry errors.

In some cases, where the manufacturer has reason to believe that local variations in board geometry could not be adequately predicted through assessment of global fiducial locations, additional "local" fiducials are incorporated in the board design. These local fiducials are located

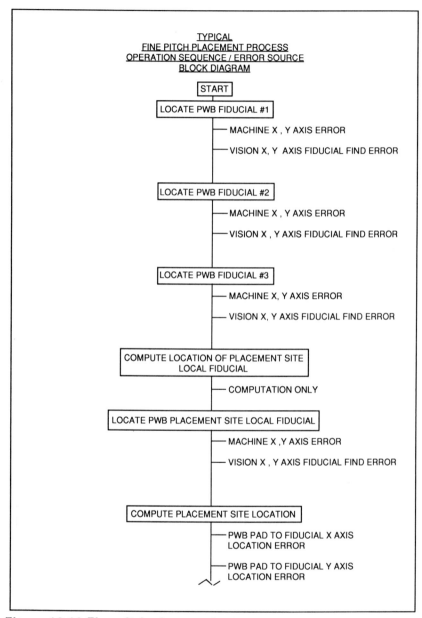

Figure 13.18 Fine pitch placement process sequence.

in close proximity to the fine pitch placement sites. Operation under this

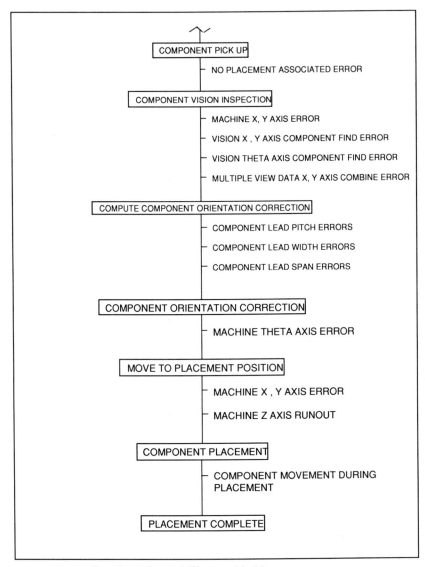

Figure 13.19 Continuation of Figure 13.18.

scenario proceeds with the use of global fiducials to locate the local fiducials, and the actual computation of the fine pitch placement site location is then carried out using the local fiducials.

In this fashion, the effect of local board geometry variations can be minimized where necessary. Typically, the use of local fiducials is

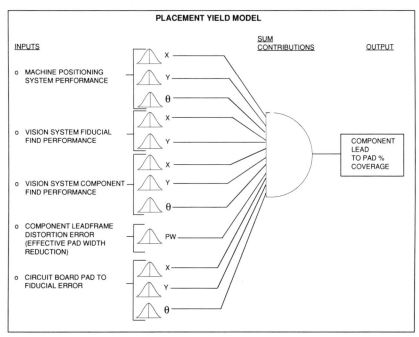

Figure 13.20 Placement error model for typical machine structure.

only used for fine pitch component placement, while the remainder of component placement utilizes global fiducials.

13.3.1 Vision system.

The ability to determine fiducial locations accurately in the machine coordinate system relies on several aspects of the placement machine structure and performance. The fiducial "finding" process is most frequently performed through the use of a "downward looking camera" attached to the placement head structure. This camera system, having a known location with relation to the machine X, Y coordinate system, is positioned over each fiducial to be observed. An automated vision system process then determines the location of the fiducial in machine coordinates.

Two elements are critical to the determination of fiducial locations:

One is the ability of the vision system to accurately recognize the fiducial and to compute its location in vision system coordinates. This ability is largely a function of the vision system design, that is, the lighting, the camera and lens optics, and the vision processor with the associated image analysis software.

The other major element is the ability of the machine positioning system to locate the camera optics accurately over the fiducials.

Examples of the type of data used to assess positioning system error, and a representative set of typical "fiducial find" data are shown in Figures 13.21, 13.22, 13.23, and 13.24.

After computation of the placement site location, the next portion of the process deals with the acquisition and inspection of the component to be placed.

For the purpose of this discussion, no component pick up errors will be considered to have an effect on final component placement. Since the predominant modality of pickup error is gross misregistration of the component on the pickup nozzle, these components are detected at the first inspection stage and culled from the placement process.

The component inspection procedures, however, have elements which play a crucial role in determining final yield to be expected from the assembly process.

Three principal elements in the inspection process affecting placement yield are the quality of the component itself, the ability of the placement machine to accurately measure the component, and the determination of suitable placement corrections.

13.3.2 Component quality.

One factor outside the direct control of the user is the component quality. The component quality attribute relevant to placement yield is the location and relative position of the component leads. Components whose lead footprint is sufficiently distorted will not give acceptable lead to pad coverage even with no other error sources considered. Figures 13.25 and 13.26 illustrate the condition described, contrasting perfect and distorted components.

Several factors can influence component leadform distortion. The manufacture of the component, etching or stamping of the leadframe, forming of the leads themselves, and removal of ESD protective "dam bars". Even the packaging and transport of the components to the final user site has a large effect on lead quality.

Even though the user can specify stringent quality and tolerance levels, the difficulty and expense of measurement to prove the components are, in fact, meeting those levels, lead many users to assume the components are of good quality.

To quantify the magnitude of the component portion of the error budget, a 37 piece sample of 208 lead PQFP (Plastic Quad Flat Pack) components in matrix trays were measured using a coordinate measuring

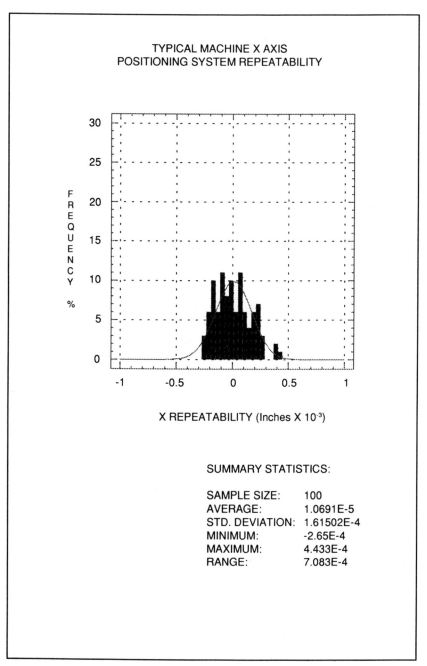

Figure 13.21 X axis repeatability.

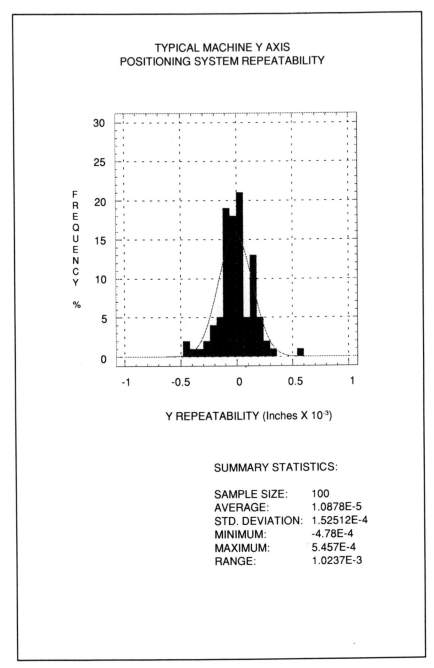

Figure 13.22 Y axis repeatability.

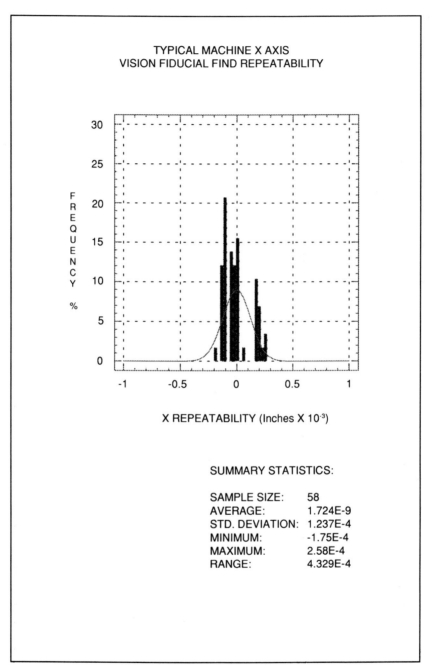

Figure 13.23 X find repeatability.

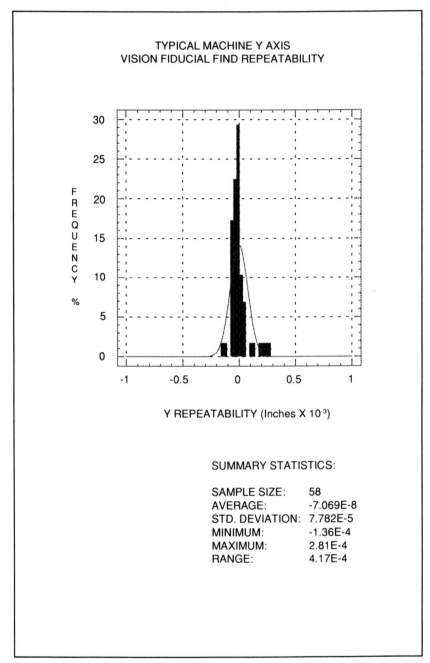

Figure 13.24 Y find repeatability.

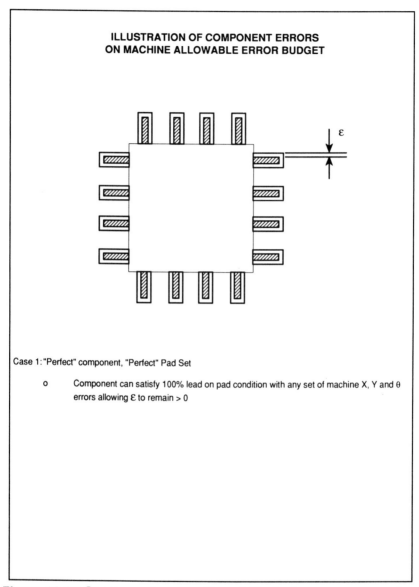

Figure 13.25 Component errors: Case 1 Perfect component, perfect pads.

machine.

      Three points on each lead on each component were measured and an additional point location was computed, as shown in Figure

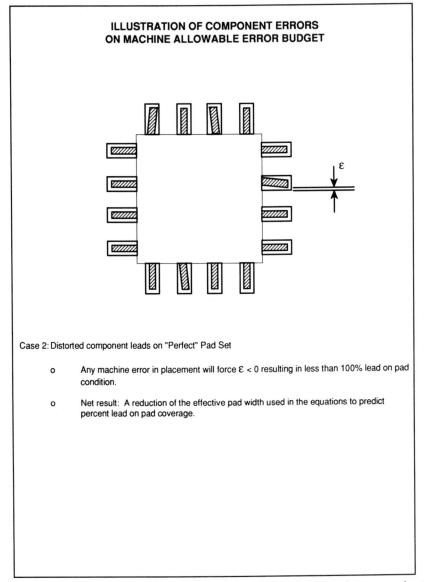

Figure 13.26 Component errors: Case 2 Distorted leads on perfect pads.

13.27.

The measured values were input into a software package (SAS Institute Incorporated) for analysis. Various attributes of the component

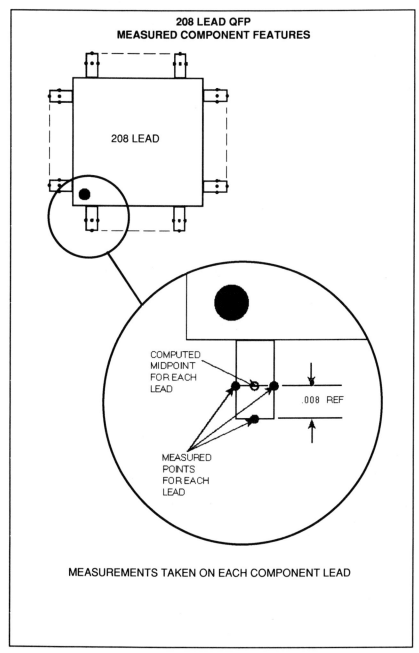

Figure 13.27 Component error measurement.

sample set were then determined.

Lead width, lead pitch and lead tip span distributions appeared as shown in Figures 13.28, 13.29, and 13.30. The lead width distribution showed two discrete distributions were involved. Two discrete lots of leadframes were apparently incorporated in the production of the sample set of components chosen for analysis.

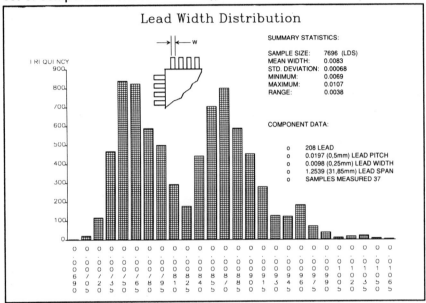

Figure 13.28 Lead width distribution.

In addition, the lead pitch distribution showed a range of over .008 inches, sufficient to cause severe problems in lead to pad coverage.

Computations were performed to show the result of overlaying the 37 piece component sample on perfectly sized and registered pad sets, corresponding to the schematic shown in Figure 13.26. With the lead width set to the mean of the measured distribution (.008 inches) and the pad width set to .010 inches, Figure 13.31 shows the distribution of all leads (208 x 37 = 7696) in terms of their individual lead to pad coverage. Approximately 98.8% of all the leads would achieve 75% or better lead to pad coverage. The remaining 1.2% would not achieve 75% pad coverage even if placed on perfect pads.

A second distribution was generated showing the dimensional deviations of the leads relative to a perfect set of pad geometry (Figure 13.32).

Finally, a computation was performed to illustrate the best possible yield from the sample set treated on a per component basis.

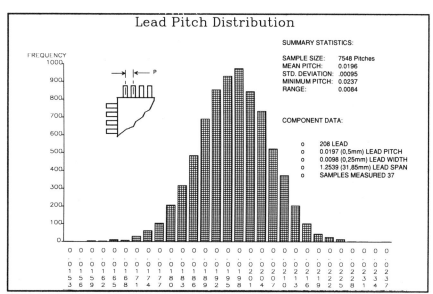

Figure 13.29 Lead pitch distribution.

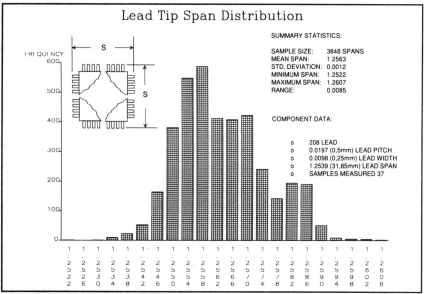

Figure 13.30 Lead tip span distribution.

Figure 13.33 shows that the best lead to pad coverage possible from the sample would be 85%, with the mean at approximately 66%.

Bearing in mind the extremely small sample size of 37 pieces, this

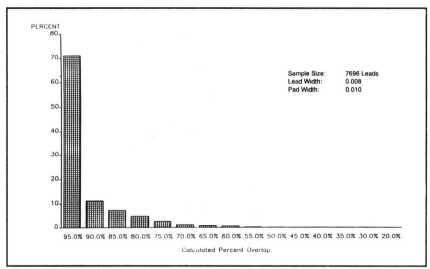

Figure 13.31 Lead displacements converted to percent lead/pad coverage: all leads in 37 piece sample.

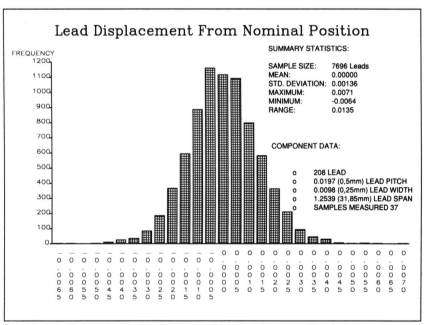

Figure 13.32 Lead displacements from perfect pad set.

data should not be considered as representative of the general case. However, it does serve to illustrate the strong coupling between possible

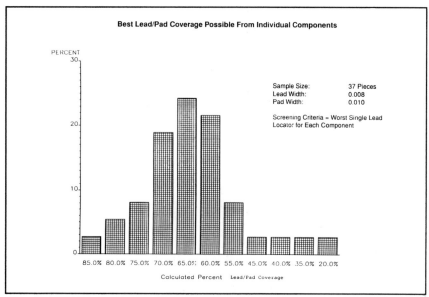

Figure 13.33 Best lead/pad coverage possible from individual components.

placement process yield and input component quality.

    The next significant factor to be considered in the inspection portion of the process is the ability of the machine to accurately measure or "find" the component.

    The component is presented over an "upward looking camera" which captures image information on each lead of the component and their relation to each other. A screening criterion can be programmed by the machine operator to cause component rejection based on the inspected component lead geometry.

    Accurate inspection depends upon the machine X, Y positioning system repeatability as well as upon the camera, lighting, optics and vision processor. A representative example of inspection system performance is shown in Figures 13.34, 13.35, and 13.36. The figures show component orientation values which would then be used to make X, Y, and Theta corrections to the component attitude prior to placement.

    The remaining part of the inspection process is the actual reorientation of the component to the appropriate angular value to be used for placement. An example of machine Theta axis repeatability is shown in Figure 13.37.

    The last elements of the placement process is to position the component over the placement site and to place it.

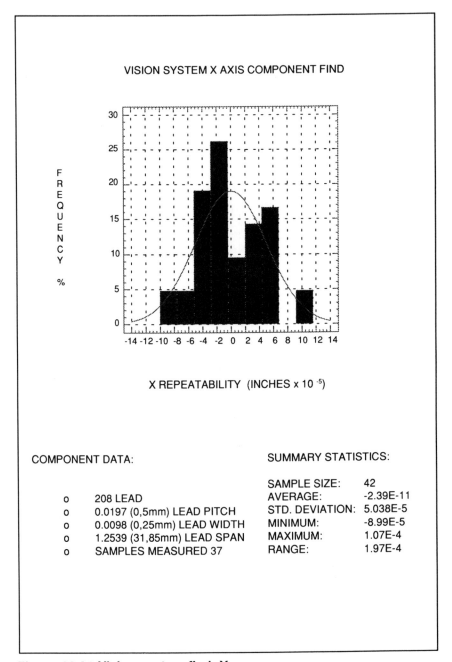

Figure 13.34 Vision system find: X.

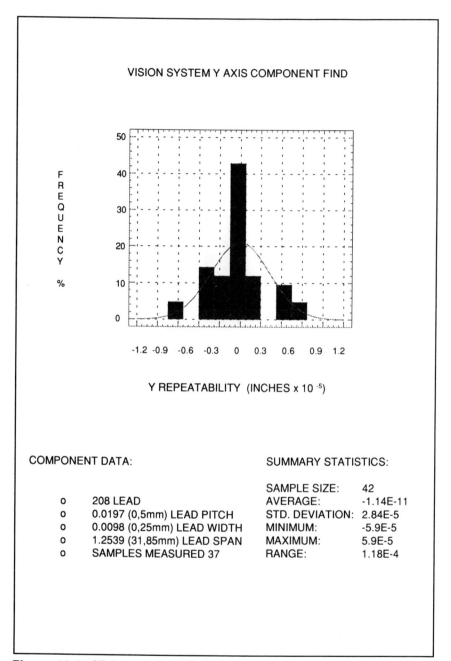

Figure 13.35 Vision system find: Y.

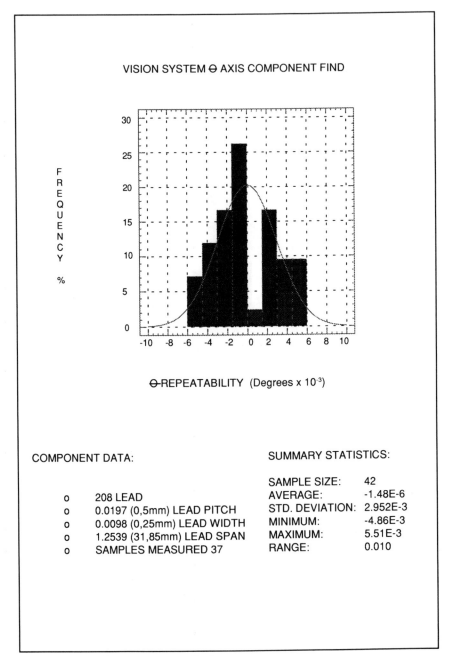

VISION SYSTEM Θ AXIS COMPONENT FIND

Θ-REPEATABILITY (Degrees x 10⁻³)

COMPONENT DATA:

- 208 LEAD
- 0.0197 (0,5mm) LEAD PITCH
- 0.0098 (0,25mm) LEAD WIDTH
- 1.2539 (31,85mm) LEAD SPAN
- SAMPLES MEASURED 37

SUMMARY STATISTICS:

SAMPLE SIZE: 42
AVERAGE: -1.48E-6
STD. DEVIATION: 2.952E-3
MINIMUM: -4.86E-3
MAXIMUM: 5.51E-3
RANGE: 0.010

Figure 13.36 Vision system find: Θ.

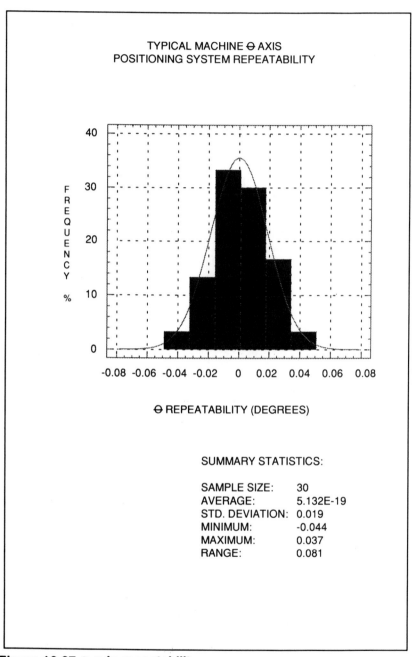

Figure 13.37 Θ axis repeatability.

Two errors can inject uncertainty into this portion of the process. The first is the ability of the machine X, Y, and Theta axes to properly position the component over the anticipated placement site, and the second concerns the actual location of the placement site pads relative to their locating fiducials.

An example of the type of error which can be observed between actual pad site location and that predicted by knowledge of the relevant local fiducials is shown in Figure 13.38.

Using measured information, and with knowledge of the particular placement machine architecture, the model shown in Figure 13.20 can be used to make an estimate of placement yield.

The variable used as the measuring to quantify placement yield is the percentage of lead to pad coverage. The relationship between the X, Y, and Theta component alignment errors and the resultant computation of lead to pad coverage percentages are shown in Figures 13.39 and 13.40 respectively.

A system, performing as discussed above, and operating on input components having the characteristics of the 37 piece sample was modeled to show the hypothetical yield.

13.3.3 Model development.

A calculation was performed in Monte Carlo fashion, in which individual representative random samples were taken from each of the error distributions shown in Figure 13.20. Those individual errors were then used as input to determine a unique X, Y, Theta gross misregistration of a particular component over a particular placement site. Superimposed on the misregistration was a leadframe distortion taken from a selection of one of the 37 component samples. Each component lead was then considered in its relation to the appropriate pad, and the amount of lead to pad coverage computed. The lead for that particular case having the poorest lead to pad percentage overlap was then selected to characterize that individual placement quality.

The entire process was repeated 740 times using new random selections from the error distributions and new selections from the 37 piece component sample.

Each of the 740 simulated placements, characterized by its worst lead to pad coverage percentage value were then combined into a 740 piece sample and the distribution plotted. The results are shown in Figure 13.41.

From the figure, the worst lead on the best placement of the 740 had 80% lead pad coverage, while the worst lead on the worst placement had 0% coverage. The median lead to pad coverage was seen to be

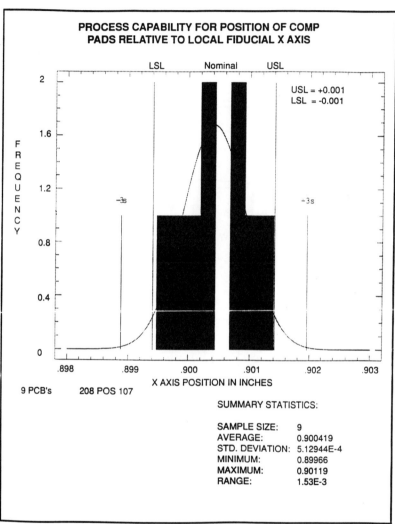

Figure 13.38 Component pad positions.

approximately 52%

13.4 Conclusions.

In summary, although there are many individual contributors to the fine

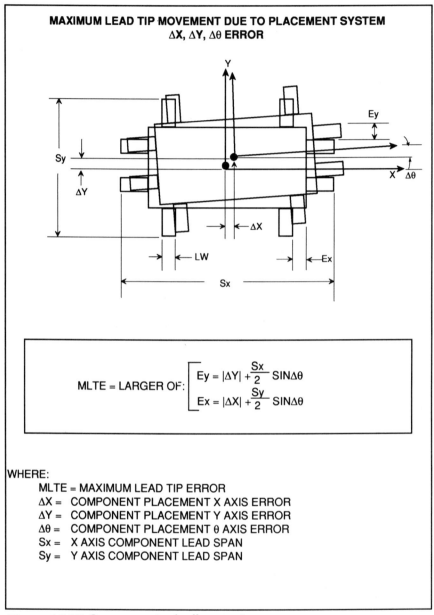

MAXIMUM LEAD TIP MOVEMENT DUE TO PLACEMENT SYSTEM
ΔX, ΔY, Δθ ERROR

$$MLTE = LARGER\ OF: \begin{bmatrix} Ey = |\Delta Y| + \dfrac{Sx}{2}\ SIN\Delta\theta \\[2ex] Ex = |\Delta X| + \dfrac{Sy}{2}\ SIN\Delta\theta \end{bmatrix}$$

WHERE:

    MLTE = MAXIMUM LEAD TIP ERROR
    ΔX = COMPONENT PLACEMENT X AXIS ERROR
    ΔY = COMPONENT PLACEMENT Y AXIS ERROR
    Δθ = COMPONENT PLACEMENT θ AXIS ERROR
    Sx = X AXIS COMPONENT LEAD SPAN
    Sy = Y AXIS COMPONENT LEAD SPAN

Figure 13.39 Component misalignment.

pitch placement error budget, the quality aspects of the individual
components themselves play a major, if not dominant role.

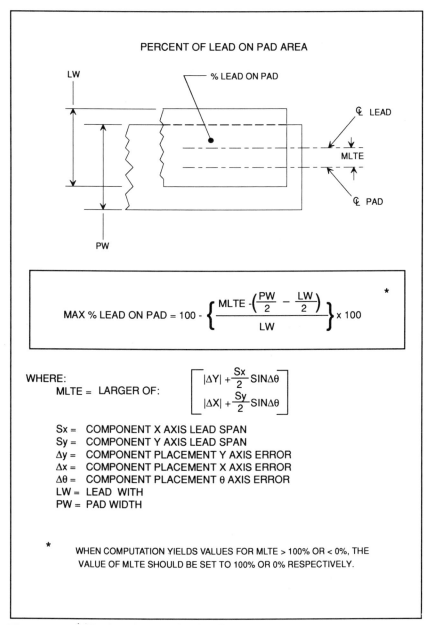

Figure 13.40 Pad coverage by lead.

An unresolved question concerns the criteria with which a

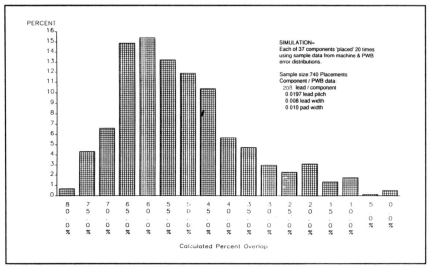

Figure 13.41 Placement simulation results with component, PWB, and machine errors included.

component placement is to be judged. Although a single numerical value is frequently stated as characterizing a "good" versus a "bad" placement (e.g. 75% lead to pad coverage is "good"), its value is questionable when considered in detail.

The real criterion for an acceptable placement is one in which, after soldering, the component leads make an adequate mechanical and electrical bond to the circuit substrate. The question then becomes twofold:

First, for a given component lead count and geometry, how many leads with a given percent of lead to pad coverage are required to provide acceptable mechanical stability?

Second, once a sufficient mechanical bond has been achieved, what degree of lead to pad coverage is necessary for the remaining leads to provide sufficient electrical connection after soldering?

The answers to these questions will have to come from test data provided by end product evaluation. Those answers will drive the selection of appropriate screening criteria, and finally help to determine placement product yield as well as the rejected component scrap rate.

13.4 Reference

1   Hadco Corporation, Solderability Issues of SMT PWB's. April 1990.

MICROELECTRONICS

PACKAGING/INTERCONNECT:

AN INDUSTRY IN TRANSITION

DANIEL C. BLAZEJ

14.1 Introduction

This presentation is adapted from a report prepared for the National Security Industrial Association (NSIA) by a Task Force formed to assess the potential impact of changes in electronics packaging/interconnect on military systems. The findings and recommendations should be of interest to all the electronics community because the current modes of doing business in military electronics are expensive and opportunities for dual-use technology applications could extend beyond the traditional military electronics industry. Some of the comments in this presentation are the author's opinion only and may not represent the consensus position of either the Task Force or Allied-Signal.

The Task Force was organized in late 1987 as part of the NSIA Automatic Testing Committee - New Technologies Subcommittee. Task Force members were drawn from the defense electronics industry and brought expertise in military packaging/interconnect (P/I) and systems technology (Appendix 1). The Task Force also arranged briefings by outside experts on the state-of-the-art and future trends in P/I technology (Appendix 2). Specific areas covered in these briefings included P/I technology in commercial and automotive electronics, collaborative P/I R&D efforts in Europe and Japan, current developments in military P/I, alternative future approaches to high density/high speed interconnect, assessments of optical and superconducting interconnect and systems considerations for year 2000 electronics.

The basis for initiating the study was a recognition that packaging/interconnect had become the limiting factor in high end system performance. In fact, much of the improvement in performance promised by rapidly advancing semiconductor device capability cannot be realized at the system level because of performance barriers imposed by today's electronics packaging/ interconnect technology. Today's P/I methods limit high speed signal integrity, component density, power distribution, thermal management, size and weight.

A mission statement was developed which formed a framework for acquiring up to date information, analyzing it, and ultimately laying out findings and recommendations. Its four parts are as follows:

- Assess role of electronics P/I technology relative to advances in semiconductor technology and military systems requirements.
- Determine applicability of commercial, consumer, and other non-military P/I approaches to military systems.
- Recommend R&D initiatives to meet future military P/I technology requirements.
- Propose methodology to accelerate and maintain active DoD

development, evaluation and insertion of evolving critical P/I technologies.

14.2 Packaging / Interconnect Today

14.2.1 Military Electronics

In terms of relative dollar value, U.S. military electronics has become a small part of the total electronics business. In the 1960s, when the Department of Defense was the major user of integrated circuits (IC), the DoD's share of the total IC market was nearly 80%. Rapid growth in the non-military sector reduced this share to 7% by 1980. According to one source, worldwide total sales of integrated circuits for 1988 are projected at $31.6 billion while the U.S. military IC market is projected to be $1.5 billion, less than 5% of the total. Other sources suggest a military share of 7% for the indefinite future [1-3].

In general, technology used in military applications is best described as mature and conservative. The standard chip carrier is ceramic, not plastic as in the commercial/consumer world. Hybrid packages are virtually always hermetic. Printed wiring boards are conventional epoxy-glass with leaded components mounted in through-holes. Although surface mount technology (SMT) is emerging and one will often see mixed technology boards, it is rare that one finds a full SMT implementation. Performance is good but not exceptional, 10 Mhz routine with an upper limit about 25 MHz, mainly because of interconnect limitations.

14.2.2 Automotive Electronics

Automotive electronics is a rapidly growing area because of the increasing electronics content of the average automobile. Production volumes are in millions of units, which helps support automated assembly and test. Cost is a major driver, but reliability is essential for both safety and customer satisfaction. Analysis of field failures is used for continual improvement of product quality. Because of low manufacturing costs, automotive electronic assemblies are throw-away items.

The dollar value of electronics in the average U.S. passenger vehicle (excluding entertainment systems) was less than $100 in 1970, about $350 in 1980 and is expected to increase to about $1,000 in 1990 and $2,000 in the year 2000, for a total market of more than $10B per year [4].

Before 1980, electronics were largely limited to the alternator rectifier, speed control and electronic ignition. Electronic engine controls, instrument clusters, trip computers, anti-lock braking, suspension control and climate control have been introduced in this decade. The next ten years will see integrated power train control, integrated electronic braking, steering and suspension, multiplexing, on-board diagnostics and communication/navigation.

Electronic engine control units, initially placed in the passenger compartment, are now located under the hood or on the engine, reducing wiring complexity and allowing for production of a complete testable engine subassembly. Electronics are also being physically integrated with electromechanical assemblies, such as anti-lock braking modulators and adaptive suspension components, thus reducing component count and cost, and increasing reliability. These trends expose the electronic assemblies to increasingly severe environmental stress.

14.2.3 Commercial Electronics

Commercial electronics seeks maximum performance at minimum cost, with reliability a secondary (but still important) concern. Volumes range from tens (super computers) to hundreds of thousands (microcomputers) per year. Performance requirements drive commercial electronics to the latest integrated circuit technology which, in turn, requires advanced and innovative P/I methods.

The commercial computer industry has been the main beneficiary of advanced P/I technology. In the early part of the decade much attention was focused on multilayer co-fired ceramic modules with flip-chip ICs and on wafer-scale integration on silicon, both approaches pioneered by U.S. companies. In recent years attention has shifted towards multilayer interconnect substrates using multilayer thin film construction and tape automated bonded (TAB) chip-on-board. It is the commercial systems companies, not device or substrate manufacturers, which are largely responsible for these advances.

Commercial systems manufacturers design for packaging/ interconnect capabilities that are expected to be available two to three years in the future, otherwise their products will not be cost-performance competitive. As a result, advanced P/I methods become part of their proprietary technology and are not readily available to competitors or to military applications.

14.2.4 Consumer Electronics

Consumer electronics (audio, Tvs, watches, games, phones and some personal computers) are characterized by very high volumes (millions), very low costs and short product cycles. The objective is maximum performance at minimum cost with extremely rapid product turn-around required to exploit rapidly developing markets. The large dollar value of sales and the major impact of P/I on manufacturing cost supports major investments in P/I R&D.

Competition in consumer electronics is intense and survival depends on satisfying consumer demands for miniaturization, performance, new capabilities and low cost. Because P/I is the biggest single contributor to manufacturing cost, consumer electronic companies are willing to invest heavily in development and implementation of potentially low cost P/I technologies.

The dominance of consumer electronics by Japanese electronics companies forms a firm economic foundation for their development of advanced packaging/interconnect technology. The Japanese lead the world in application of surface mount, TAB and chip-on-board. The emphasis is on miniaturization and low cost, and product targets are to an increasing degree commercial rather than just consumer.

The Japanese electronics industry, based on consumer applications, has moved aggressively into the commercial area, establishing strong positions in computers (minis, mainframes and supers), test equipment and broadcast studio systems. The multimarket presence which characterizes the larger Japanese electronic companies assists the orderly and effective transfer of P/I technology developed for consumer applications into the commercial area. Japan's growing ability to compete in military/ aerospace electronics is shown by vigorous competition from indigenous Japanese companies for avionics on Japanese aircraft programs.

14.2.5 Lessons for Military Electronics

It is commonly perceived that military electronics operate in environments dramatically more strenuous than all other applications and that packaging techniques are therefore necessarily more complex and costly. Inspection of Table 14.1 shows that automotive operating environments are not all that different from military. In fact, test procedures and reliability requirements for the automotive and trucking industry are often more stringent than in the military.

Part of the reason for vastly lower costs in the automotive industry is the larger production volume, but more importantly is the fact that cost is factored into all phases of the design cycle. The result is that reliability and manufacturability are designed into a product right from the start. Low cost components are assembled using technologies

Table 14.1 Typical environmental exposure conditions for electronic modules.

Environment	Military	Automotive	Commercial	Consumer
Temperature	-55 to 125°C	-40 to 125°C	0 to 70°C	0 to 40°C
Humidity	85% RH (85°C)	85% RH (85°C)	Controlled	Normal ambient
Shock	1,000 to 100,000g	150g	Minimal	Minimal
Vibration	20-100g, 20 to 20,000Hz	20g, 20 to 20,000Hz	Minimal	Minimal
Chemical Resistance	Salt spray	Salt spray, automotive fluids	Generally not resistant	Not resistant

developed in the commercial/consumer industries. Modules containing plastic packages, polymer encapsulants and gasket seals are expected to last 2000 hours (100,000 miles) in the engine compartment of a car. This is done quite successfully without the use of expensive hermetic packages or even specially selected components (capacitors and resistors) which are five to ten times more costly than commercial devices.

The lesson for military electronics is that significant opportunities for cost reduction can be obtained by tapping the know-how extant in the automotive and commercial arenas. Not only are advanced technologies readily available, but so are low cost manufacturing techniques and design approaches.

14.3 Packaging / Interconnect in the Year 2000

14.3 1 Eliminating a Level of Interconnect

The major thrust in advanced P/I over the next decade will be to "eliminate a level of interconnect." This will be done through development of the multichip module (MCM). Usual practice today is to mount single chips into individual packages with the chip input/output (I/O) pads connected to the package lead frame or metallizations by wire bonding. The single chip packages are, in turn, connected to a board (polymer-based PWB or ceramic) by external leads or metallizations. Thus, each chip I/O goes through an intermediate chip-to-package connection before being connected to the board. For each individual chip integrated into a multichip module, the number of connections is reduced by half.

Reducing the number of interconnects by going to multichip modules will improve reliability, as interconnects are the most common cause of system failure. Migrating functionality from a printed wiring board to multichip module will also increase component density and shorten interconnect distance, thus reducing signal delay and improving signal integrity, resulting in improved performance.

14.3.2 State-of-the-Industry in the Year 2000

By the year 2000, single chip packages, multichip packages and multichip modules will all be used, with the more demanding applications using hybrid wafer scale integration or chip on board (COB) multichip modules and the most demanding applications probably employing monolithic wafer scale integration.

For those applications still using printed wiring boards most attachments will be by surface mount with compliant leads, so thermal expansion constraint of PWBs will not be a requirement. Leaded components for plated through-holes will have virtually disappeared and leadless surface mount will be on the decline because of reliability problems resulting from thermal expansion mismatch.

Multichip packages, comprising largely chips and interconnect structure, will be mounted to a PWB for mechanical support, power and ground. Multichip modules will include more passive components, and an integral foundation structure which includes power, ground and a third level interconnect means, for example, a multilayer ceramic substrate with an area array connector to a backplane. Both multichip packages and modules will use multilayer thin film interconnects fabricated by sequential deposition of polymer dielectric with buried vias, having two to three times the interconnect density attainable with plated through-hole PWBs. For many military applications (active decoys, smart weapons/ munitions and SDI kinetic energy weapons) the entire digital electronics system will be a single MCM.

Polymer dielectrics will include new materials with lower dielectric constant and lower moisture absorption than current polyimides. Non-hermetic chip protection will dominate and give reliability comparable to present-day hermetic packages.

The key to successful manufacturing implementation of advanced P/I will be process control; first pass yield in the high ninety percent range will be essential and will require a dedication to quality at every step of the process.   Quality will be built-in, not inspected-in, as advanced P/I will not be compatible with rework.

A new electronics manufacturing infrastructure will develop because the new P/I technologies are not evolving from traditional PWB or hybrid circuit manufacturing practices but will require fundamentally different manufacturing methods and cultures.   Many of the companies currently involved in manufacture of military electronics will be unable to make the transition.

14.4 Conclusions

Electronics P/I is a Critical Limiting Technology, But Not Widely Recognized

Packaging/Interconnect is clearly a second sister to the glamorous world of integrated circuits and truthfully, until now, there has been little need to be concerned with the impact of P/I on system performance.   But the situation is changing, and rapidly.

Opportunities Exist for the Insertion of Dual-Use Technology in Military Applications

Military electronics are separated from the mainstream of commercial and consumer electronics by perceived conflicts between military and commercial/consumer performance, reliability and cost requirements. This separation is exacerbated by military specifications and standards as well as by separate developmental, engineering and manufacturing cultures.   Insertion of commercially available packaging/interconnect technology into military applications could be encouraged through policies which favor dual-use technology, military P/I standards which are compatible with commercial practices, more flexible military qualification procedures and joint military-commercial industry development efforts.

Developments in Japanese Commercial/Consumer Electronics Are Potentially Applicable to Military Systems and Pose a Threat to Domestic Suppliers

Japanese electronics industry developments in packaging/interconnect technology are threatening to overtake the capabilities of U.S. commercial system houses and may soon lead the world.   This capability is being built upon Japanese dominance of consumer electronics and growing strength in commercial applications.   Of particular concern is that development and use of multichip module technology in Japan appears to be ahead of U.S. capabilities.   The dropping of barriers for insertion of dual-use technology in military applications may be seen as an opportunity for U.S. industry but it is also a threat since Japanese companies could, and would, respond even more aggressively.

Development of Key Support Technologies is Required for Advanced P/I

One of the unique aspects of packaging/interconnect is that it is highly interdisciplinary. Integrated design tools are needed which address mechanical, thermal, and electrical aspects simultaneously. Similarly, a wide variety of materials are utilized (metals, polymers, and ceramics) in composite systems which don't always behave in a predictable fashion. Basic R&D in these generic technologies, along with others such as test methodologies and reliability science, are opportunities for joint industry, government, and/or university cooperation.

14.5 Summary

Electronics is an increasingly important aspect of the world economy. In 1980 electronic hardware represented 2.3% of the world's gross domestic product. In 1990, this grew to 3.3% and by year 2000, electronic hardware is expected to be 4.2% of the world's gross domestic product [5].

Packaging/Interconnect is a stepchild of the electronics industry which is not widely recognized and yet represents the single largest cost item in an assembly. Increasingly, P/I is becoming more critical to system performance and cost. This is exemplified in the current emergence of the multichip module, seen by many as the single most important change in P/I since the invention of the printed wiring board. As such, the multichip module also represents a major opportunity because those who can produce a low cost module will have control of the most powerful component of the most advanced electronic assemblies, including workstations, personal computers, communications systems including high definition television (HDTV), and advanced automotive guidance systems. The multichip module, and other advanced electronics technologies, also represent a threat to U.S. industry because implementation requires that current business practices undergo change.

The most ominous threat to domestic industry is not that adaptation can't be done, but that it can't happen quickly enough. In the old days (twenty years ago) when the only market of real concern was domestic and the playing field was U.S. only (and the same rules applied to all players) it was relatively easy for companies old and new to bring out new products and technologies and prosper. Nowadays, the worldwide market is essential for survival because the real competition is not domestic. Particularly in electronics the competition is in Japan, Korea, the other Far Eastern countries and Europe. With the large differences in cost of capital from country to country, unfavorable trade practices for the U.S. and enormous differences in government support of industry (both moral and financial) the playing field is not even anymore and it becomes very difficult for domestic manufacturers to be cost competitive and timely. Given the total loss of the consumer electronics market and the rapid erosion of the automotive market, one wonders if the U.S. will be able to effectively compete in the advanced electronics arena.

The multichip module is representative of a technology that is capital intensive, requires the best materials available, and requires a broad spectrum of specialized talents (design, materials handling, very high yield manufacturing, test, etc.). There is no doubt that the technical ability exists in the U.S. The question is whether a company can pull together the talent and the finances long enough to survive. Domestic industry puts great emphasis on technology innovation and clearly that is a crucial aspect to success in multichip module development. More important though is the manufacturing philosophy which insists on building it right the first time. This is probably the single most important differentiator between the U.S. and Japanese industries and one of the keys to Japanese success. (It is most unfortunate that in the military electronics business the exact opposite holds true. Electronics are not expected to work right in assembly or for very long in use and, in fact, repairability is _designed in_ from the start!)

The business dislocation mentioned earlier is therefore multifaceted. Introduction of multichip module technology will require a multitude of skills and a healthy financial foundation. (It is interesting to speculate on what industry will succeed in multichip module technology: system houses, IC manufacturers, hybrid industry, left field?) Success in the new technology will require advanced manufacturing techniques, a highly integrated technical and material support system (close partnerships?), and ready access to the market. The market aspect is interesting because, when all is said and done, an electronic module is useless unless it goes into something. If it turns out that the major market is in HDTV and they are all built offshore, there will be a problem for the domestic supplier.

What to do? The first thing is simply to recognize the problem. Specifically, the critical importance of packaging/interconnect needs to be widely appreciated. It is an area that has never appeared on the list of national critical technologies, and maybe it's time that it did. It is not enough that just the technical community be made aware; top business executives and government representatives also need to know.

Secondly, the industry needs to take a closer look at how it does business and how well it's doing relative to the rest of the world. The Task Force recommends that R&D consortia be formed to pool resources. I would suggest that even more aggressive approaches need to be looked at. Historically, the R&D consortium is relatively ineffective in actually transferring and implementing technology onto the factory floor. Much more rapid means of insertion need to be devised. U.S. industry is the best in the world in R&D and simply carrying on that tradition isn't enough anymore. Making products that sell in today's world requires low cost, high quality manufacturing techniques. The transition from laboratory to factory needs to be emphasized far more than it is today. In packaging/interconnect this can be a very difficult thing to do because of the wide variety of skills required and the capital investment required. Partnering with other companies can relieve this burden and provide opportunities for rapid market entry. A creative example of this is a vertically integrated partnership of material supplier, partsmaker, and end user recently organized to market superplastic steel [6].

And finally, the government could be more supportive of industry's efforts to be competitive worldwide. Specific recommendations for DoD have been mentioned already with regard to insertion of dual-use approaches and generic supporting technologies. It is also time that our legislators helped industry out of the low spot in the worldwide economic field with more far-sighted approaches to budget control and foreign trade policy. The alternative is continued erosion of key industries, loss of technical and economic leadership positions, and ultimately a deterioration in the nation's defensive posture.

14.6 References

1   "The VHSIC Program: Its Impact on the Commercial IC Industry", The Information Network (1988).

2   "Microelectronics to the Year 2000", SRI International, Business Intelligence Program, Report No.739.

3   "The World of Silicon: It's Dog Eat Dog", IEEE Spectrum, Sept 1988.

4   Ford Motor Company estimates and the University of Michigan Delphi III market research study.

5   Ralph Anavy "Strategy 2000", Electronic Outlook Corporation, 1990.

6   "Swords Into Plowshares", Forbes, July 23 1990.

Appendix 1
Task Force Members

Allied-Signal Inc.	Bruce E. Kurtz (Chairman)
	Daniel C. Blazej (Secretary)
DARPA/ISTO	John C. Toole
Eaton AIL	Paul Heller
General Electric	Raymond Fillion
Honeywell	Bernie E. Goblish
IBM Corporation	Roy R. Stettinius, Jr.
Martin Marietta Corporation	Ronald J. Cybulski / Charles A. Hall (Vice Chairman)
McDonnell Douglas	Thomas E. Schulte / Emery E. Griffin
Raytheon	Frank J. Cheriff
TRW Inc.	Robert Smolley
Unisys Corporation	Victor A. Wells

Appendix 2
Task Force Briefings

Advanced P/I & Manufacturing Technology at Motorola
Mauro Walker, Motorola

VHSIC P/I at IBM
Roy Stettinius, IBM

Issues in High Density Interconnect
Robert Smolley, TRW

P/I Lessons from Semiconductor Experience
Paul Hart, Hughes Aircraft Company

Electronics P/I Technology in Japan and Implications for Military Electronics
Don Brown, Information Incorporated

Optical Interconnect for Advanced Systems
Robert Burmeister, Saratoga Technology Assc.

Electronics P/I for the Year 2000
Richard Landis, Martin Marietta Corporation

Electronics P/I R&D at AT&T
John Segelken, AT&T Bell Labs

P/I Systems Considerations in Electronics P/I
Dennis Herrel, MCC

P/I Research in the EEC
Ted Ralston, MCC

Superconducting Interconnect
Harry Kroger, MCC

P/I Research in Japan
Atsushi Akera, MCC

Photonic Interconnect for Computing
Anis Husain, Honeywell

P/I and the Automotive Electronics Environment
Roel Hellemans, Allied-Signal Inc. - Bendix Electronics

High Speed Digital P/I
Daniel Schwab, Mayo Foundation

Electronics P/I Technology at Honeywell
Jerry Loy, Honeywell

INDEX

direct
air cooling 210
chip attach 168
liquid immersion 190
directional couplers 286, 287
discretization 108
dissipation factor 248
distributed power 162, 172, 178
DMOS 166
dual-use technology 444
Duhamel-Neumann law 42
dynamic response 119, 127, 129, 135

E

E-glass 269
edge connectors 119, 127
effective dielectric constant 256
efficiency 143, 144, 148, 156, 169
elastic constants 88
elastic pin deformation 95
electric dipoles 251
electrical
bonding resistance 370
noise 154
overstress 299
resistivity 240
electro-magnetic interference 149
electro-optic effect 291
electrodeposited nickel 389
electroless nickel plate 372, 380
electrolytic capacitors 160 161 162
electromagnetic
compatibility 154, 171
interference 365
pulses 299
electron injection 311, 316
electronics manufacturing 444
electroplated nickel 372
connector 391
stainless steel connectors 365
electrostatic discharge 56, 299
damage 308
emitter follower 143
EMPs 299
encapsulants 264
encapsulation 28
energy storage elements 146
engineering expenses 155
ENIAC 184
entropy 44

environmental
exposure 377
stress 440
stress screening 365
EOS 299, 311, 315
epoxy 168
encapsulants 30
/fiberglass laminates 268
resin 263
equivalent CTE 90
equivalent series inductance 162
equivalent series resistance 162
ESD 299, 309, 311, 315, 316, 326
sensitivity 306
eutectic 23
excising 32
expanded PTFE 270
experimental modal analysis 104
external thermal resistance 188, 192, 212

F

failure mechanisms 300
failure modes 52
fan-out on level 33
fan-out on slope 33
fatigue 86
cracking 55
fault isolation 175
FEM 107
ferrite 164
fibers 269
filter capacitors 146
filtering 141
fin density 218
fin tip clearance 215
fine pitch 59, 396, 406
assembly 409
component assembly 409
placement 433-434
TAB 56
finite element 49, 86, 98
methods 52, 96, 104, 106, 114
model 131, 133
modeling 61
finite width heat sinks 220
finned air-cooled heat sinks 210
fins 215
flame hydrolysis deposition 283
flammability hazard 357
flange mount connector 375, 382

457